W9-AXO-827

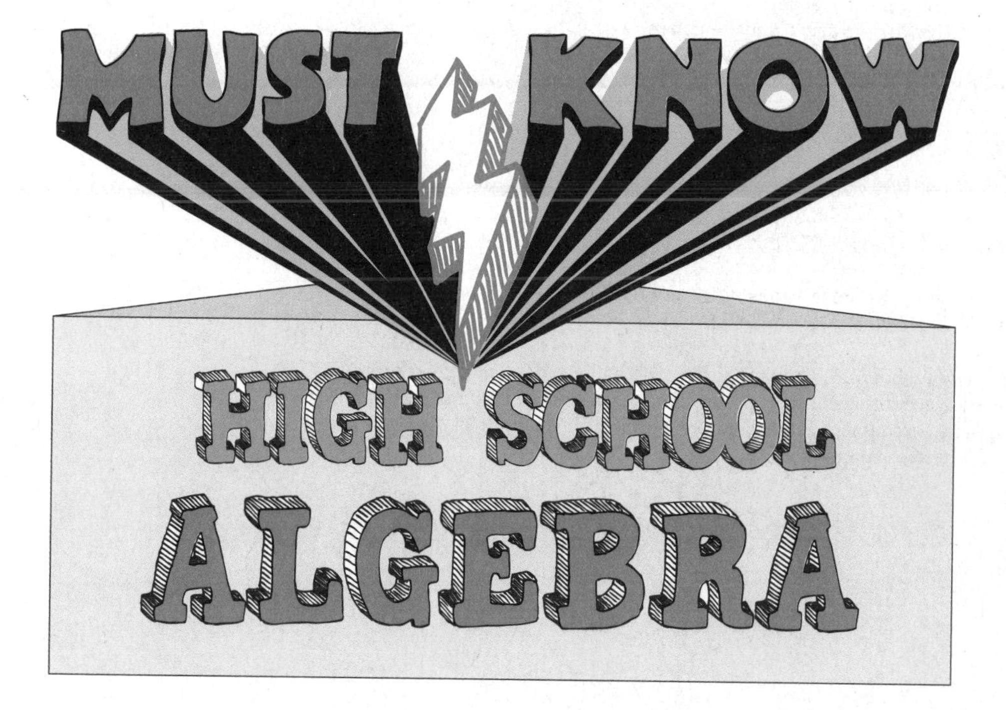

Chris Monahan

New York Chicago San Francisco Athens London Madrid
Mexico City Milan New Delhi Singapore Sydney Toronto

Copyright © 2019 by McGraw-Hill Education. All rights reserved. Printed in the United States of America. Except as permitted under the United States Copyright Act of 1976, no part of this publication may be reproduced or distributed in any form or by any means, or stored in a database or retrieval system, without the prior written permission of the publisher.

1 2 3 4 5 6 7 8 9 LCR 24 23 22 21 20 19

ISBN 978-1-260-45292-1
MHID 1-260-45292-1

e-ISBN 978-1-260-45293-8
e-MHID 1-260-45293-X

Interior design by Steve Straus of Think Book Works.
Cover and letter art by Kate Rutter.

McGraw-Hill Education books are available at special quantity discounts to use as premiums and sales promotions or for use in corporate training programs. To contact a representative, please visit the Contact Us pages at www.mhprofessional.com.

Dedication

I want to thank my wife, Diane, for her support while I was writing this book, and I would like to acknowledge the work done by Garret Lemoi of McGraw-Hill. The questions he asked me during the writing of this book were invaluable in making sure that we concentrated on the concepts and skills that you ***must know!***

And to all of you who have been in my classes, you've taught me so much. Thank you!

Author's Note

Must Know High School Algebra continues to build on the skills and concepts learned in your middle school math classes. Much of the work done in this book concentrates on a formal process for solving equations and inequalities (including graphical approaches). In addition, you will learn about the arithmetic of polynomials, functions and function notation, solving quadratic equations, solving exponential equations, and dig a little deeper into the study of probability and statistics.

As a student, I used to hate the questions that started with "Kathy has 60 coins in her purse" and then go on to ask how many coins of each type she has in her purse. I always wanted to tell Kathy to dump the contents of her purse on the table and count the coins. As a teacher, I understood the importance of learning the process of clearly defining the meaning of the variables used and the relationships described in each of the equations and inequalities written about those variables. I hope you will be patient as you work through the book's problems and remember that they will help you understand how to analyze problems in real life.

Mathematics, in general, and this book in particular are not meant for you to read only. You should always have a pencil, paper, and your calculator with you when doing math. There are plenty of examples in the book with explanations that will guide you through the steps involved. I encourage you, though, to try to do the problems before you read the explanations. This will give you a reasonable measure of how much you understand when you first read the problem. Did you get stuck at some point? That's fine. The reason you bought this book was to help you get over whatever may have been causing you trouble. Read through the explanations. Make sure they seem logical to you. Go back and try the problem again. There are more problems for you to try in the exercise sections of each chapter, with the answers and explanations available to you in the back of the book. There are also electronic flash cards for you in each of these sections.

This book begins each chapter with statements about what it is you **must know** when you have completed the chapter. If you do not feel that you have a good grasp of the items described at the beginning of the chapter, please take the time to go back through the chapter until you do. You will be glad that you did.

Contents

Introduction

Welcome to your new algebra book! Let us try to explain why we believe you've made the right choice. This probably isn't your first rodeo with either a textbook or other kind of guide to a school subject. You've probably had your fill of books asking you to memorize lots of terms (such is school). This book isn't going to do that—although you're welcome to memorize anything you take an interest in. You may also have found that a lot of books jump the gun and make a lot of promises about all the things you'll be able to accomplish by the time you reach the end of a given chapter. In the process, those books can make you feel as though you missed out on the building blocks that you actually need to master those goals.

With *Must Know High School Algebra*, we've taken a different approach. When you start a new chapter, right off the bat you will immediately see one or more **must know** ideas. These are the essential concepts behind what you are going to study, and they will form the foundation of what you will learn throughout the chapter. With these **must know** ideas, you will have what you need to hold it together as you study, and they will be your guide as you make your way through each chapter.

To build on this foundation you will find easy-to-follow discussions of the topic at hand, accompanied by comprehensive examples that show you how to apply what you're learning to solving typical algebra questions. Each chapter ends with review questions—more than 400 throughout the book—designed to instill confidence as you practice your new skills.

This book as other features that will help you on this algebra journey of yours. It has a number of "sidebars" that will both help provide helpful information or just serve as a quick break from your studies. (As good a book as we think this is, we know that you may need a little respite once in a while to pace yourself.) The BTW sidebars ("by the way") point out important information as well as tell you what to be careful about algebra-wise. Every once in a while, an IRL sidebar ("in real life") will tell you what you're studying has to do with the real world; other IRLs may just be interesting factoids.

In addition, this book is accompanied by a flashcard app that will give you the ability to test yourself at any time. The app includes more than 100 "flashcards" with a review question on one "side" and the answer on the other. You can either work through the flashcards by themselves or use them alongside the book. To find out where to get the app and how to use it, go to the next section, The Flashcard App.

We also wanted to introduce you to your guide throughout this book. Chris Monahan has more than 30 years' experience teaching students at the high school and college levels. He knows what you should get out of an algebra course and his strategies will help get you there. Chris has seen the typical kinds of problems that students can have with algebra, and he is experienced at solving those difficulties. In this book, he uses that experience to show you not only the most effective way to learn a given concept but how to get yourself out of traps you might have fallen into. We've had the pleasure of working with Chris before are confident that we're leaving you in good hands.

Before we leave you to Chris's surefooted guidance, let us give you one piece of advice. While we know that something "is the *worst*" is a cliché, we might be inclined to say that the quadratic formula is the worst. Don't try to see if you can get away with going through the motions; instead, let the author introduce you to the concept and show you how to apply it confidently to your algebra work. If you'll take our word for it, mastering the quadratic formula will leave you in good stead for the rest of your math career.

Good luck with your studies!

The Editors at McGraw-Hill

The Flashcard App

This book features a bonus flashcard app. It will help you test yourself on what you've learned as you make your way through the book (or in and out). It includes 100-plus "flashcards," both "front" and "back." It gives you two options as to how to use it. You can jump right into the app and start from any point that you want. Or you can take advantage of the handy QR Codes near the end of each chapter in the book; they will take you directly to the flashcards related to what you're studying at the moment.

To take advantage of this bonus feature, follow these easy steps:

Search for ***Must Know High School*** App from either Google Play or the App Store.

↓

Download the app to your smartphone or tablet.

↓

Once you've got the app, you can use it in either of two ways.

↙ ↘

Just open the app and you're ready to go.	Use your phone's QR Code reader to scan any of the book's QR codes.
You can start at the beginning, or select any of the chapters listed.	You'll be taken directly to the flashcards that match your chapter of choice.

↘ ↙

Be ready to test your algebra knowledge!

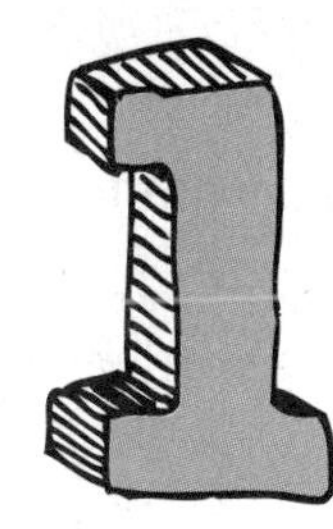

Putting Numbers Together and Taking Them Apart

The order of operations is our guide to working clearly and accurately—in algebra as well as arithmetic.

The **order of operations** dictates how we combine numbers when computing a total from a set of numbers including evaluating an algebraic expression. You've heard this in earlier grades. The order of operations also dictates how we solve equations.

The Order of Operations

You have already learned that Please Excuse My Dear Aunt Sally (PEMDAS) has been the mnemonic that most students have learned for remembering the order of operations. Parentheses (or any grouping symbols) have the top priority followed by exponents, multiplication and division as they occur from left to right, and addition and subtraction as they occur from left to right. The order of operations is important in the evaluation of algebraic expressions as well as in determining the method with which equations are solved.

EXAMPLE

- Compute: $87 + 9(27 - 5^2)^3 \div 12 \times 3$
- As we look at the problem, we are directed to work within the parentheses first.
- Compute 5^2: $87 + 9(27 - 25)^3 \div 12 \times 3$
- Finish the arithmetic inside the parentheses: $87 + 9(2)^3 \div 12 \times 3$
- Since there are no more grouped expressions, evaluate terms with exponents: $87 + 9(8) \div 12 \times 3$
- Multiplication and division have equal priority and are addressed from left to right as they occur in the problem:

  Multiply 9 and 8: $87 + 72 \div 12 \times 3$

  Divide 72 by 12: $87 + 6 \times 3$

  Multiply 6 and 3: $87 + 18$

  Add to finish the problem: $87 + 18 = 105$

A computational problem that is a bit more complex is

EXAMPLE

▶ Compute: $\dfrac{87 + 9(27 - 5^2)^3}{12 \times 3}$

▶ This almost looks like the last problem, with the exception that the denominator contains the product of 12 and 3.

$$\frac{87 + 9(27 - 5^2)^3}{12 \times 3} = \frac{87 + 9(27 - 25)^3}{12 \times 3} = \frac{87 + 9(2)^3}{12 \times 3}$$

$$= \frac{87 + 9(8)}{12 \times 3} = \frac{87 + 72}{12 \times 3} = \frac{159}{36} = \frac{3(53)}{3(12)} = \frac{53}{12}$$

▶ If you were to enter this into your calculator, you would enter $(87 + 9(27 - 5^2)^3 \div (12 \times 3)$.

This next example involves a complex fraction. A **complex fraction** is a fraction in which the numerator or denominator contains fractions.

EXAMPLE

▶ Calculate: $\dfrac{\frac{3}{4} + \frac{5}{8}}{\frac{11}{12} - \frac{7}{24}}$

▶ Most people would argue that this problem is just plain ugly. I am going to disagree with that opinion because I am going to make the problem easier to compute. We all know that when you add or subtract fractions, you must first find a common denominator. So, in complex fractions, you first multiply the numerator and denominator of the complex fraction by the common denominator of ALL of the fractions in the problem. In this case, that number is 24.

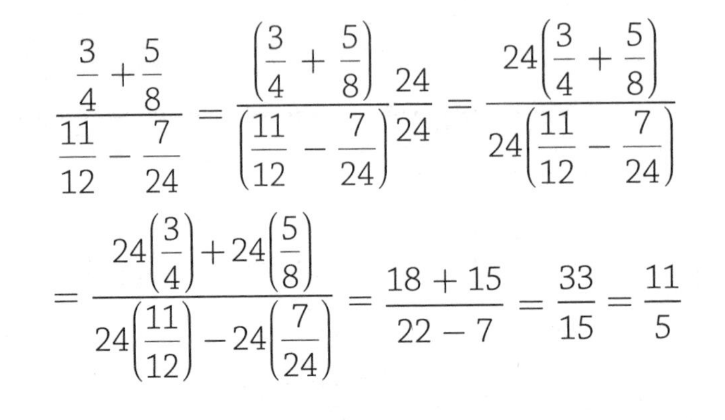

Addition and multiplication allow for some flexibility in computing that subtraction and division do not. Addition and multiplication allow for **commutativity**, meaning that terms can be rearranged. For example, $3 + 5 = 5 + 3$ and $3 \times 5 = 5 \times 3$. These two operations also allow for regrouping, or **associativity**. You probably use this more often than you think. Because the mind (yours and a computer's CPU) can only process two numbers at a time (though the computer processes much faster), a problem such as $11 + 23 + 29$ is treated as $(11 + 23) + 29$. Many of you, without being aware of it, will change this problem to $(23 + 11) + 29 = 23 + (11 + 29) = (11 + 29) + 23 = 40 + 23 = 63$ because you know that gathering 10s is a quick way to do mental arithmetic. As was said earlier, this combination of the associative and commutative properties is a common practice. An application to multiplication is to compute $4 \times 19 \times 25$. (You got the answer 1900 quickly, didn't you?)

A third property of real numbers is the **distributive property of multiplication over addition**, most often referred to as simply the distributive property. The expression $5(4 + 6)$ can be evaluated as $5(10) = 50$ by first doing the addition and then the multiplication, or by first applying the distributive property: $5(4 + 6) = 5(4) + 5(6) = 20 + 30 = 50$. You used the distributive property to justify why the product of two negative numbers is a positive number. For example, you know that $-8(4 + (-4)) = 8(0) = 0$ when the addition is done first. When the distributive property is applied, $-8(4 + (-4))$ becomes $-8(4) + (-8)(-4)$ and this must equal 0. We know $-8(4) = -32$, so $-32 + (-8)(-4) = 0$. Therefore, $(-8)(-4)$ must equal positive 32, and we can state that when two numbers with the same sign are multiplied, the product is positive.

EXAMPLE

- Compute: $23(12) + 23(8)$
- The order of operations tells us to do the multiplication first and add second. Mental computation of the product of 23 and 12 can be done, but not all that quickly. Noticing that 23 is a factor of both terms in the problem, you can rewrite the problem as $23(12 + 8)$. You can now work inside the grouping symbols: $12 + 8 = 20$. So the problem is now $23(20)$, and that equals 460.

IRL **Many businesspeople will take a problem such as 23×12 and rewrite it as $23(10 + 2)$. They know that $23 \times 10 = 230$ and that $230 + 46$ is $230 + 40 + 6 = 270 + 6 = 276$. In fact, it took much longer for me to type this and you to read it than it takes a typical businessperson to do the computation.**

Try this.

EXAMPLE

- Mentally compute: 32×19.
- Take a moment to do this before you read the material in the next paragraph.
- Rewrite 19 as $20 - 1$. The problem now becomes $32 \times (20 - 1)$. Apply the distributive property so the problem now reads $(32 \times 20) - (32 \times 1)$. Compute 32×20 to be 640, and, of course, $32 \times 1 = 32$. $640 - 32$ can be written as $640 - 30 - 2$. $640 - 30$ is 610 and $610 - 2 = 608$.

There are many applications of algebraic formulas to get shortcuts for mental computation. I'll point them out as we develop the formulas.

Mentally compute: $\frac{36}{3 \times 4}$

Now use a calculator to compute the value. Did you get the same answer with the calculator as you did mentally and is that answer 3?

The problem asks that 36 be divided by the product of 3 and 4. That is, $\frac{36}{3 \times 4}$ means $36 \div (3 \times 4)$. You know that 3 times 4 is 12 and 36 divided by 12 is 3. However, a large number of people will get the answer 48 when using the calculator because they will enter the problem as $36 \div (3 \times 4)$ and the order of operations dictates that multiplication and division, with equal priority, are done from left to right. $36 \div 3 \times 4$ becomes 12×4, and this, in turn, equals 48. For this reason, you must understand the order of operations; then when entering expressions into calculator, you will be entering a true representation of the original problem.

EXAMPLE

Compute, both with paper and pencil and with a calculator: $\frac{28(15 + 9) + 128}{25(7) - 15}$.

Make sure that you understand that this problem really reads as

$$(28(15 + 9) + 128) \div (25(7) - 15)$$

This helps you understand that you can work within the two sets of parentheses to get to the answer.

$(28(15 + 9) + 128) \div (25(7) - 15)$ becomes

$(28(24) + 128) \div (175 - 15)$ after you apply the rules

BTW

Like written and spoken language, the grammar of mathematics is very specific. Unlike the written language in which the punctuation marks must be included to ensure the correct concept is being conveyed (e.g., "I'm sorry I like you." and "I'm sorry, I like you."), the grammar of mathematics is implied within the order of operations.

for the order of operations to each of the parentheses.
$(28(24) + 128) \div (175 - 15)$ now becomes $(672+128) \div (160)$
$= (800) \div (160) = 5$.

- Understanding the rules of operation will allow you to get the correct result when using a calculator.

Variables and Expressions

You have been using letters to represent numbers for a few years. In your early education, you used boxes or blanks to represent numbers. Do you remember having to fill in the blank for problems like $3 \pm = 5$? This is really algebra. The boxes and blanks were used so that you would not be distracted. A **variable** is a symbol that is used to represent a quantity. You'll see that the most popular choices for variables come from the end of the alphabet, primarily x, y, and z. You will also see w, t, and n on occasion.

Variables can be combined with other variables or constants to form mathematical **expressions**. (Expressions are different from equations in that expressions do not have an equal sign.) You'll see expressions like $4x$, $8xyz$, and $3x + 4y$. Following are some terms you will need to understand:

- **Monomials** are the products of variables and constants. You will not find any addition, subtraction, or division in a monomial.
- **Polynomials** are made up of the sum and difference of monomials (not division). Polynomials are broken down to **binomials** (two terms) and **trinomials** (three terms). Anything more than three terms is simply referred to as a polynomial.

Combining monomials and polynomials is just an application of the distributive property. While the majority of the applications in your experience have been to change the expression $a(b + c)$ to the form $ab + ac$, you will use the distributive property to change $ab + ac$ to $a(b + c)$.

EXAMPLE

▶ Combine $4x + 8x$.

▶ Use the distributive property to rewrite $4x + 8x$ as $(4 + 8)x$. Clearly this is equal to $12x$.

You will need to be careful as the examples get more complicated. The bottom line is that in order for you to combine two or more monomials through addition and subtraction, the variable piece of the monomial must be an **exact** match so that the distributive property can be applied.

EXAMPLE

▶ Combine each of the following terms:

(1) $12x + 9y + 10x - 4y$

(2) $12x + 9xy + 10xy - 4y$

▶ Solutions:

(1) We can apply the commutative property to rearrange the problem to read $12x + 10x + 9y - 4y$. Apply the distributive property twice to get $(12 + 10)x + (9 - 4)y$. Do the arithmetic to get the answer $22x + 5y$.

(2) While this problem might look similar to item (1), it is very different. The distributive property applies just once: $12x + 9xy + 10xy - 4y = 12x + (9 + 10)xy - 4y$, and this equals $12x + 19xy - 4y$. Do you see that the three monomials in this answer are all different? That is why they cannot be combined.

Bases and Exponents

The exponential expression b^n, in which b is a number other than 0 or 1 and n is a counting number, is an indication of repeated multiplication. Following are some results of this definition.

$(a^n)(a^m)=a^{n+m}$	When terms with a common base are multiplied, add the exponents.
$\frac{a^n}{a^m}=a^{n-m}$	When terms with a common base are divided, subtract the exponents.
$(ab)^n=a^nb^n$	A product raised to a power yields the product of each factor raised to the power.
$\left(\frac{a}{b}\right)^n=\frac{a^n}{b^n}$	A quotient raised to a power is the quotient of each term raised to the power.
$(a^m)^n=a^{mn}$	When a term raised to a power is then raised to a power, the result is the term raised to the product of the powers.

Two applications of these rules are

$$(-4p)^5=(-4p)\times(-4p)\times(-4p)\times(-4p)\times(-4p)=1024p^5$$

$$\left(\frac{2a^4}{3b^5}\right)^3=\frac{2a^4}{3b^5}\times\frac{2a^4}{3b^5}\times\frac{2a^4}{3b^5}=\frac{8a^{12}}{27b^{15}}$$

We exclude 0 and 1 from being the base because the results are basic. Zero raised to any positive power will be 0, and 1 raised to any power will be 1.

EXAMPLE

▶ Simplify: (a) $(6z)^3$ (b) $(6q^4)^3$ (c) $\left(\frac{2t}{3p^2}\right)^5$

(a) $(6z)^3 = (6)^3(z)^3 = 216z^3$

(b) $(6q^4)^3 = (6)^3(q^4)^3 = 216q^{4\times3} = 216q^{12}$

(c) $\frac{(2t)^5}{(3p^2)^5} = \frac{(2t)^5}{(3p^2)^5} = \frac{(2)^5(t)^5}{(3)^5(p^2)^5} = \frac{32t^5}{243p^{10}}$

Since expressions can also contain terms with exponents, you have to be even more careful when combining two or more expressions.

EXAMPLE

▶ Simplify $4x^2 + 5x + 8 + 6x - 3x^2 - 9$.

▶ As usual, gather common terms together: $4x^2 - 3x^2 + 5x + 6x + 8 - 9$ $= (4-3)x^2 + (5+6)x + (8-9) = x^2 + 11x - 1$.

Subtracting polynomials gets tricky because you need to be sure to attach the subtraction to the entire polynomial that follows it.

EXAMPLE

▶ Simplify: $(4x^2 + 5x + 8) - (6x - 3x^2 - 9)$.

▶ While you never have to write it, technically there is a 1 in front of each of these parentheses. Multiplying $4x^2 + 5x + 8$ does not change anything, but multiplying $6x - 3x^2 - 9$ by –1 changes all the signs. So, $(4x^2 + 5x + 8) - (6x - 3x^2 - 9) = (4x^2 + 5x + 8) + (-1)(6x - 3x^2 - 9) =$ $4x^2 + 5x + 8 - 6x + 3x^2 + 9 = 7x^2 - x + 17$.

Evaluating Algebraic Expressions

When being directed to substitute constants for the variables in algebraic expressions, you might find it easier to first simplify the expression rather than make the substitutions. The last example from the previous section is a good example of this.

BTW

Algebraic expressions that are the product of numbers and variables (such as $6x$ and $5xyz^2$) are called ***monomials****. Terms that are added together (such as $3x$, $4x^2 + 5x - 2$, and $5x^2y - 4xy^2$) are called* ***polynomials****.*

EXAMPLE

Given $z = 6$, $p = -2$, $q = -3$, and $t = 4$, evaluate (a) $(6z)^3$ (b) $(6q^4)^3$ (c) $\left(\frac{2t}{3p^2}\right)^5$

(a) $(6z)^3 = 216z^3$. Substituting $z = 6$, $216z^3$
$= 216(6)^3 = 216 \times 216 = 46{,}656$

(b) $(6q^4)^3 = 216q^{12}$.
Substituting $q = -3$, $216q^{12} = 216(-3)^{12} = 114{,}791{,}256$

(c) $\left(\frac{2t}{3p^2}\right)^5 = \frac{32t^5}{243p^{10}}$.
Substituting $p = 2$ and $t = -4$, $\frac{32t^5}{243p^{10}}$
$= \frac{32(4)^5}{243(-2)^{10}} = \frac{32(1024)}{243(1024)} = \frac{32}{243}$

Some of those values are very large. These are good examples of why you should have your calculator with you when working on your material AND know how to enter the material correctly.

EXAMPLE

Simplify $\dfrac{12x + 9y}{5x - 4y}$ when

(a) $x = 5$ and $y = 2$;

(b) $x = 5000$ and $y = 2000$;

(c) $x = 45$ and $y = 23$.

(a) The expression cannot be simplified, so make the substitution immediately: $\dfrac{12x + 9y}{4x - 3y} = \dfrac{12(5) + 9(2)}{5(5) - 4(2)} = \dfrac{78}{17}$.

(b) These numbers are much larger than those in part (a). However, did you notice that these values are each 1000 times larger than the values from part (a)? Are you not surprised that the answer is again $\dfrac{78}{17}$? $\dfrac{12x + 9y}{4x - 3y} = \dfrac{12(5000) + 9(2000)}{5(5000) - 4(2000)} = \dfrac{78000}{17000} = \dfrac{78}{17}$. Please observe that it is to your advantage to examine the problem before you begin to "plug and chug," as there might be a shortcut to doing the problem.

(c) There are no common factors for 45 and 23, and some of these computations are tedious, so use your calculator. Enter the keystrokes $(12(45) + 9(23)) \div (5(45) - 4(23))$ to get the answer $\dfrac{747}{133}$.

Here is another problem.

EXAMPLE

- Simplify $\frac{20x+15y}{4x+3y}$ when $x=5$ and $y=2$.
- If you make the substitution, you get $\frac{20(5)+15(2)}{4(5)+3(2)}=\frac{130}{26}=5$.

Perhaps you noticed that $\frac{20x+15y}{4x+3y}=\frac{5(4x+3y)}{4x+3y}$ and reduced the fraction by cancelling the common factor $4x + 3y$ to get the answer 5. There will be times that you will see the shorter approach. Keep in mind, the important piece here is that you are able to get the correct answer!

There is one more item that is important for us to talk about at this point. When asked to evaluate $-x^2$ when $x = -5$, you must see that there a subtle, but very important, difference between $-x^2$ and $(-x^2)$. The first item tells us to square x and then multiply by –1, while the second tells us to multiply x by –1 and then square the product. The value of $-x^2$ when $x = -5$ is –25.

Solving Two-Step linear Equations

You read earlier that the order of operations is key to solving equations. Solving an equation such as $12x = 84$ goes back to elementary school. "12 times what number is 84?" Divide 84 by 12 to get the answer 7. However, solving an equation such as $12x + 19 = 103$ is a bit different. "When 19 is added to the product of 12 and some number, the answer is 103" leads to the thought process: "I multiplied 12 and the unknown number and then added 19, getting the result 103. I need to work the order of operations from lowest priority back to highest priority. Addition is lower in the priority system than multiplication, so I'll remove the 19 by subtracting. This gives me the new problem that 12 times some number is 84, and I just saw that the answer to this question is 7. The unknown number is 7."

The mathematical work corresponding to this dialogue is

$$12x + 19 = 103$$

Horizontal Method

$$12x + 19 - 19 = 103 - 19$$
$$12x = 84$$
$$\frac{12x}{12} = \frac{84}{12}$$

Vertical Method

$$\begin{aligned} 12x + 19 &= 103 \\ -19 \quad &\; -19 \\ 12x &= 84 \\ \frac{12x}{12} &= \frac{84}{12} \\ x &= 7 \end{aligned}$$

Both procedures require the same work. It is a matter of preference as to which method you choose. I will use both methods depending on the amount of writing that is needed. Eventually, we will not show all the steps (such as writing –19 on both sides of the equation) but will provide a dialogue of what is being done to solve the equation.

EXAMPLE

▶ Solve: $45n - 129 = 1086$

Solution: We begin by removing the 129. Since it was subtracted from the product of 45 and the unknown number, we'll add it to both sides of the equation. We'll then divide the new result by 45.

Add 129: $45n = 1215$

Divide by 45: $\frac{45n}{45} = \frac{1215}{45}$

$n = 27$

This next question is similar to the previous. The big difference is that the coefficient of x is a fraction. Wait! Who cares what set of numbers the coefficient comes from? The process does not change!

EXAMPLE

Solve: $\frac{x}{4}+13=27$

Solution:

Subtract 13: $\frac{x}{4}=14$

Since x was divided by 4, multiply by 4: $4\left(\frac{x}{4}\right)=4(14)$

Compute: $x=56$

This problem shows, initially, a different look than the previous two equations.

EXAMPLE

Solve: $123-5x=138$

Although this looks different with the $5x$ following the subtraction, it really is exactly the same type of problem.

Subtract 123 from both sides of the equation: $-5x=15$

Divide by -5: $x=-3$

An alternative approach would be to rewrite the problem as $123+(-5x)=138$ or that $-5x=123=138$. Do you see how this equation is solved as the first two examples were solved?

Don't let the signs, size, or type of number throw you for a loop when solving equations. First remove whatever quantity is added or subtracted from the variable and then divide by the coefficient of the variable. Does the problem contain ugly arithmetic? Use your calculator.

EXERCISES

EXERCISE 1-1

Evaluate the expressions with x = −4, y = 7, and z = 12.

1. $5x^2 + 6y - 5z$
2. $(4x + 3y)(4z + 9x)$
3. $\dfrac{-3x^2 + 7y + 2z}{x(y - z)}$
4. $6xyz$
5. $3y(2x - 3z)^2$

EXERCISE 1-2

Evaluate the expressions with $p = \frac{2}{3}$, $r = \frac{-3}{4}$, ***and*** $q = \frac{5}{6}$.

1. $18p - 12q$
2. $72p + 64r$
3. pqr
4. $12(p + r) - 72q^2$
5. $\dfrac{5p + 3r}{4p - q}$

EXERCISE 1-3

Simplify each of the following terms.

1. $3y + 4z - 5w + 9w - 5y + 2z$
2. $12ab - 9bc + 8ab - 6bc$
3. $9p^2 - 8p + 4 + 10p^2 - 11p - 5$
4. $(9p^2 - 8p + 4) - (10p^2 - 11p - 5)$
5. $(3x^3 + 5x^2 - 4x + 6) + (4x^3 - 7x^2 - 8x + 9)$
6. $(3x^3 + 5x^2 - 4x + 6) - (4x^3 - 7x^2 - 8x + 9)$
7. $4(3x^3 + 5x^2 - 4x + 6) + 5(4x^3 - 7x^2 - 8x + 9)$
8. $5(3x^3 + 5x^2 - 4x + 6) - 2(4x^3 - 7x^2 - 8x + 9)$
9. $12ab^2c - 9abc^2 + 8ab^2c - 6a^2bc$
10. $3xyz + 4xyz - 5wxy + 9wxy - 5wxyz + 2wxyz$

EXERCISE 1-4

Solve each of the following equations.

1. $4x - 3 = 21$
2. $4x + 3 = 21$
3. $82 - 7y = 54$
4. $0.4z + 2.9 = 8.6$
5. $\frac{2}{3}w + 19 = 27$

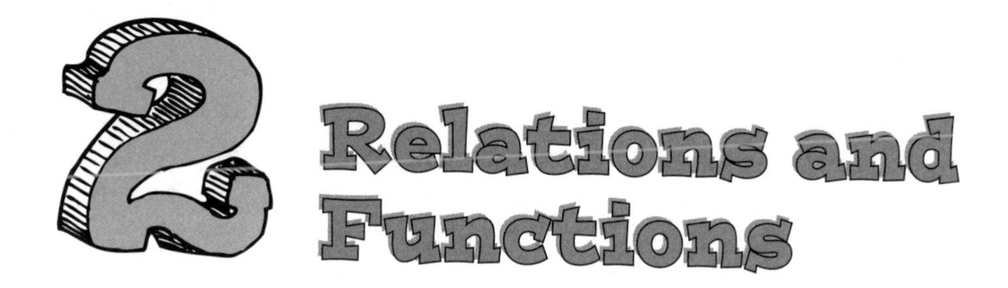

2 Relations and Functions

MUST KNOW

- The set of any ordered pairs is a relation. An important subset of relations is the set of functions. Knowing how to tell them apart is important when solving equations and examining applications.

- Functions are relations that guarantee for each legitimate input value there will be exactly one output value.

picture is worth a thousand words. While many think that algebra is the branch of mathematics in which letters are used to resemble numbers, visual and verbal descriptions are equally important.

Cartesian Coordinate Plane

There is a legend that says when René Descartes (1596–1650) was in the army, he had permission from the doctor to lie in bed later than the other soldiers because of poor health. While in bed, he observed a bug on the ceiling (some legends say it was a fly, others a spider). He attempted to describe the location of the bug by drawing to lines that were perpendicular to each other through the point that was in the center of the ceiling. He then divided each line with equally spaced markings. The process required that the horizontal position be determined by moving left or right and then the vertical position by moving up or down to get to the bug's location. Many of us know this process as the coordinate system, but its full name is the Cartesian coordinate system in honor of the man who created it. (The Cartesian coordinate system can even be extended into three dimensions to identify the location of an object in space.)

The center point of the coordinate system is called the **origin**, while the lines are called the **axes**. Note that although in the vast majority of applications of the coordinate system, these lines are referred to as the ***x*-axis** and ***y*-axis**, they are generically known as the horizontal and vertical axes. We'll use the common convention of calling them x and y axes in our general discussion and will identify them differently when the application involved requires us to do so. These axes divide the coordinate plane into four regions, each called a **quadrant**. The quadrants are numbered sequentially counterclockwise as shown in the diagram.

BTW

Descartes is as famous for his work in philosophy as mathematics. Known in particular for his famous proposition "Cogito, ergo sum," or "I think, therefore I am," he is considered a pioneer of rationalism and modern thought.

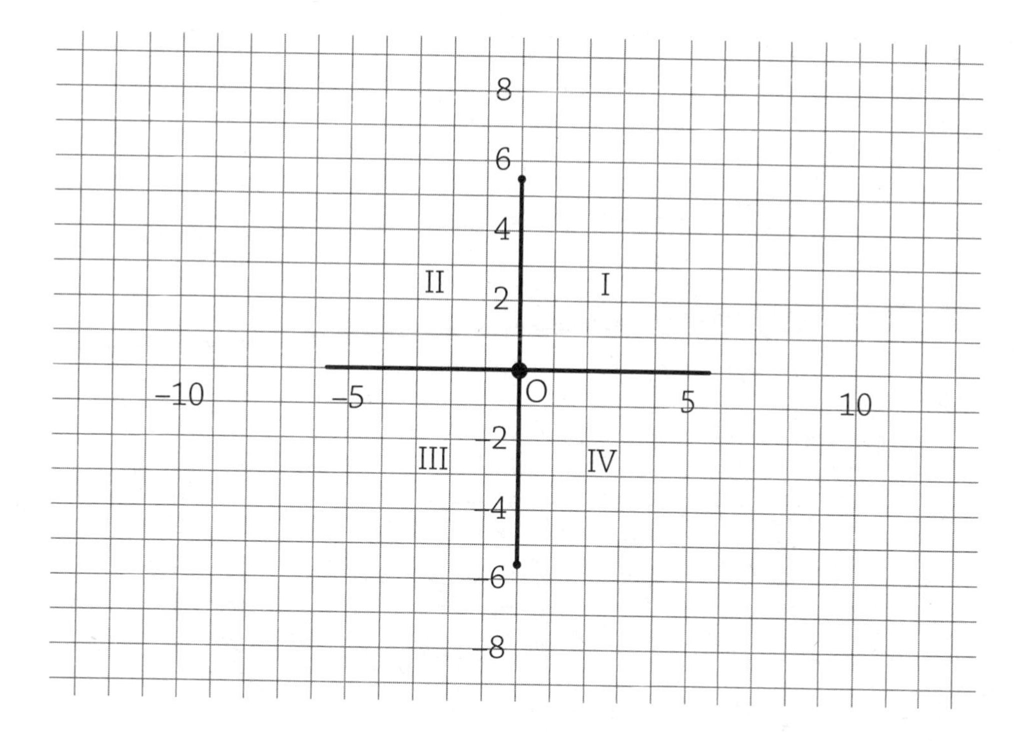

Point O is the origin and is given the coordinates (0, 0). The points to the right of the origin are assigned a positive first coordinate, and the values to the left of the origin are assigned a negative value. The coordinates above the origin are assigned a positive value and the points below a negative value. The first value in an ordered pair is called the **abscissa**, and the second value in the ordered pair is called the **ordinate**.

EXAMPLE

▶ Give the ordered pairs for each of the points shown in the diagram.

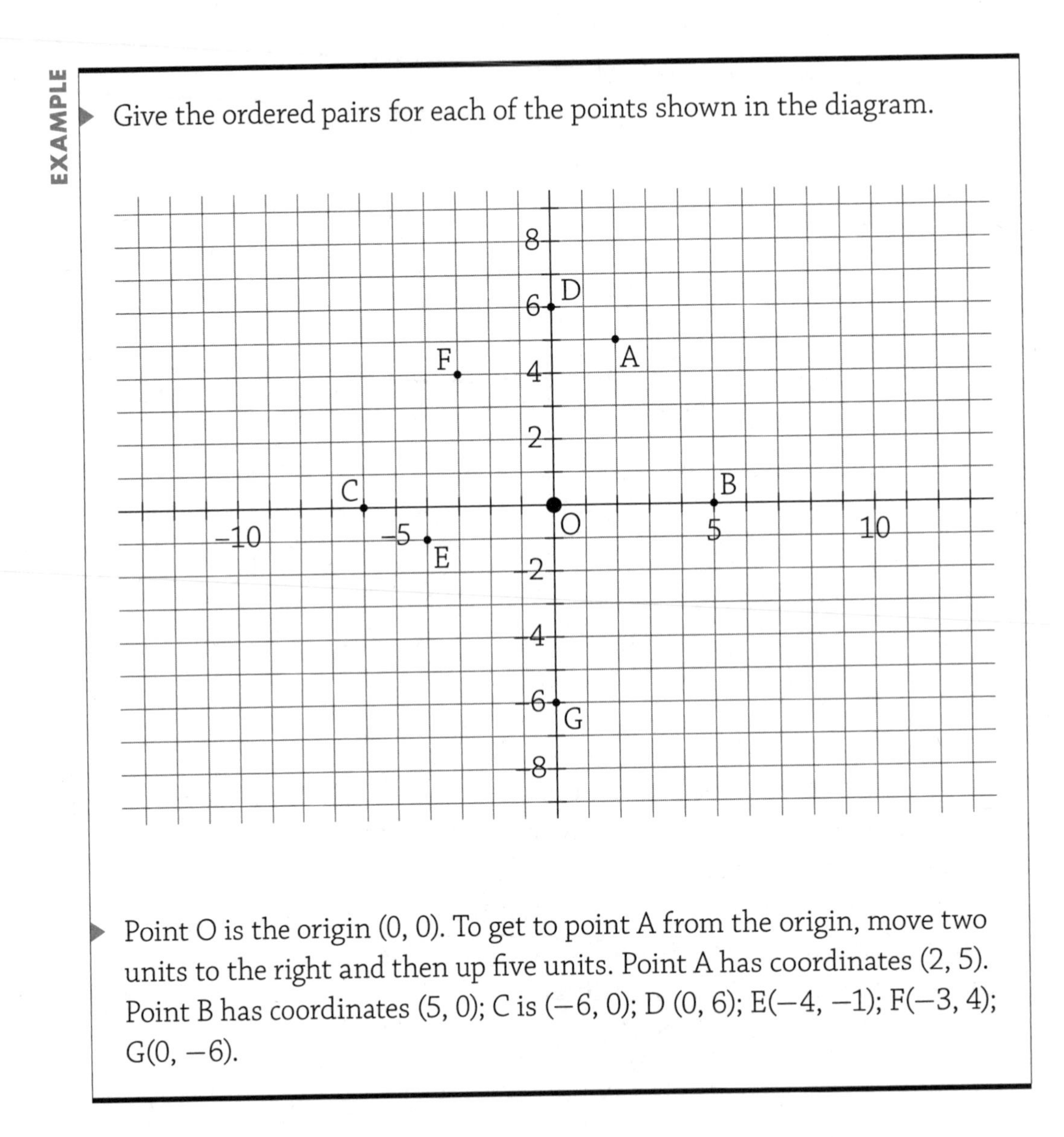

▶ Point O is the origin (0, 0). To get to point A from the origin, move two units to the right and then up five units. Point A has coordinates (2, 5). Point B has coordinates (5, 0); C is (−6, 0); D (0, 6); E(−4, −1); F(−3, 4); G(0, −6).

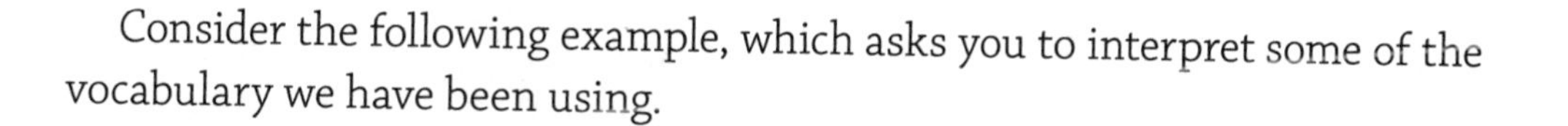

Consider the following example, which asks you to interpret some of the vocabulary we have been using.

EXAMPLE

▶ Determine the coordinates of five points for which the ordinate is 3 less than twice the abscissa.

▶ We're being asked to pick an abscissa, double it, and then subtract 3. The directions do not tell which values to pick for the abscissa. Let's choose −3, 2, 0, 5, and 7. When the abscissa is −3, the ordinate is

$$2(-3) - 3 = -6 - 3 = -9$$

▶ Our first ordered pair is (−3, −9). The other ordered pairs are

abscissa		ordinate		ordered pair
2		$2(2) - 3 = 4 - 3 = 1$		(2, 1)
0	➡	$2(0) - 3 = 0 - 3 = -3$	➡	(0, −3)
5		$2(5) - 3 = 10 - 3 = 7$		(5, 7)
7		$2(7) - 3 = 14 - 3 = 11$		(7, 11)

▶ Grab a pencil and let's plot these points on a coordinate axis, which will give us a sense of what this looks like.

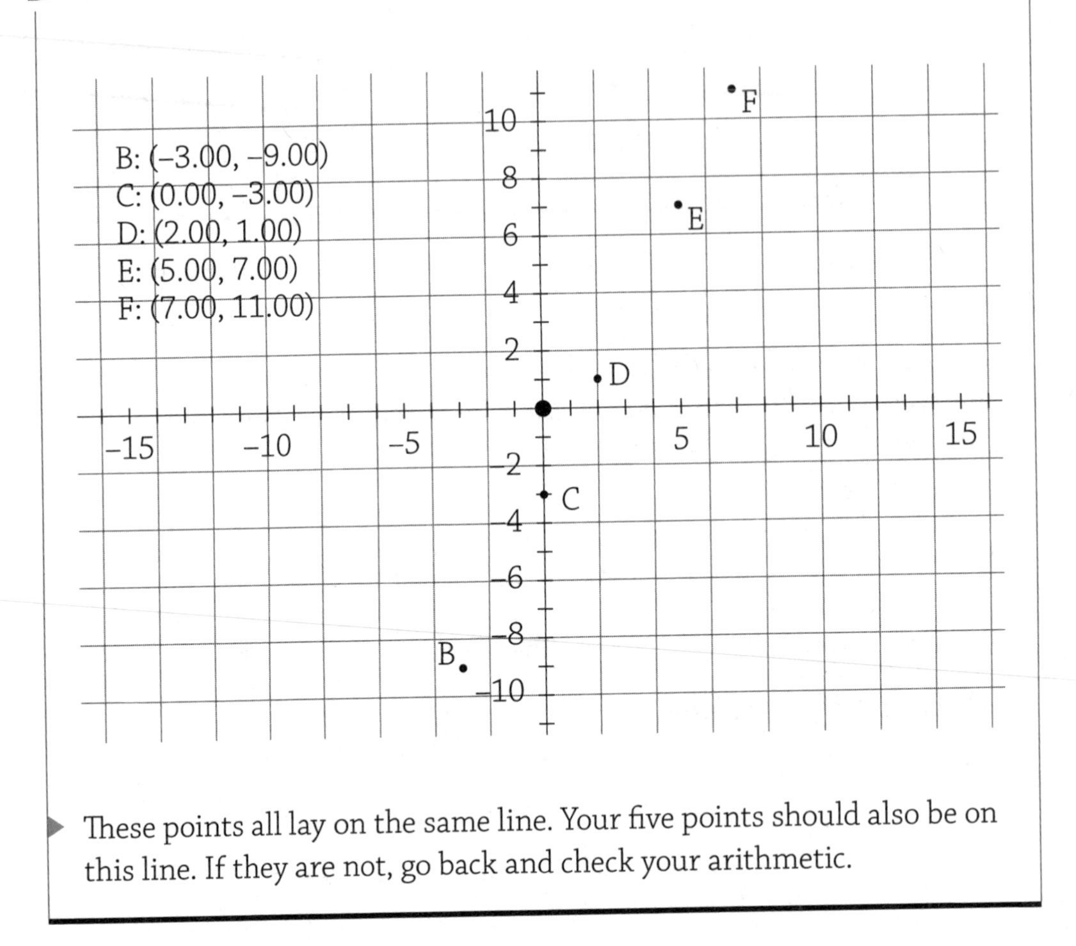

▶ These points all lay on the same line. Your five points should also be on this line. If they are not, go back and check your arithmetic.

We can memorize rules and definitions. That is important. More important than the memorization is the application of that knowledge. Reading graphs is a skill that puts knowledge to use. In the next sections, we will use graphical representations to classify connections between variables.

Relations

The most important aspect of mathematics is its ability to identify the connections among different characteristics. The work you have done in math so far might not seem like you were making connections, nor will some of the work you will do in this course. You will have to trust that they will come. Here is a simple example: You learned how to multiply 98 × 8. What is the price of the 8 markers you purchased at $0.98 each? The arithmetic now has a context.

One of the first connections we make in mathematics without concerning ourselves with computation is that of a relation. By definition, a **relation** is any set of ordered pairs. An example of a relation is (the members of your mathematics class, the email accounts at which they can be reached). The set of all the first elements in the ordered pairs is called the **domain**, and the set of all the second elements is called the **range**. Relations are usually identified by a capital letter. We'll choose E, since this is a relation about emails. A partial listing of what this relation might look like is

E = (Mike A., mikea@aol.com), (Mike A., mikea@gmail.com), (Janet B., janet1@yahoo.com), (Charlie K., thatsus@hotmail.com), (Nancy K, thatsus@hotmail.com)

Let's look at this list. Mike A. has two email accounts, while Charlie K. and Nancy K. share an email account. The domain of E is {Mike A., Janet B., Charlie K., Nancy K.}. Mike A. appears twice in the relation E but only needs to be listed once in the statement of the domain. The range of E is {mikea@aol.com, mikea@gmail.com, janet1@yahoo.com, thatsus@hotmail.com}. The e-mail address thatsus@hotmail.com is used twice in the relation but only needs to be noted once in the statement of the range.

Here is an example for domain and range that involve a more traditional mathematical approach.

EXAMPLE

Determine the domain and range of the relation:

$$A = \{(2, 3), (4, 6), (2, -1), (5, 8), (0, -2), (9,7)\}$$

The domain is {2, 4, 5, 0, 9}, while the range is {3, 6, –1, 8, –2, 7}.

Although this next problem might look different from the last problem, it really is exactly the same.

EXAMPLE

Determine the domain and range of the relation represented by the graph:

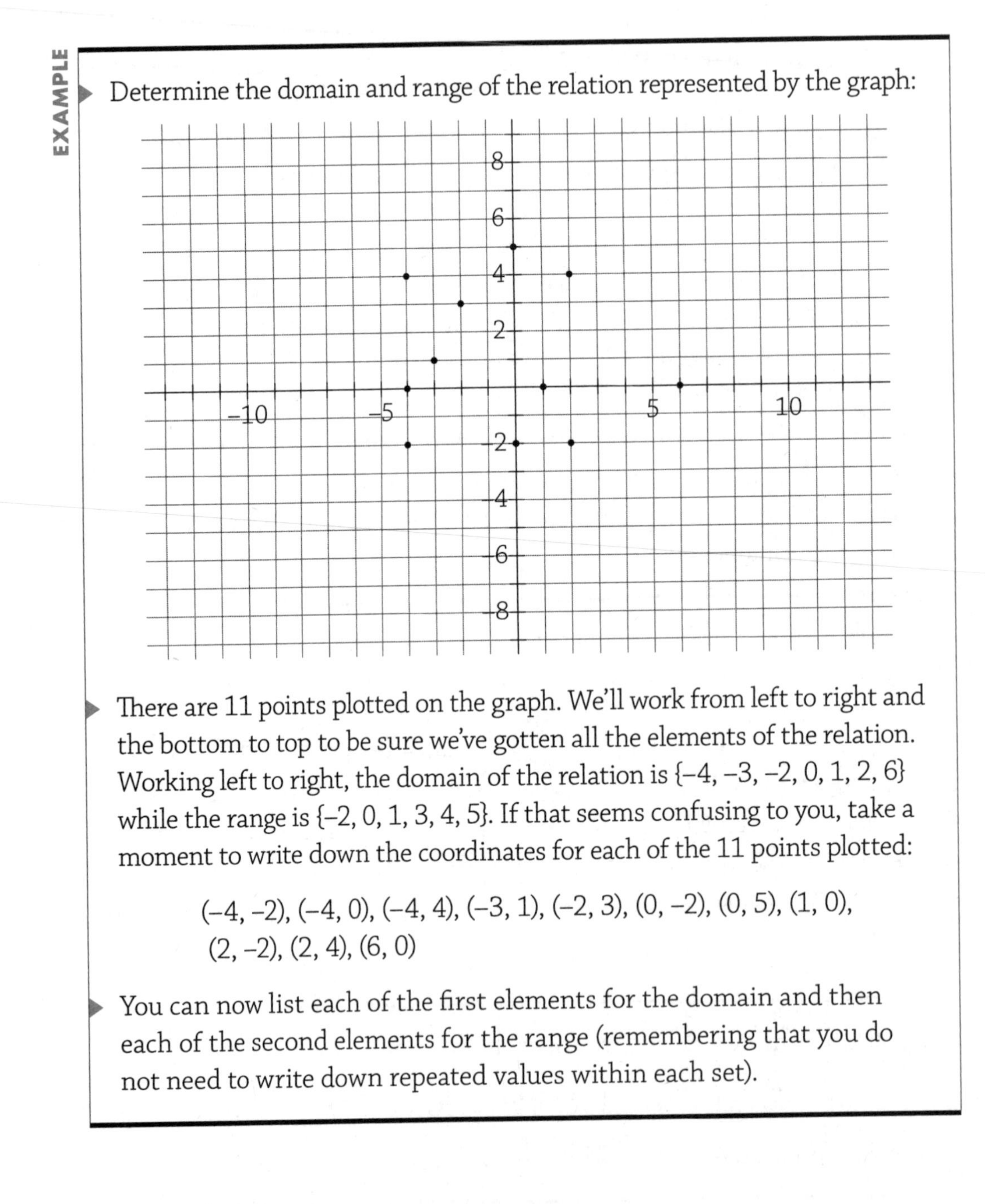

There are 11 points plotted on the graph. We'll work from left to right and the bottom to top to be sure we've gotten all the elements of the relation. Working left to right, the domain of the relation is {−4, −3, −2, 0, 1, 2, 6} while the range is {−2, 0, 1, 3, 4, 5}. If that seems confusing to you, take a moment to write down the coordinates for each of the 11 points plotted:

(−4, −2), (−4, 0), (−4, 4), (−3, 1), (−2, 3), (0, −2), (0, 5), (1, 0), (2, −2), (2, 4), (6, 0)

You can now list each of the first elements for the domain and then each of the second elements for the range (remembering that you do not need to write down repeated values within each set).

Functions

Functions are one of the most important concepts in all of mathematics. To understand this, let's first look at exactly what a function is.. A **function** is a

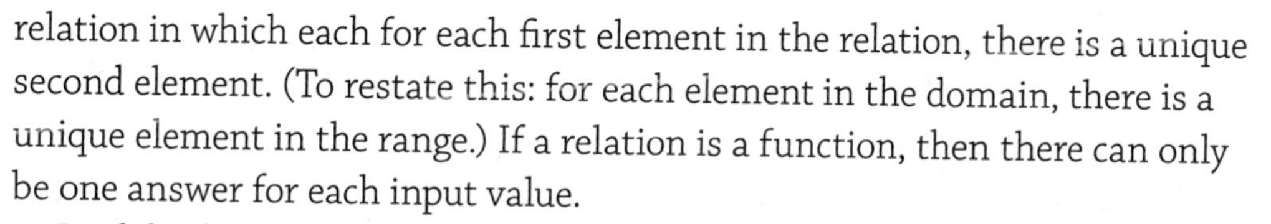

relation in which each for each first element in the relation, there is a unique second element. (To restate this: for each element in the domain, there is a unique element in the range.) If a relation is a function, then there can only be one answer for each input value.

Look back in the last section to the example of the relation concerning emails. If you want to be sure that Mike A. gets your email, you need to realize that there are multiple accounts and you cannot just send the message to one account. However, if you want to send a message to Nancy K., you know there is only one account that she uses.

Verbal descriptions of examples identifying functions require us to give thought to all the possibilities of what is included. A graphical representation covers all possibilities.

EXAMPLE

▶ Is the relation represented by the graph below a function?

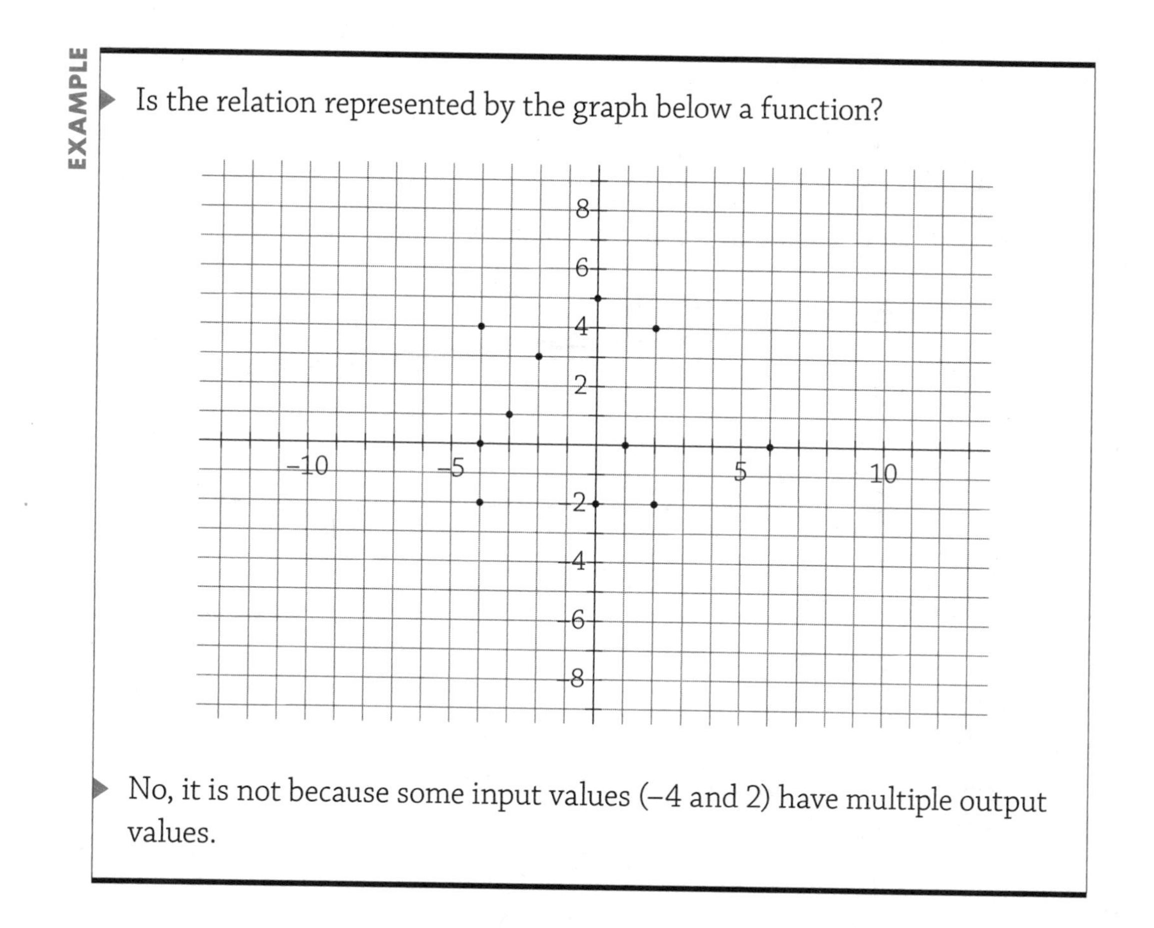

▶ No, it is not because some input values (−4 and 2) have multiple output values.

BTW

A very useful guideline for determining functions is the **vertical line test**. If a vertical passes through multiple points anywhere on a graph, then the relation determined by those points is not a function.

Let's apply the vertical line test to a graph. Does this graphical representation represent a function?

The graph does not satisfy the vertical line test, so it does not represent a function.

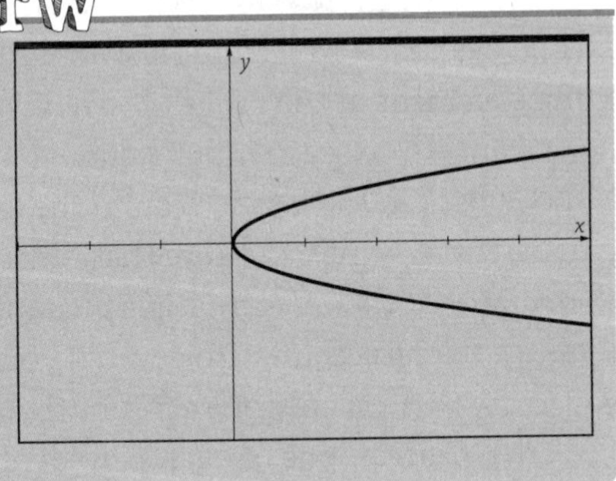

Functions can be defined in any number of different manners. They can be given by equation, by graph, or by verbal description.

EXAMPLE

▶ Does the following represent a relation that is a function?

Domain: all the students in your school

Range: the student's social security number

▶ It does represent a function because each person has exactly one social security number.

Function Notation

If asked to determine the value of y when $x = 4$ in the equation $y = 3x + 7$, you would substitute 4 for x and get $y = 3(4) + 7 = 19$. You have been doing that for a while now.

There is a special notation for functions that work in the same way. The $f(x)$ notation (read as "f of x") is a substitution-based process. If $f(x) = 3x + 7$, then $f(4) = 3(4) + 7 = 19$ and is associated with the ordered pair (4, 19). **To say it another way, $f(4)$ tells us to find the y-value for $3x + 7$ when $x = 4$.** What is particularly useful about the function notation is that each function can be given its own name so that there will be no ambiguity as to which function is being referenced. Naming the functions $f(x) = 3(4) + 7$, $g(x) = 7x - 3$, and $h(x) = 7x + 3$ allows me to tell you that $f(4) = 19$, $g(4) = 25$, and $h(4) = 31$.

IRL

Many disciplines use functions and they do call them $f(x)$. Economists might use $R(p)$ to represent the revenue earned when the price of an item sold is \$$p$. A chemist might use the notation $V(t)$ to represent the volume of a gas when the temperature of the gas is t degrees. It is commonplace to use $f(x)$ when first learning about functions because f is the first letter of function and x is frequently used to represent the input variable.

EXAMPLE

Given that $f(x) = 2x^2 + 7x - 5$, determine the value of $f(3) - f(2)$.

$$f(3) = 2(3)^2 + 7(3) - 5 = 2(9) + 21 - 5 = 18 + 16 = 34$$

$$f(-2) = 2(-2)^2 + 7(-2) - 5 = 2(4) - 14 - 5 = 8 - 19 = -11$$

↓

$$f(3) - f(-2) = 34 - (-11) = 45$$

We aren't always given the rule by which the function is defined. There will be times when the function will be defined by a table of values or by a graph.

EXAMPLE

▶ Given the function $y = f(x)$ as defined by the accompanying graph. Determine the value of $f(2)$.

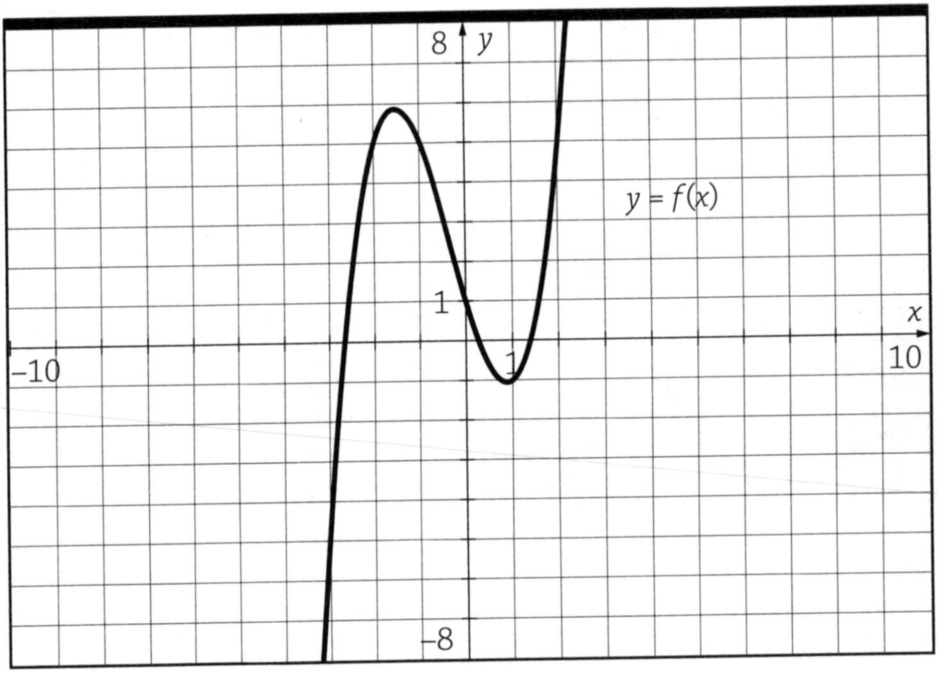

8 y
$y = f(x)$
1
x
−10
1
10
−8

▶ When $x = 2$, the graph has a y value of 5. Therefore, $f(2) = 5$.

EXERCISES

EXERCISE 2-1

Identify the coordinates for each of the labeled points on the accompanying coordinate system.

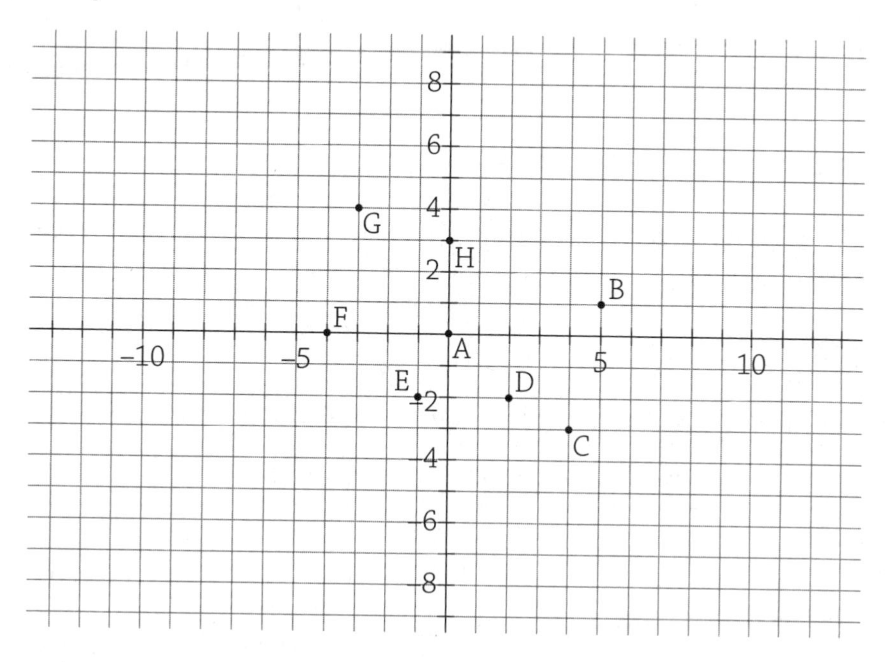

EXERCISE 2-2

Let the relation defined in Exercises 2-1 be labeled as G.

1. Determine the domain of G.
2. Determine the range of G.

EXERCISE 2-3

Let P be the relationship defined by (Host City of the Winter Olympics in the 21st Century, Country with the Most Gold Medals at those Olympic Games).

3. Give a list of all the elements of *P*.
4. Determine the domain of *P*.
5. Determine the range of *P*.
6. Explain why *P* represents a function.
7. Let Q be the relation defined by (Members of your math class, Telephone Numbers where they can be reached). Is Q a function? Explain your answer.

EXERCISE 2-4

Let $f(x) = x^2 + 9$, $g(x) = \dfrac{x+4}{x-2}$, ***and*** $h(x) = 11 - 7x$. ***Evaluate each of the following. Use these definitions to answer questions 8 to 16.***

1. $f(3)$
2. $f(-3)$
3. $g(8)$
4. $g(-4)$
5. $h(2)$
6. $h(-3)$
7. $f(2) + g(3)$
8. $\dfrac{f(4) \times g(5)}{h(2)}$
9. Alex claims that $f(x)$ is not a function because $f(3) = f(-3)$. Is Alex correct? Explain your reasoning.

3 Linear Equations

MUST KNOW

- A linear equation is an equation between two variables that gives a straight line when plotted on a graph.
- Solving linear equations is the first step in learning how to solve all equations.
- The ability to translate sentences into equations is essential in mathematics, in particular algebra.

A linear equation is one in which the variable of the equation has an exponent of 1. In this chapter, you will learn how to solve equations whose coefficients (the number used to multiply the variable) are either constants or literal (letter) values.

Solving Multistep Linear Equations

You saw in Chapter 1 how to solve one-step equations. We can extend this process to solve multistep equations. Let's begin with a simple two-step example.

EXAMPLE

- Solve: $3x + 5 = 26$
- As you can see, two things happened to the variable. First, it was multiplied by 3, and then 5 was added to the product.
- The last operation performed on the variable expression was to add 5. Remove it.

$$\begin{array}{r} 3x + 5 = 26 \\ -5 \quad -5 \\ \hline \end{array} \quad \text{or} \quad \begin{array}{c} 3x + 5 - 5 = 26 - 5 \\ 3x = 21 \end{array}$$

- Divide by 3 to get $x = 7$.

BTW

We solve equations by implementing the order of operations in reverse order.

We can show the work vertically or horizontally. For the straightforward problems, the work will be shown horizontally. If the problem is more complex, the work will be shown vertically to make the explanation clearer.

EXAMPLE

- Solve: $23x + 19 - 15x + 8 = 77$
- The variable is multiplied by two different factors and then two constants are added to these products. Let's clean up the left side of the equation by combining common terms. We can apply the commutative property to rewrite the equation to read $23x - 15x + 19 + 8 = 77$. Combine the variable terms and the constants to get

$$8x + 27 = 77.$$

- As with the last example, subtract the value added:

$$8x + 27 - 27 = 77 - 27$$

- This leaves $8x = 50$.
- Divide by 8 to determine that $x = 6.25$ or $6\frac{1}{4}$.

There are times when you will have a choice as to what the first step in the solution should be.

EXAMPLE

- Solve: $15(3x - 7) = 390$
- You have two choices of how to solve this problem. You can first divide by 15 (keeping with the notion that the last operation performed in the first to be removed), or you can apply the distributive property and make this problem look like the first problem from this section. Which way should you do the problem? A good rule of thumb is if the division does not leave you with fractions, do that first:
- Divide by 15: $\frac{15(3x-7)}{15} = \frac{390}{15} \Rightarrow 3x - 7 = 26$
- Add 7: $3x - 7 + 7 = 26 + 7 \Rightarrow 3x = 33$
- Divide by 3: $x = 11$.

Now, here is a really ugly problem.

EXAMPLE

▶ Solve: $\frac{2}{3}\left(\frac{5}{8}x+\frac{8}{9}\right)=\frac{27}{10}$

▶ This is truly as ugly a problem as it can possibly get. What you need to do when you see this is take a deep breath and realize that you need to clean up the way it looks before you get to the business of solving the equation. Clean up? Getting rid of the fractions would go a long way to make this problem look easier to do. We can multiply both sides of the equation by $\frac{3}{2}$ to help get rid of the parentheses, or we can multiply by 360, the common denominator of all the fractions, to remove all of the fractions. Let's do it both ways, and then you can decide which way you prefer.

▶ Multiply by $\frac{3}{2}$:

$$\frac{3}{2}\left(\frac{2}{3}\right)\left(\frac{5}{8}x+\frac{8}{9}\right)=\left(\frac{27}{10}\right)\frac{3}{2}$$

$$\frac{5}{8}x+\frac{8}{9}=\frac{81}{20}$$

▶ Multiply by 360.

$$360\left(\frac{5}{8}x\right)+360\left(\frac{8}{9}\right)=\left(\frac{81}{20}\right)360$$

$$225x+320=1458$$

▶ Apply the distributive property first:

$$\frac{2}{3}\left(\frac{5}{8}x\right)+\frac{2}{3}\left(\frac{8}{9}\right)=\frac{27}{10}$$

▶ Perform the calculations:

$$\frac{5}{12}x+\frac{16}{27}=\frac{27}{10}$$

▶ Multiply both sides of the equation by 540, the least common denominator of 12, 27, and 10:

$$540\left(\frac{5}{12}x+\frac{16}{27}\right)=\left(\frac{27}{10}\right)540$$

$$225x+320=1458$$

- Each approach will leave us with the same equation to solve.
- Subtract 320: $225x + 320 - 320 = 1458 - 320 \Rightarrow 225x = 1138.$
- Divide by 225: $\frac{225x}{225} = \frac{1138}{225} \Rightarrow x = \frac{1138}{225}$

The key to solve the more ugly equations is not to get bothered by the style of the number (fractions, decimals, negative numbers). Concentrate on the process. You can use your calculator to do the arithmetic. Just be sure you are aware of how you are "undoing" the order of operations.

Solving Linear Equations with Variables on Each Side of the Equation

We have practiced solving equations that have all variable expressions on one side of the equation. We can now take on the case where there is a variable expression on each side of the equation. The guiding principle will be to gather common terms with the variable terms on one side of the equation and the constants on the other side.

EXAMPLE

- Solve: $9x - 17 = 5x + 19$
- The phrase "gather common terms on one side of the equation" tells you to move the $5x$ to the left and 17 to the right in this problem. (Why in that order? Because it keeps the coefficients positive.)

$$9x - 5x - 17 + 17 = 5x - 5x + 19 + 17$$
$$\Rightarrow 4x = 36$$

- Divide by 4 to get $x = 9$.

BTW

I haven't mentioned this yet but now seems an appropriate time. It is always a good idea to check your answer after solving an equation—especially when multiple steps are involved. You can check the problem $9x - 17 = 5x + 19$ mentally. $9(9) - 17 = 81 - 17 = 64$ and $5(9) + 19 = 45 + 19 = 64$. Both sides of the equation have the same value when $x = 9$.

In algebra, you will frequently encounter problems involving the distributive property, fractions and decimals, and multiple variables on each side of the equation. What to do?

- Gather common terms on each side of the equation.
- Eliminate fractions when you can but multiplying both sides of the equation by a common denominator.
- Variables on each side of the equation? Move similar terms to one side of the equation.

EXAMPLE

- Solve for t: $4t-3(18-2t)=5(3t+8)-7t$.
- Gather common terms on each side of the equation requires the distributive property.

$$4t-54+6t=15t+40-7t$$

$$10t-54=8t+40$$

- Move similar terms: $10t-8t-54+54=8t-8t+40+54 \Rightarrow 2t=94$
- Divide by 2 to get $t=47$.
- Check your work:

  Substitute 47 for t: $4(47)-3(18-2(47))=5(3(47)+8)-7(47)$
- All you need to do now is arithmetic. Do not move items from one side of the equation to the other.

$$188-3(18-94)=5(141+8)-329$$

$$188-3(-76)=5(149)-329$$

$$188+228=745-329$$

$$416=416$$

- We've verified that $t=47$.

Here are a couple of problems with fractions and decimals.

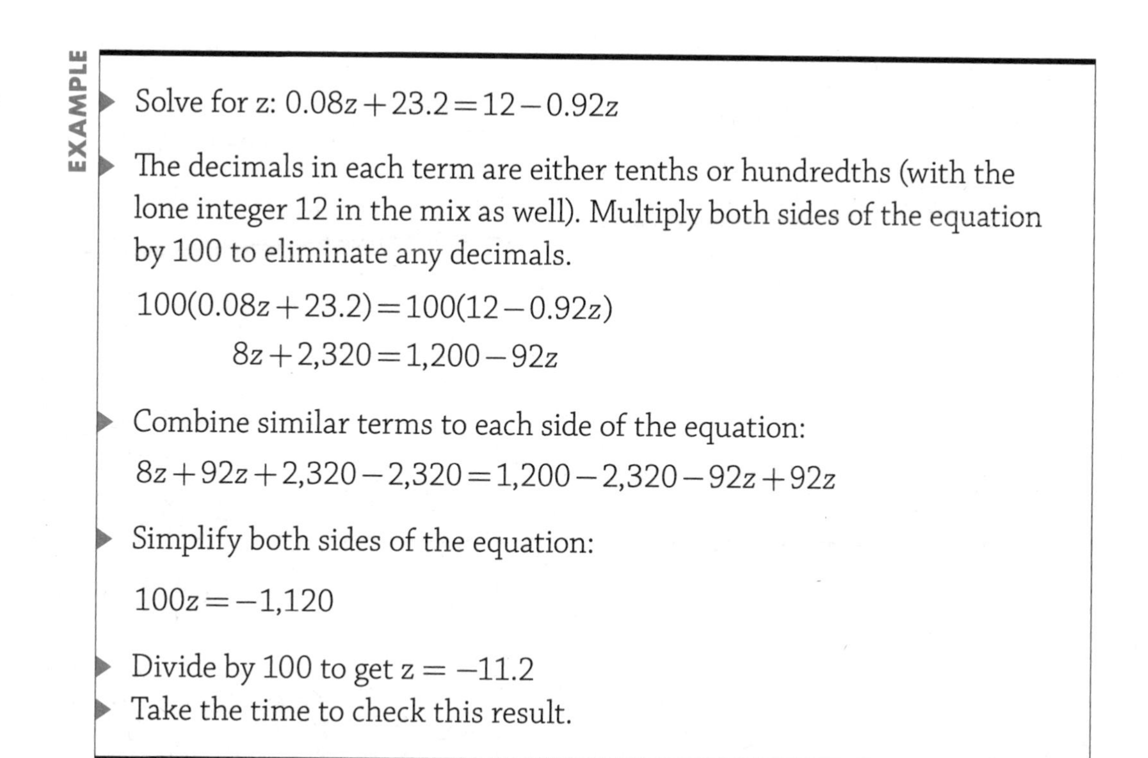

EXAMPLE

- Solve for z: $0.08z + 23.2 = 12 - 0.92z$
- The decimals in each term are either tenths or hundredths (with the lone integer 12 in the mix as well). Multiply both sides of the equation by 100 to eliminate any decimals.

$$100(0.08z + 23.2) = 100(12 - 0.92z)$$
$$8z + 2{,}320 = 1{,}200 - 92z$$

- Combine similar terms to each side of the equation:

$$8z + 92z + 2{,}320 - 2{,}320 = 1{,}200 - 2{,}320 - 92z + 92z$$

- Simplify both sides of the equation:

$$100z = -1{,}120$$

- Divide by 100 to get $z = -11.2$
- Take the time to check this result.

Now, let's take a look at a problem with fractions.

EXAMPLE

- Solve for v: $\frac{2}{5}(v-4) + \frac{2}{3}v = \frac{7}{15}v + \frac{3}{4}$
- The common denominator for 3, 4, 5, and 15 is 60. Multiply both sides of the equation by 60.

$$60\left(\frac{2}{5}(v-4) + \frac{2}{3}v\right) = 60\left(\frac{7}{15}v + \frac{3}{4}\right)$$

$$60\left(\frac{2}{5}(v-4)\right) + 60\left(\frac{2}{3}v\right) = 60\left(\frac{7}{15}v\right) + 60\left(\frac{3}{4}\right)$$

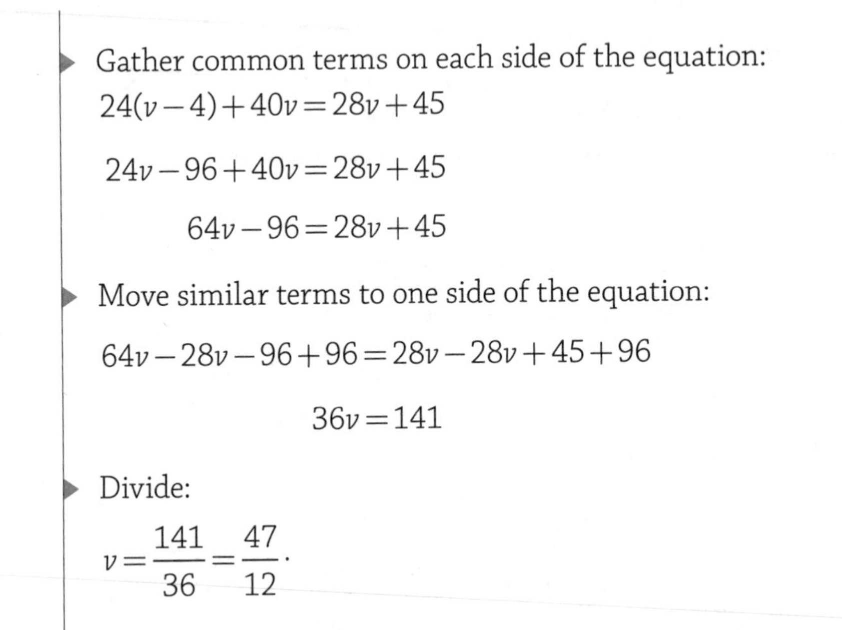

- Gather common terms on each side of the equation:

$$24(v-4)+40v=28v+45$$

$$24v-96+40v=28v+45$$

$$64v-96=28v+45$$

- Move similar terms to one side of the equation:

$$64v-28v-96+96=28v-28v+45+96$$

$$36v=141$$

- Divide:

$$v=\frac{141}{36}=\frac{47}{12}.$$

Literal Equations

You probably know that not all equations have numeric coefficients. But if you don't, consider this. What is the formula for the area of a rectangle? Area = length times width (or base times height, if you prefer). As an equation, this is $A = lw$ or $A = bh$.

EXAMPLE

- Solve the equation $A = lw$ for the width of the rectangle.
- All you need to do is to divide the area by the length to get $w=\frac{A}{l}$.

Equations such as $A = lw$ are called **literal equations** because the coefficients of whatever is determined to be the variable of choice are letters rather than numeric values. The way we deal with literal equations is exactly the same as we deal with all equations, we look at the order of operations.

EXAMPLE

- Solve for x: $ax + b = c$
- This is your basic two-step linear equation. First, subtract b from both sides of the equation to get $ax = c - b$. Second, divide by a to get $x = \frac{c-b}{a}$.

Let's look at a couple of formulas with which you may be familiar.

EXAMPLE

- The formula for the perimeter of a rectangle is $P = 2l + 2w$. Solve the equation for the length of the rectangle.
- First, subtract $2w$ to get $P - 2w = 2l$. We'll rewrite this as $2l = P - 2w$ to have the variable on the left side of the equation. (This is simply a personal preference and not a requirement; you will develop your own preferences.) Divide both sides of the equation by 2 to get $l = \frac{P-2w}{2}$.

The next problem is the same problem but is written after the distributive property is applied.

EXAMPLE

- The formula for the perimeter of a rectangle is $P = 2(l + w)$. Solve the equation for the length of the rectangle.
- You could multiply through by 2 and get the same answer as the last example. Let's deal with the formula as it is written.
- First, divide by 2: $\frac{P}{2} = l + w$.

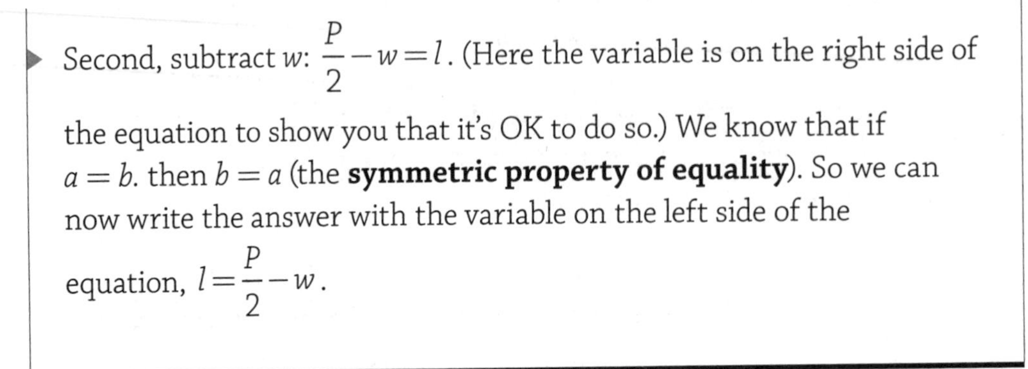

- Second, subtract w: $\frac{P}{2} - w = l$. (Here the variable is on the right side of the equation to show you that it's OK to do so.) We know that if $a = b$. then $b = a$ (the **symmetric property of equality**). So we can now write the answer with the variable on the left side of the equation, $l = \frac{P}{2} - w$.

The formula for simple interest is $I = Prt$, where I is the amount of interest, P is the principal (the amount of money borrowed), r is the rate of interest, and t is the amount of time.

EXAMPLE

- Solve the interest equation for t.
- Divide both sides of the equation by the product of P and r: $t = \frac{I}{Pr}$.

Let's take a look at an equation that is not tied to any particular formula but does look like equations that you have solved.

EXAMPLE

- Solve the equation $ax + b = cx + d$ for x.
- The rule is to get all terms with the variable on one side of the equation and all terms without the variable on the other side. Consequently, we'll need to subtract b and cx from both sides of the equation:
- Subtract: $ax - cx + b - b = cx - cx + d - b$
- Simplify: $ax - cx = d - b$
- Apply the distributive property: $(a - c)x = d - b$
- Divide by $a - c$: $x = \frac{d-b}{a-c}$

Let's extend our sampling to include variables on both sides of the equation with the distributive property included.

EXAMPLE

- Solve for v: $pv + q - rv = z(v + r) + w$

That looks tricky enough. We'll need to apply the distributive property to the right hand side of the equation:

$pv + q - rv = zv + rz + w$

- Bring all terms involving v to one side: $pv - rv - zv = rz + w - q$
- (Did you notice not all the subtraction steps are included? If you prefer, you can write down each the subtraction steps to see the process in full and verify it was done correctly.)
- Apply the distributive property to all terms with v:

$(p - r - z)\, v = rz + w - q$

- Divide by the coefficient of v: $v = \dfrac{rz + w - q}{p - r - z}$.

Writing Equations

When you are studying mathematics, translating written statement into mathematical equations and equations into written statements are important skills. A key piece to the translation is clearly defining the variables involved. When discussing money, for instance, one must be sure to define whether the variable expression represents the amount of money in a given denomination or the number of pieces of currency with a given value. For example, if n represents the number of nickels in a piggy bank and d represents the number of dimes, the equation $n + d = 60$ tells us that there are a total of 60 dimes and nickels in the piggy bank, whereas the equation $5n + 10d = 420$ indicates that the value of all the nickels and dimes is $4.20.

In the last chapter, we saw "Find 5 points whose ordinates are 3 less than twice the abscissa." Given that the abscissa is usually represented by the variable x and the ordinate is represented by the variable y, we can write a more general relation between the variables to be $y = 2x - 3$.

EXAMPLE

▶ Translate the statement "the number of quarters in the piggy bank is equal to the sum of the number of nickels and the number of dimes in the piggy bank."

▶ Letting q represent the number of quarters, n represent the number of nickels, and d represent the number of dimes, the statement translates to $q = n + d$.

You need to carefully read the directions when translating statements into equations.

EXAMPLE

▶ There are a total of 60 nickels and dimes in a piggy bank. Represent the number of dimes in the piggy bank in terms of the number of nickels.

▶ If we let d represent the number of dimes, then the directions "Represent the number of dimes" tells us our equation will begin $d =$. With n of the coins being nickels, the rest need to be dimes so the equation for this statement is $d = 60 - n$.

Let's take this problem a step further.

EXAMPLE

▶ A piggy bank contains a total of 60 nickels and dimes. Represent the total value of the coins in the bank in terms of the number of nickels in the piggy bank.

▶ We just determined that the number of dimes, d, in the piggy bank can represented by $d = 60 - n$. Each nickel is worth 5 cents (which is easier to work with than \$0.05), and each dime is worth 10 cents. To get the total value, v, of the money in the piggy bank, you need to multiply the value of each coin with the number of such coins in the piggy bank. Therefore, the value of the nickels will be $5n$, and the value of the dimes will be $10d$, or $10(60 - n)$ because the request was to represent the value in terms of the number of nickels. Finally, we can write that $v = 5n + 10(60 - n)$.

IRL

Most people do not like to be told to break things down to such simple steps. However, the more complicated the problem gets, the more important it is to be able to have this skill. In my experience, this process has proven most useful when writing formulas for spreadsheets. Labeling data (defining your variable) is a critical piece of getting useful results from the spreadsheet.

You are probably familiar with the following type of problem.

EXAMPLE

▶ Whippet Tours has buses with a capacity of 75 passengers. They offer a tour of the Blue Ridge Mountains at a price of \$150 per person. A nearby travel agent has worked out a deal with the tour company. The travel agent guarantees a minimum of 60 travelers and for each person in excess of 60, the tour company will reduce the price by \$3 for each person taking the tour. Write an equation for the total revenue, R, the tour company will earn in terms of the number of people in excess of 60 who take the trip.

- The revenue is the amount of money the company collects from all the passengers, so we will need to multiply the number of people on the tour with the price paid by each person who takes the tour. You might find it helpful to pick variable names that remind you of what the variable represents. The problem reads "for each person in excess of 60" so it's more helpful to choose p to represent the number of extra people who sign up for the tour than a generic variable like x.

- So, how many people are going on the trip? The travel agent guaranteed 60 people, so that is the minimum. There could be an additional people, p, taking the trip, so the number of people taking the trip is $60 + p$. What is the cost per person for this trip? If 60 or fewer people take the trip, the price is \$150 per person. Every time an extra person signs up for the trip, the price goes down by \$3. If one extra person goes, the price drops by \$3, and if 2 extra people go, the price drops by 2(\$3). If p additional people go, the price drops by \$3p. Remember, the price drops from a beginning value of \$150, so if an additional p people sign up for the tour, the price will be \$150 − 3p. Therefore, the revenue is given by the equation $R = (60 + p)(150 - 3p)$.

BTW

This is a good time to talk about sets of numbers. In a problem such as this, the value of p must be a whole number. You cannot have $\frac{2}{3}$ of a person attending, nor can you have π people signing up for this trip. When solving applications, you need to remember to take into account the types of numbers that are acceptable as answers.

We need to ask one more important question about the Whippet Tour to the Blue Ridge Mountains.

EXAMPLE

- What is the domain of the revenue equation?
- We've already discussed that fractional and irrational numbers are not acceptable values. Can there be negative numbers? No, there cannot be a negative amount of people. What does a negative person look like? (Not their attitude, but their physical being.) Is zero an acceptable value for the number of additional people who sign for the tour? Yes. Can 10 extra people sign for the tour? Yes. Can 20? No. The bus has a seating capacity of 75 people, so the largest number of extra people who can join the tour is 15. Therefore, the domain for the revenue statement is {0, 1, 2, 3, 4, ..., 15}.

Ratios and Proportions

This section examines ratios and proportions and some applications. A **ratio** is a comparison of two integers. For example, if a jar contains 7 red marbles and 4 blue marbles, the ratio of red marbles to blue marbles is 7 to 4. There are two different ways this is represented. Sometimes colons are used: the ratio of red marbles to blue marbles is 7:4. Other times, fractions are used. The ratio of red marbles to blue marbles is $\frac{7}{4}$.

When two ratios are set equal to each other, the result is called a **proportion**. Proportions are used to solve for quantities that exist in the same ratio. For example, if 5 oranges cost $5.25, what will 7 oranges cost? You may have already determined that if 5 oranges cost $5.25, then 1 orange costs $1.05 so 7 oranges will cost $7.35. Awesome!

BTW

*Numbers are called **rational** if they can be written as the ratio of two integers. The numbers $\frac{2}{3}$, 5, and $-\frac{3}{5}$ are rational. The number $\frac{3}{2}/\frac{5}{6}$ is rational because when you evaluate $\frac{3}{2} \div \frac{5}{6} = \frac{10}{9}$, the result meets the definition of a rational number.*

Now let's expand the solution out a little bit more so you can use it as a model to solve all proportions.

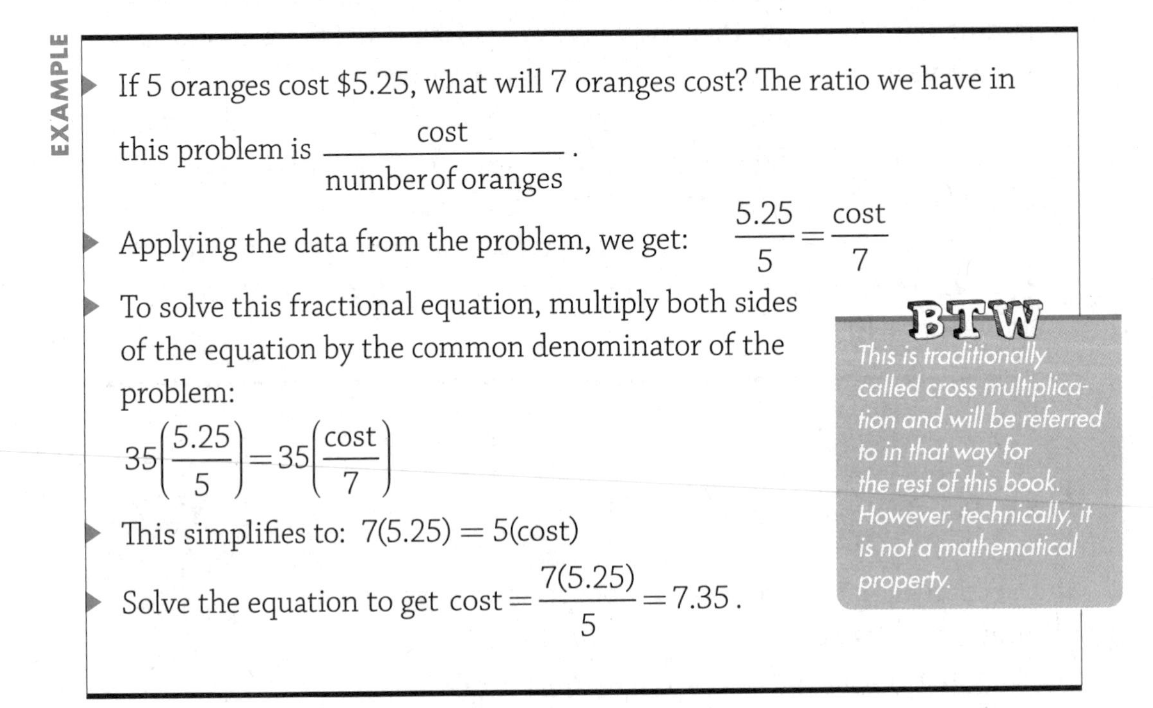

EXAMPLE

- If 5 oranges cost \$5.25, what will 7 oranges cost? The ratio we have in this problem is $\frac{\text{cost}}{\text{number of oranges}}$.
- Applying the data from the problem, we get: $\frac{5.25}{5} = \frac{\text{cost}}{7}$
- To solve this fractional equation, multiply both sides of the equation by the common denominator of the problem:

  $35\left(\frac{5.25}{5}\right) = 35\left(\frac{\text{cost}}{7}\right)$
- This simplifies to: $7(5.25) = 5(\text{cost})$
- Solve the equation to get $\text{cost} = \frac{7(5.25)}{5} = 7.35$.

BTW

This is traditionally called cross multiplication and will be referred to in that way for the rest of this book. However, technically, it is not a mathematical property.

The next problem is a little more challenging.

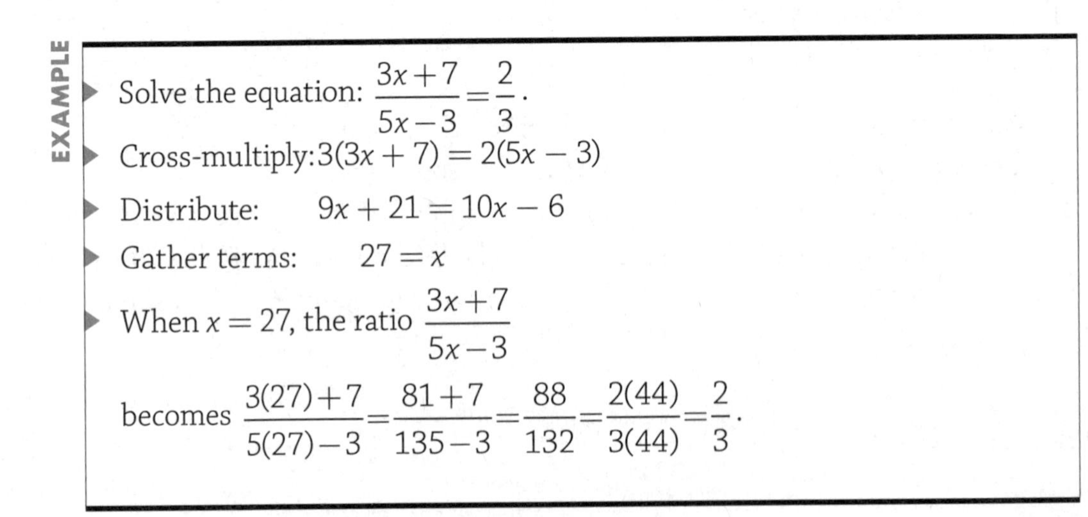

EXAMPLE

- Solve the equation: $\frac{3x+7}{5x-3} = \frac{2}{3}$.
- Cross-multiply: $3(3x + 7) = 2(5x - 3)$
- Distribute: $9x + 21 = 10x - 6$
- Gather terms: $27 = x$
- When $x = 27$, the ratio $\frac{3x+7}{5x-3}$

  becomes $\frac{3(27)+7}{5(27)-3} = \frac{81+7}{135-3} = \frac{88}{132} = \frac{2(44)}{3(44)} = \frac{2}{3}$.

The most used application of proportions is percent. The words "per cent," as you know, mean out of 100. When solving percent problems, you learned quickly to write the denominator of one fraction as 100 and the numerator of that fraction as the indicated percentage.

You might also have learned that all percentage problems can be fit into the sentence "x% of base is part," and this taught you that the proportion is always $\frac{x}{100} = \frac{\text{part}}{\text{base}}$.

EXAMPLE

What number is 25% of 84?

"Of 84" indicates that the base is 84 so the proportion is $\frac{25}{100} = \frac{p}{84}$. When you solve the equation, your answer, of course, is 21.

The proportion used to solve percent questions can be modified by multiplying both sides of the equation by the base. That is, $\frac{x}{100} = \frac{\text{part}}{\text{base}}$ becomes $\frac{x}{100} \times \text{base} = \text{part}$. The advantage to this equation is that it allows us to use decimals more naturally in solving percent problems.

EXAMPLE

"84 is 25% of what number?" becomes the equation: $0.25(\text{base}) = 84$ so $\text{base} = 84 \div 0.25 = 336$.

You will see that this second form of the equation is more practical to use.

Percent of Change

Computing the percentage of the money saved at a sale, the sales tax rate for a particular county, or the amount of a raise earned are just three examples of percent of change.

EXAMPLE

- Determine the percent raise given to an employee whose hourly wage went from \$8.25 to \$8.50.
- The change in the hourly wage is \$0.25 per hour, so the question being asked is "0.25 is what percent of 8.25?"
- Using r for the rate of percent, the equation is $r \times 8.25 = 0.25$.
- Solving this equation, we get that $r = \frac{0.25}{8.25} = 0.030303$, or roughly 3.03%.

As you can see from the title of this section and from the example, the key is to compute the change from the base.

EXAMPLE

- Determine the percent raise given to an employee whose hourly wage went from \$8.25 to \$8.50.
- Charlie solve this problem with the equation $r \times 8.25 = 8.50$ and found that $r = 1.0303$. He reasoned that the base wage, \$8.25 per hour, corresponded to 100% of what the working was earning, so that the worker received a 3.03% wage increase.
- Charlie did not concern himself with the percent change until the end of the problem.

As you can see, there are a few ways to interpret how to do these questions. Here is another question that is answered in two different ways.

EXAMPLE

- While shopping, Arlene saw a blouse that was on sale for \$19. The tag on the blouse showed that the original price for the blouse was \$25. By what percent was the price of the blouse changed?
- The price is dropped by \$6, so the equation for this problem is $r \times 25 = 6$, so that $r = 0.24$. The price of the blouse is reduced by 24%.
- The second way to look at this problem is to work with the sale price: $r \times 25 = 19$. Solve the equation to determine that $r = 0.76$. This tells us that the sale price of the blouse is 76% of the original price, so the percent discount is $100\% - 76\% = 24\%$.

More often than not, you know what the rate of change is going to be. In solving such problems, you can either compute the amount of change and then determine the final answer or you can adjust the percentage and then compute the final answer.

EXAMPLE

- A jacket with a retail price of \$229 is put on sale at a 30% discount. What is the sale price of the jacket?
- Compute the change first: 30% of \$229 is $0.3 \times \$229 = \$68.70$. Therefore, the sale price is $\$229 - \$68.70 = \$160.30$.
- Change the percentage first: If the jacket has a 30% discount, then sale price is 70% of the retail price, or $0.7 \times \$229 = \$160.30$.

You have to be very careful when more than one change occurs to a base value. Examples are successive discounts and the change in a labor force. In situations like these, it is best to work the altered percentages.

EXAMPLE

▶ Diane has been looking for a particular coffee maker. She read the ads for a department store and saw that it had the model she wanted for $89. When she went to the store, she was pleasantly to see the model on sale with a 15% discount. Diane had a coupon for an additional 20% discount. What was the price of the coffee maker (before taxes)?

▶ The sale price of the coffee maker was 85% of $89. When the coupon was applied, Diane paid 80% of the store sale price. That is, Diane paid $0.80 \times 0.85 \times \$89 = \$60.52$ for the coffee maker.

This next problem pertains to employment rates.

EXAMPLE

▶ The local manufacturing company had a workforce of 300 people when it announced that it had to lay off 20% of its workforce. Six months later, the company proudly announced that it was increasing its workforce by 20%. What percent of the original workforce are now employed?

▶ The percentage of the workforce employed after the layoffs is 80%. The percentage of the workers employed after the increase is 120%. Therefore, the number of employed people is now $300 \times 0.80 \times 1.20 = 288$, which is 4% ($0.8 \times 1.2 = 0.96$) less than the number originally employed.

The importance here is to be aware. A company that lays off 20% of its workforce one month and later that they are restoring the workforce to full strength by hiring an additional 20% is not telling the truth. As you can see from the explanation above, the workforce is still 4% less than what is was before the layoffs were made.

Weighted Averages

There are instances when some aspects are given a higher priority than others. For instance, when magazines announce the 10 best places to live in a given area, some of the characteristics that might be considered are population, tax rates, employment percentages, crime rate, number of acres of green space, and access to public transportation. It is not unreasonable for people to value employment percentages and access to public transportation more highly than population and the number of acres of green space. In order to create a scaled score, a number to emphasize its importance might multiply the more highly valued aspects.

EXAMPLE

What would be the score for a town's popularity based on the following criteria?

Characteristic	Population	Tax Rate	Employment	Public Transportation	Green Space
Scale Factor	1	1	2	2	1
Score	7	8	9	8	4

The score for this town would be $1(7) + 1(8) + 2(9) + 2(8) + 1(4) = 53$.

The notion of weights also applies to frequency charts. Using weights reduces the amount of work that needs to be done.

EXAMPLE

The scores for the students who took the Algebra I final exam in Piedmont School District last year are shown in the following table. Determine the average score for the students who took the test.

Score	Frequency
100	7
97	15
93	12
86	45
80	72
75	64
70	24
60	15

We need to add the numbers in the frequency column to determine the number of students who took this test.

$7 + 15 + 12 + 45 + 72 + 64 + 24 + 15 = 254$

Multiply each score by the number of students who earned that score and put that number in the total column.

Score	Frequency	Total
100	7	700
97	15	1455
93	12	1116
86	45	3870
80	72	5760
75	64	4800
70	24	1680
60	15	900

▸ The average score is the sum of the total column divided by 254. The weighted average is $\frac{20{,}281}{254} = 79.85$ (answer rounded to the nearest hundredth).

Two-Way Tables

Interpreting data that is presented in a table is an important skill. You must pay particular attention to the specifics of the problem. The accompanying scenario is to be used to answer the three examples that follow.

Students are asked to name their favorite sport from the list of basketball, soccer, tennis, or lacrosse. The results are shown.

	Basketball	Soccer	Tennis	Lacrosse
Boys	25	40	10	30
Girls	20	45	30	10

EXAMPLE

▸ What percentage of the students chose basketball as their favorite sport?

▸ There are 105 boys and 105 girls involved in this survey. Of the 210 students, 45 chose basketball as their favorite sport. This represents $\frac{45}{210} = 21.4\%$ of the students. (Answer rounded to the nearest tenth.)

"What percentage of the students" tells us that all students are involved in the conversation.

EXAMPLE

▶ What percentage of the girls chose basketball as their favorite sport?

▶ The only people to consider in this problem are the girls. Of the 105 girls, 20 chose basketball as their favorite sport. This number represents $\frac{20}{105} = 19.0\%$ of the girls. (Answer rounded to the nearest tenth.)

The population being considered in this problem was just the girls. None of the information about the boys were needed to answer the question.

EXAMPLE

▶ What percentage of those who chose basketball as their favorite sport are girls?

▶ The only people to consider in this problem are those who chose basketball as their favorite sport. Of the 45 students who chose basketball, 20 girls. This number represents $\frac{20}{45} = 44.4\%$ of the basketball players. (Answer rounded to the nearest tenth.)

EXERCISES

EXERCISE 3-1

Solve each of the following equations.

1. $13x - 19 = 9x + 37$
2. $8w + 27 = 5w - 15$
3. $10t + 42 = 86 - 12t$
4. $35 - 7p = 5p + 11$
6. $92w + 149 = 47w - 26$
7. $14 + 3(5x - 10) = 4(x - 9) + 3x$
8. $74 - 2(17 - 7x) = 11x + 82$
9. $45v - 7(6v + 5) = 13v - 2(4v + 9)$
10. $120p - 1340 = 480 - 3(95 - 20p)$
11. $\frac{2}{3}x + \frac{5}{8} = \frac{5}{12}x - \frac{3}{4}$
12. $\frac{5}{9}\left(x + \frac{3}{10}\right) = \frac{1}{2}\left(x + \frac{3}{8}\right)$
13. $\frac{5}{3}(2x - 5) + 4 = \frac{7}{6}(3x + 7) - 5$
14. $0.05(7x - 19) = 0.3x + 1.5$
15. $\frac{12}{35} = \frac{x}{105}$
16. $\frac{2x + 5}{10} = \frac{3x - 4}{8}$
17. $\frac{20}{x - 4} = \frac{25}{x + 2}$

EXERCISE 3-2

Solve each equation for the indicated variable.

1. Solve $5x + 4y = 30$ for y.
2. Solve $A = \frac{h}{2}(a+b)$ for h.
3. Solve $ax + by = cx + dy$ for x.

EXERCISE 3-3

The newspaper advertised that all items at Garret's Emporium is on sale today at a 15% discount. Use this information to answer the questions in this exercise.

1. Grace found a bookcase at the emporium with a retail price tag of \$240. Grace has a coupon for an additional 25% discount. What is the sale price of the bookcase?
2. The tax rate in the county is 7%. What is the final price for the bookcase?
3. George told Grace that he found a recliner at the Emporium and that after the discount and tax, he paid \$582.08 for the recliner. What was the retail of the recliner?

EXERCISE 3-4

Use the given tables to answer these questions.

1. What is the class average for the test with scores shown?

Score	Frequency
98	5
94	7
91	10
85	16
78	18
74	7
68	3

2. The class average for the scores shown is 85.25. How many students scored 87 on the test?

Score	Frequency
95	4
89	6
87	x
83	10
75	5

EXERCISE 3-5

Students are asked to name their favorite subject from the list of math, English, science, or social studies. The results are shown.

	Math	English	Science	Social Studies
Boys	35	30	40	30
Girls	40	55	30	30

Use the accompanying table to answer the following questions. Answer to the nearest tenth of a percent.

1. What percent of the students surveyed picked math as their favorite subject?
2. What percent of the girls surveyed picked math as their favorite subject?
3. What percent of the students who picked science as their favorite subject are boys?
4. What percent of the students who picked social studies or English as their favorite subjects are boys?
5. What percent of the boys picked social studies or English as their favorite subjects?

4 Equations of Linear Functions

MUST KNOW

- Lines are the most basic mathematical graphs. They can be described by two points or by one point and directions on how to reach the next point.

- The slope-intercept form for a line, $y = mx + b$, is the most useful form for the equation of a line because one of the variables stands alone on one side of the equation.

There are many applications of mathematics that behave in a linear fashion. Some of these are:

- Revenue earned from sales
- Earnings for an employee who is paid by the hour
- The amount of money paid to fill up the tank of an automobile

We will look at the equations of a line in this chapter and some of the special features we need to understand: slope, y-intercept, and the x-intercept.

Rate of Change and Slope

Statements such as \$10.50 per hour, 65 miles per hour, \$2.59 per gallon, and 1 tablespoon per quart all indicate the rate of how one quantity will change in relation to another. In particular, if these rates of change remain constant, we will notice that the graphs of the relationships will be linear.

Computing a rate of change is as simple as computing the ratio between the difference in one quantity to the difference of a second quantity. For example:

EXAMPLE

▸ Alice noticed that her friend Janice earned \$78 for working 5 hours last week and the week before had earned \$106 for working 7 hours. She computed the rate of change with the ratio $\frac{106-78}{7-5} = \frac{\$28}{2 \text{ hours}} = \$14$ per hour. Alice said that she understood the notion of earning \$14 per hour, but that would mean Janice should earn \$70 for working 5 hours and \$98 for working 7 hours. Why was she paid \$8 more than that each week?

▶ Janice explained that she is given \$8 per week to cover her transportation costs to get to work. "Aha," Alice said. "So, if you work 8 hours next week, you'll earn \$120." "That's right," Janice replied.

In analyzing the problem, Alice determined that the amount of money earned depended upon the number of hours worked. In writing the ordered pairs, she chose to write the facts she knew as (5, 78) and (7, 106). The first element in the ordered pair (the abscissa) represents the **independent variable**, while the second element (the ordinate) represents the **dependent variable**. That gives two ways of interpreting the ordered pair (a, b). We can call it (abscissa, ordinate) or we can call it (independent variable, dependent variable). A third way to look it is (input, output).

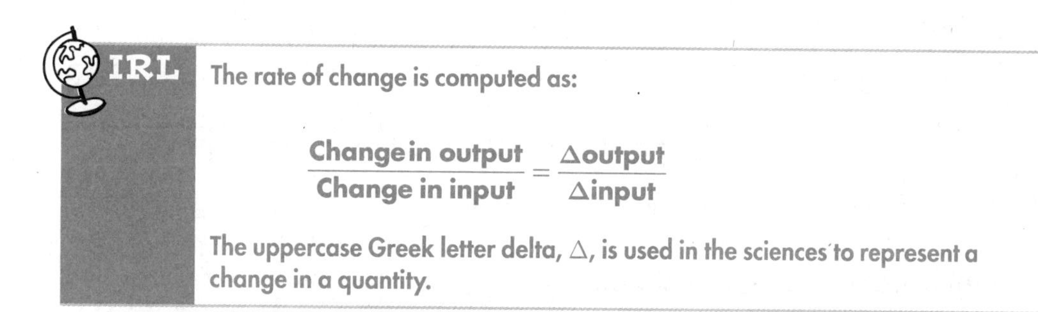

IRL

The rate of change is computed as:

$$\frac{\text{Change in output}}{\text{Change in input}} = \frac{\Delta\text{output}}{\Delta\text{input}}$$

The uppercase Greek letter delta, Δ, is used in the sciences to represent a change in a quantity.

Let's practice computing the rate of change for a few ordered pairs.

EXAMPLE

▶ Compute the range of change for the ordered pairs (19, 210) and (27, 288).

▶ Using the ratio $\frac{\Delta\text{output}}{\Delta\text{input}}$, we get $\frac{288-210}{27-19}=\frac{72}{8}=9$.

It is important that when computing the differences, you recognize that the elements of the same ordered pair are written in the same order. That is, we wrote the elements from the ordered pair (27, 288) first in both the numerator and denominator. Could we have written the elements from (19, 210) first? Yes!

$$\frac{210-288}{19-27}=\frac{-72}{-8}=9$$

We **cannot** mix the ordered pairs like this: $\frac{288-210}{19-27}=\frac{72}{-8}=-9$, because it does not give a true representation of the data.

EXAMPLE

▶ While using his car's cruise control, Alex notices that she passes mile marker 122 at exactly 10 a.m. and then passes mile marker 138 at 10:15 a.m. At what speed does Alex have the cruise control set?

▶ The change in the number of miles driven is 16, while the change in time is 15 minutes. The speed the car is traveling is $\frac{16 \text{ miles}}{15 \text{ minutes}}$.

Multiply numerator and denominator by 4 so that the time unit is 1 hour (60 minutes) to determine that Alex has the cruise control set for 64 miles per hour.

IRL

I'll take a moment here to for tell a story about slope. Descartes (remember him from the Cartesian coordinate plane?), as the story goes, was trying to explain the notion of slope to his audience. (The story is unclear as to whether they were students or other professionals.) You and I take for granted that we can put graphs on the coordinate plane and do some measuring. This was brand-new, never-before-seen material to them. In relating the notion of slope, Descartes referred to something everyone could relate to: the mountains. You can go up a mountain. The elevation gets higher (bigger) as you climb up and lower (smaller) as you go down. This helps us to understand positive and negative slope. If you are on flat ground, the elevation does not change.

We now understand a slope equal to zero. Lastly, there is the precipice that is a vertical climb. You cannot walk up or down this section of the mountain because there is no horizontal change. What value do we assign to this slope? More on that in a minute. There is some disagreement among mathematical historians about the next statement the statement that follows, but it is interesting because I will make, but I like this version and it is not inconceivable that it is true. Since Descartes was using the mountains to explain his notion of slope, he used the letter *m*, the first letter in the word for mountain (or *monte*, in French) to indicate slope.

The following examples have you compute slopes. Does the interpretation for the meaning of each slope make sense to you?

EXAMPLE

▶ Compute the slope of the line that passes through (3, 7) and (9, 21).

▶ $m=\frac{21-7}{9-3}=\frac{14}{6}=\frac{7}{3}$. The positive slope tells us the graph of the line rises from left to right.

Beware the negative numbers in the next problem.

EXAMPLE

▶ Compute the slope of the line that passes through (–4, 10) and (–1, 16).

▶ $m=\frac{16-10}{-1-(-4)}=\frac{6}{3}=2$. The positive slope tells us the graph of the line rises from left to right.

This next example is just slightly different.

EXAMPLE

▶ Compute the slope of the line that passes through (−4, 10) and (−1, −16).

▶ $m=\frac{-16-10}{-1-(-4)}=\frac{-26}{3}$. The negative slope tells us the graph of the line falls from left to right.

The next two examples involve zero in the problem.

EXAMPLE

▶ Compute the slope of the line that passes through (2, 6) and (11, 6). Can you picture the lines in your mind?

▶ $m=\frac{6-6}{11-2}=\frac{0}{9}=0$. A slope of zero tells us the line is horizontal.

EXAMPLE

▶ Compute the slope of the line that passes through (2, 6) and (2, 11). Can you picture the lines in your mind?

▶ $m=\frac{11-6}{2-2}=\frac{5}{0}$. Division by 0 is undefined. A vertical line has a slope that is undefined. This is the answer to the precipice that Descartes needed to explain.

The notion of slope is also explained from a graphical point of view. The line connecting the points (3, 7) and (9, 21) looks like this:

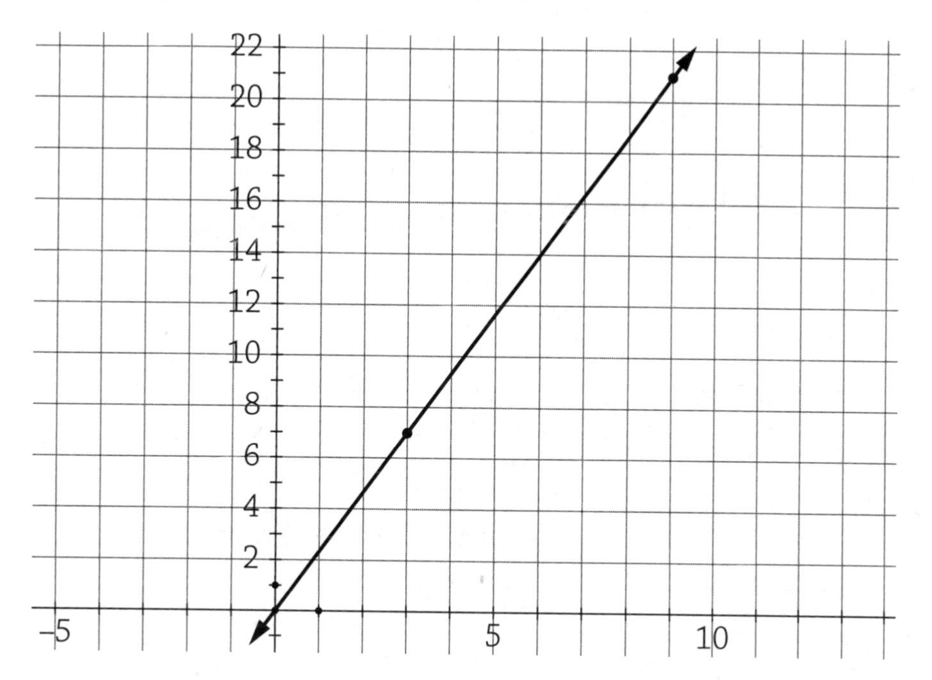

The concept of rise over run, $\frac{\text{rise}}{\text{run}} = \frac{\Delta\text{vertical}}{\Delta\text{horizontal}} = \frac{\Delta y}{\Delta x}$, is used to compute slope.

The slope of the line passing through two points (x_1, y_1) and (x_2, y_2) is given by the equation:

$$m = \frac{\Delta y}{\Delta x} = \frac{y_2 - y_1}{x_2 - x_1}$$

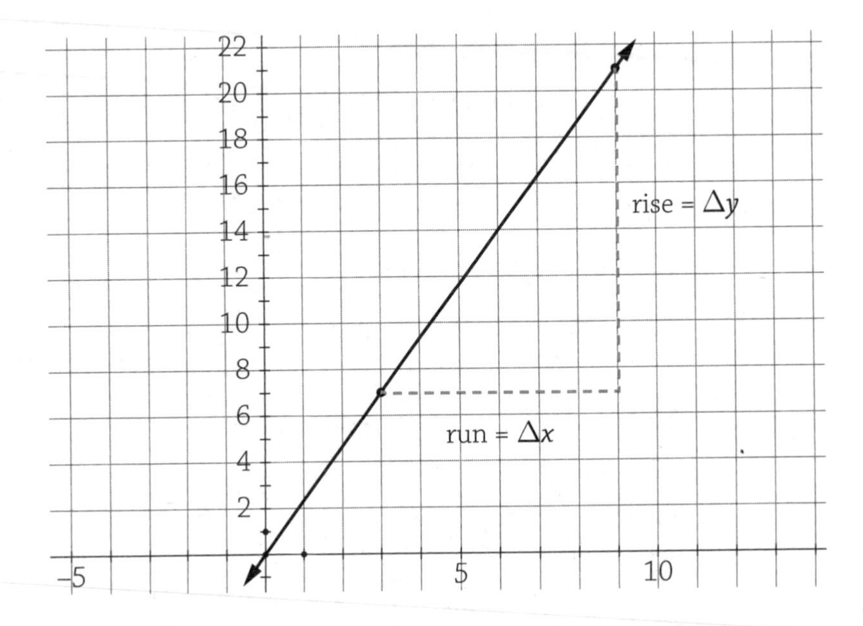

When using this approach, you must make sure to compute the run from left to right. That is, with the exception of a vertical line, the run will always be computed as a positive value. The rise will be positive if the line rises, will be negative if the line falls, and will be zero if the line is horizontal.

EXAMPLE

▶ Determine the slope for each of the following lines.

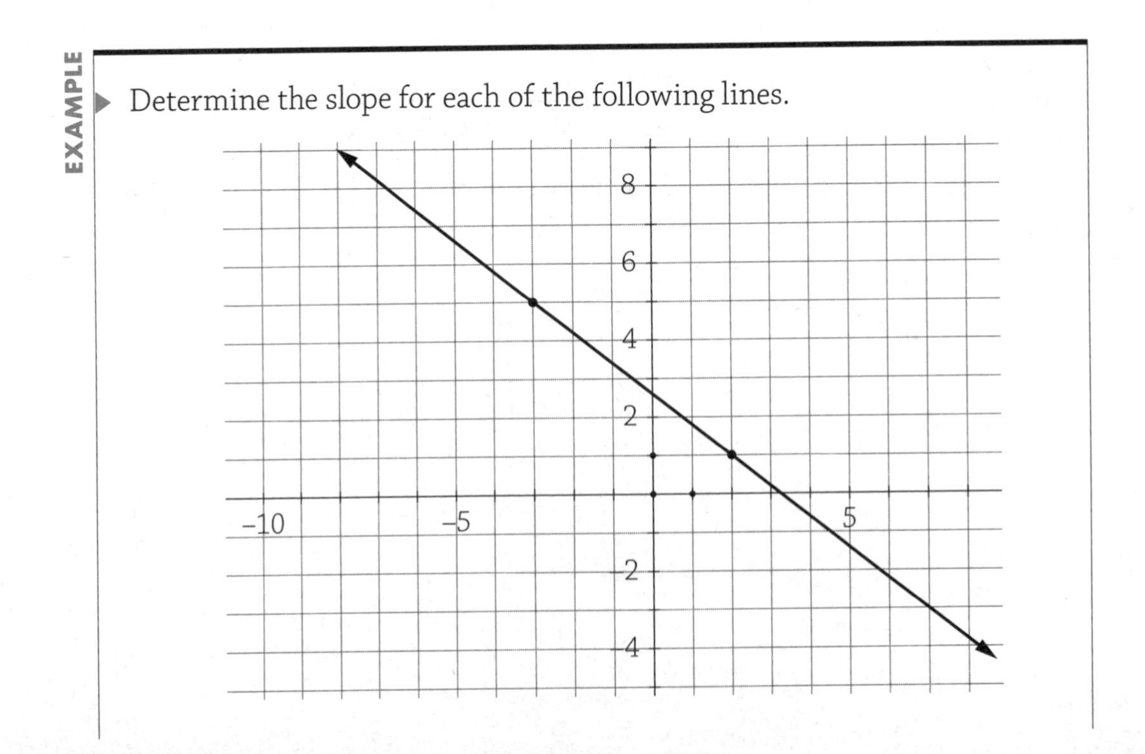

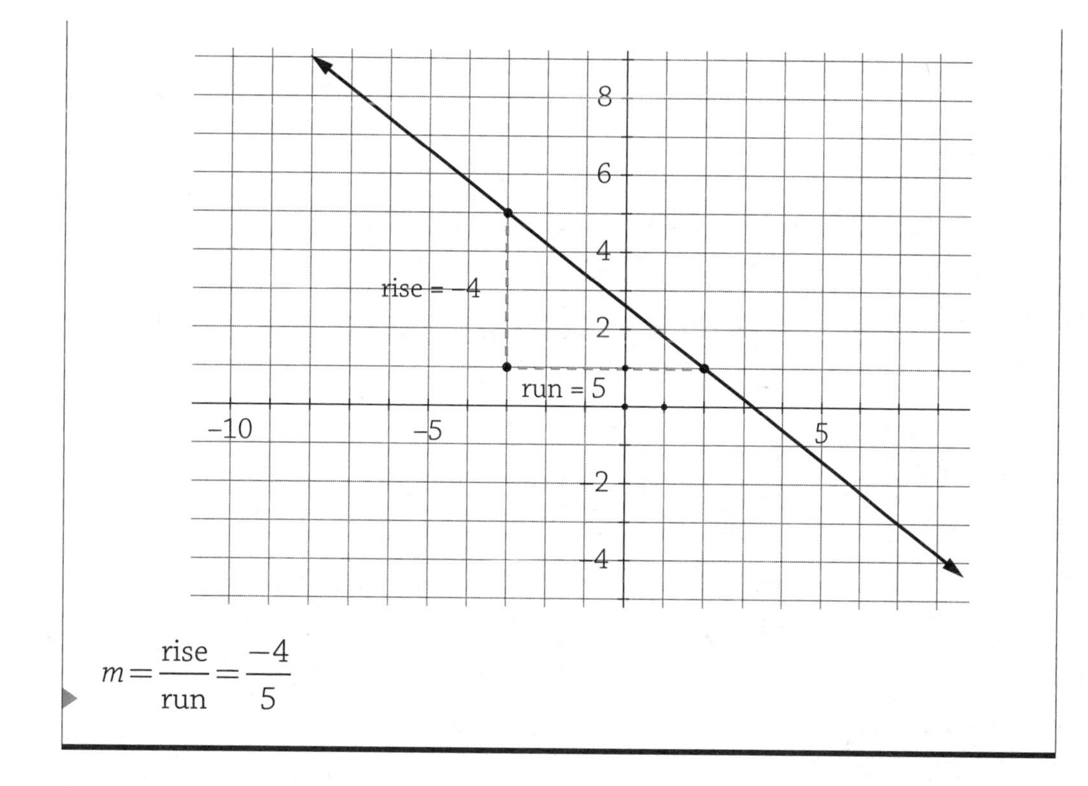

$$m = \frac{\text{rise}}{\text{run}} = \frac{-4}{5}$$

Here is another problem for you to try.

EXAMPLE

Determine the slope of the line shown.

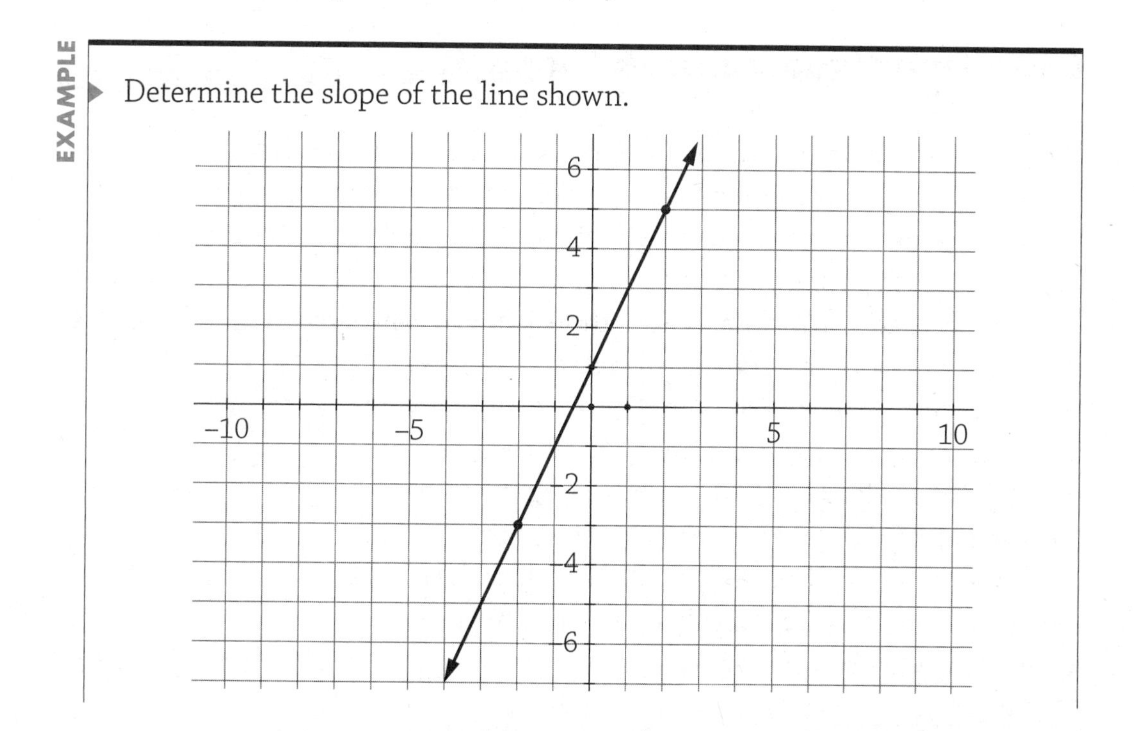

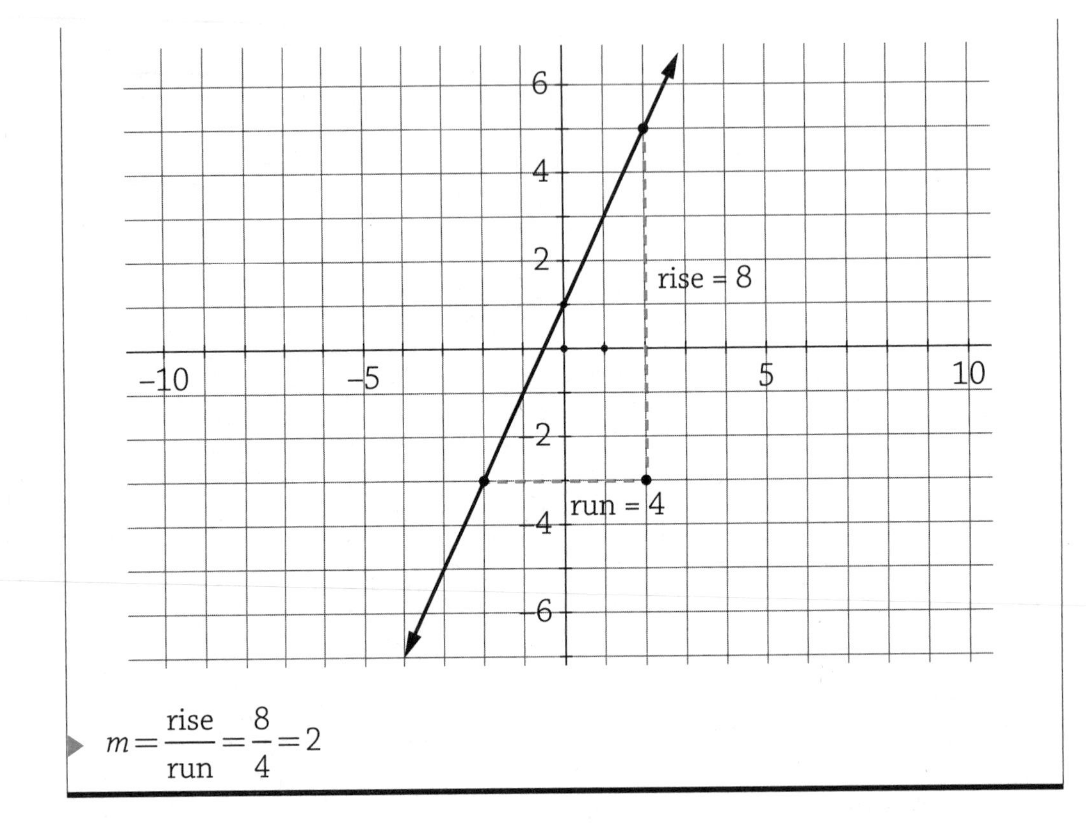

$m = \frac{\text{rise}}{\text{run}} = \frac{8}{4} = 2$

Graphing Equations in Slope-Intercept Form

The points where a graph crosses the axes are called the **intercepts**. It will ALWAYS be true that the abscissa for a point on the *y*-axis will be 0, while the ordinate for a point on the *x*-axis will be 0. You will take advantage of this many times during your study of mathematics.

While there are many variations for the equation of a line, the first form that is traditionally studied is $y = mx + b$. We know the m represents the slope. What does b represent? If we let $x = 0$, the equation becomes $y = m(0) + b$, or, simply, $y = b$. Looking back at the last paragraph, we see that the *y*-intercept for the line $y = mx + b$ is the point $(0, b)$. This is why the equation $y = mx + b$ is called the **slope-intercept** form for the line. It is not difficult to remember that the intercept in question is the *y*-intercept.

BTW

Whenever we write an equation of the form $y = f(x)$, that is, y depends on some expression in x, the purpose of the x and y in the equation is to represent the coordinates any arbitrary point that lies on the graph.

EXAMPLE

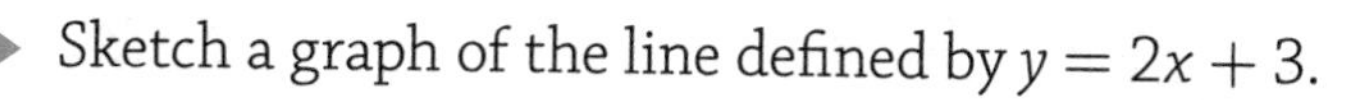

▶ Sketch a graph of the line defined by $y = 2x + 3$.

▶ We know two key pieces of information. We know a point on the line, the y-intercept is (0, 3), and we know the slope is 2. Since this is our first example, let me write the slope as $\frac{2}{1}$ so that we go back to the notion of $\frac{\text{rise}}{\text{run}}$. A slope of 2 tells us that for each movement of 1 to the right, the graph rises 2 units. We can sketch the graph by plotting the y-intercept and use the slope to plot a few more points:

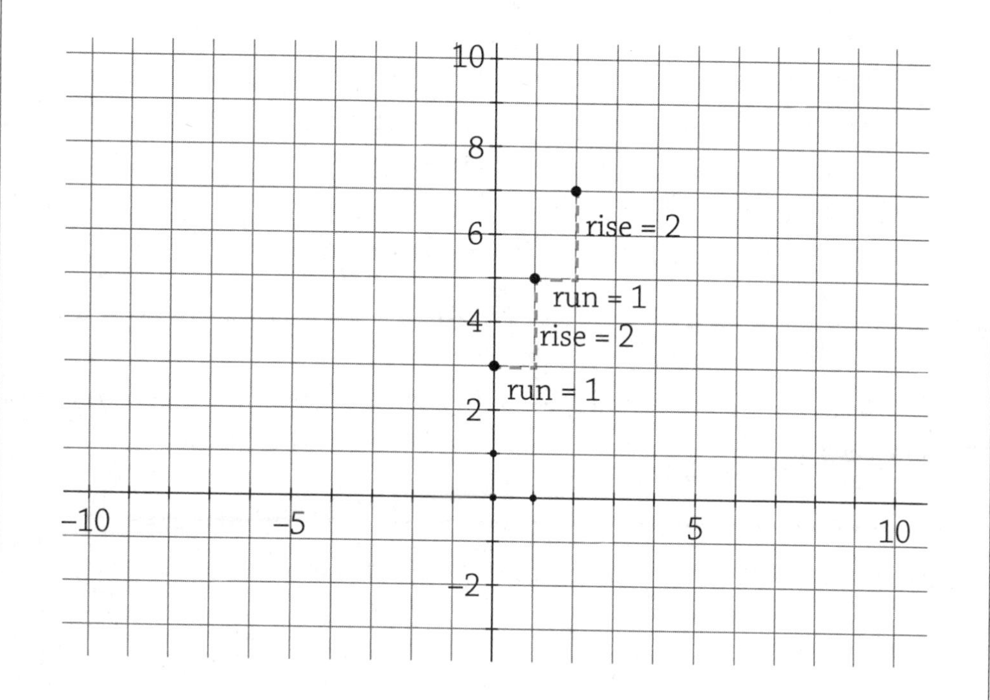

▶ Connect these points to get the line:

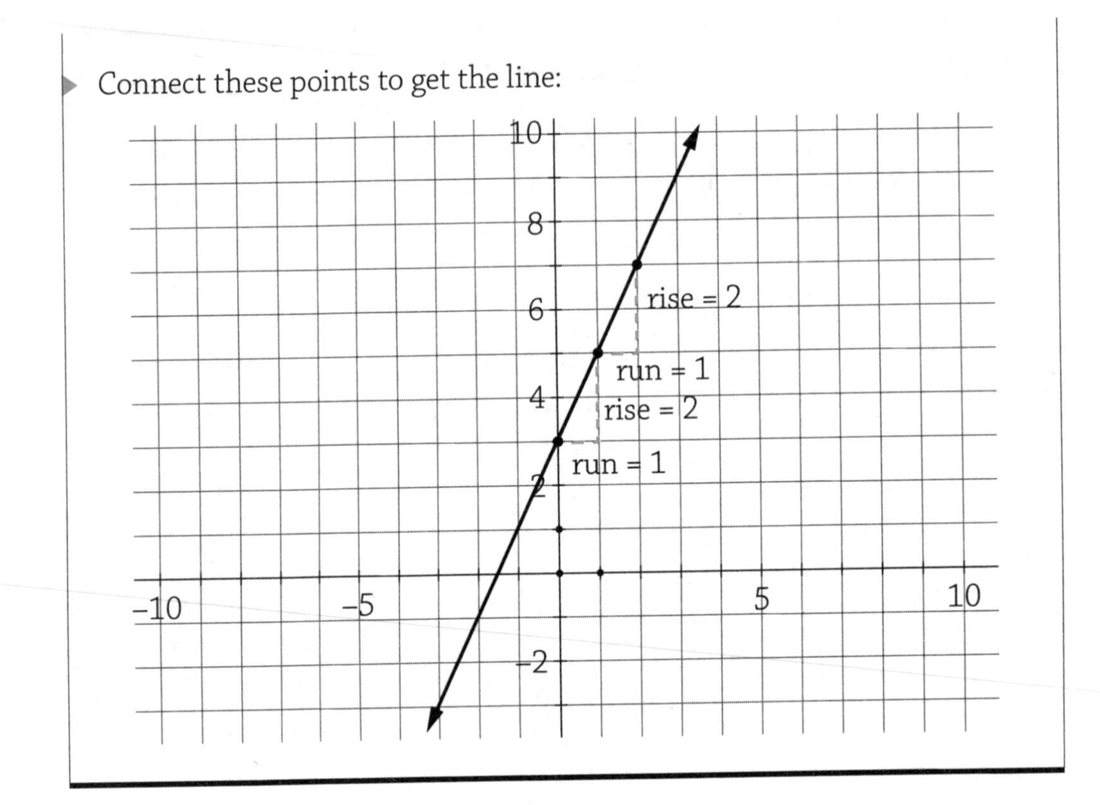

This process for graphing lines actually works well when the slope is a fraction. You can easily see the run and rise pieces of the fraction.

EXAMPLE

▶ Sketch a graph of the line defined by $y=\frac{-2}{3}x+5$.

▶ The y-intercept is 5 and the slope is $\frac{-2}{3}$. Plot the y-intercept, move right 3 units, and down 2 units to get to the next point. Repeat to get the third point. Sketch the graph.

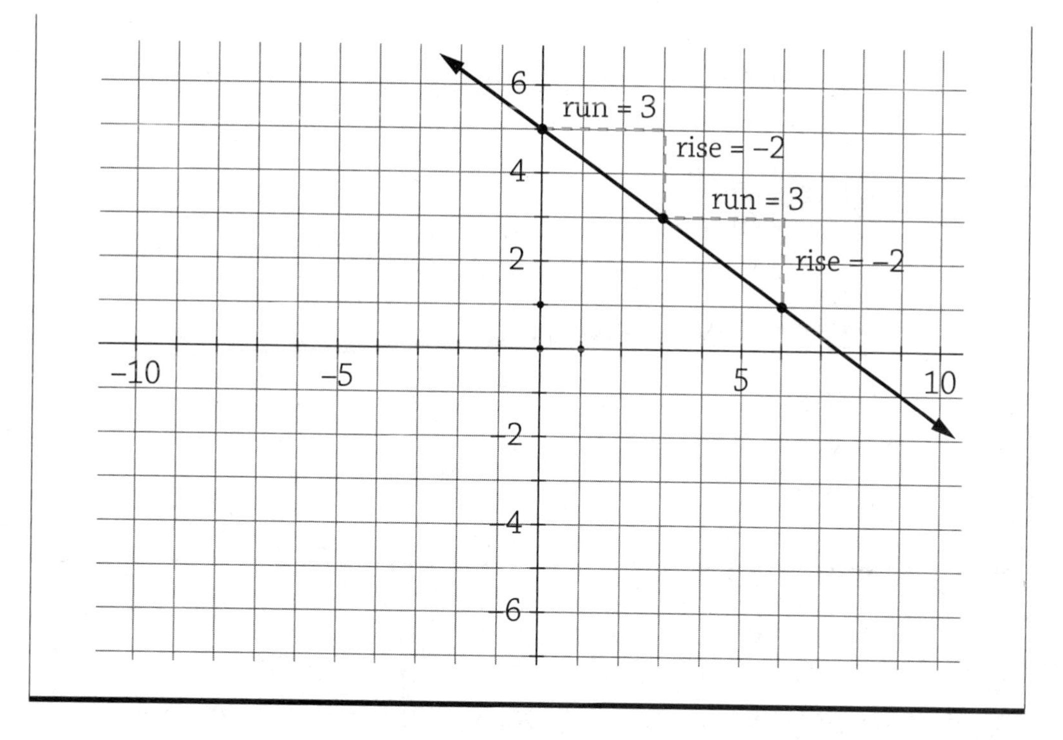

Do you see how this process can be used to create a table of values for a graph that has a fractional slope? Instead of creating a table of values with x equal to consecutive integers, you can choose values for which the coordinates are guaranteed to be integers.

EXAMPLE

▶ Create a table with the coordinates of 5 points that will lie on the line with equation $y = \frac{3}{7}x - 4$.

▶ The slope is $\frac{3}{7}$ and the y-intercept is -4. We know that the point $(0, -4)$ is on the graph. The slope has a run of 7 and a rise of 3. The next point we can plot has coordinates $(0 + 7, -4 + 3) = (7, -1)$. After

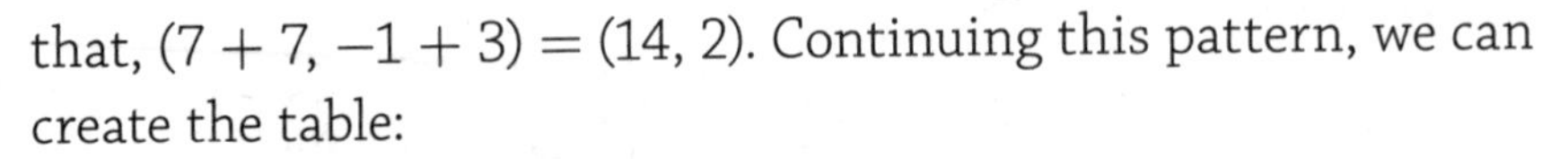

that, $(7 + 7, -1 + 3) = (14, 2)$. Continuing this pattern, we can create the table:

x	y
0	−4
7	−1
14	2
21	5
28	7

Writing Equations in Slope-Intercept Form

When asked to write an equation for a line, you have to be given a minimum of two pieces of information:

- You are either told the coordinates of a point and the slope of the line.
- You are told the coordinates of two points.

An easy problem would be if you were told the slope of the line and the y-intercept.

EXAMPLE

▶ Write an equation for the line with slope -3 and y-intercept 7.

▶ The equation for the line would be $y = -3x + 7$.

Problems do not tend to be that easy. We need to know a general plan that will allow us to determine the equation for the information given to us.

EXAMPLE

- Write an equation for the line with slope –3 and that passes through the point (2, 5).
- We know the slope so we are half way to getting the equation. Right now we have the equation to be $y = -3x + b$. We substitute the coordinates of the given point for x and y to determine the value of b.

$$\begin{aligned} & y = -3x + b \\ \text{becomes} \quad & 5 = -3(2) + b \\ \text{solve} \quad & 5 = -6 + b \\ & 11 = b \end{aligned}$$

- The equation is $y = -3x + 11$.

What do we do when we are given the coordinates of two points that are on the line?

EXAMPLE

- Write an equation for the line that passes through the points (6, 9) and (–2, 5).
- The first order of business is to find the slope of the line:

$$\frac{9-5}{6-(-2)} = \frac{4}{8} = \frac{1}{2}$$

- You can use either of the given points to find the y-intercept. I'll choose the point (6, 9) because we will not have to deal with negative numbers.

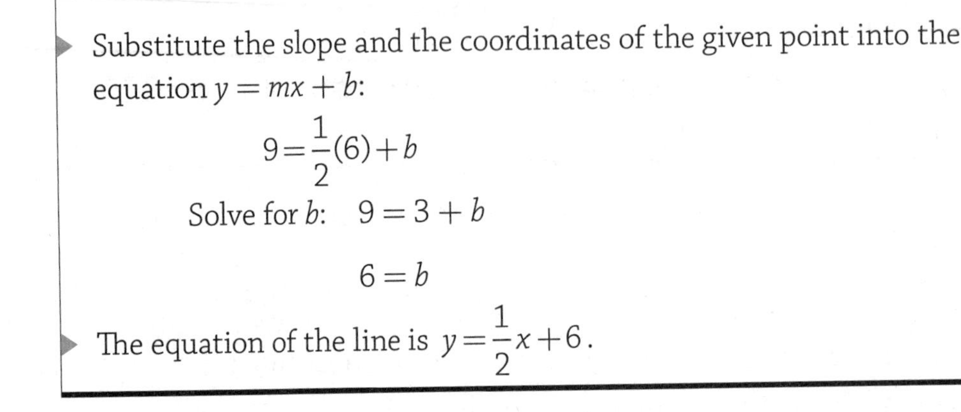

▶ Substitute the slope and the coordinates of the given point into the equation $y = mx + b$:

$$9 = \frac{1}{2}(6) + b$$

Solve for b: $9 = 3 + b$

$$6 = b$$

▶ The equation of the line is $y = \frac{1}{2}x + 6$.

Unfortunately, the problems you'll run into will not always have arithmetic that is that clean.

EXAMPLE

▶ Write an equation for the line passing through the points (−5, 11) and (−2, 7).

▶ Determine the value of the slope:

$$\frac{7 - 11}{-2 - (-5)} = \frac{-4}{3}$$

▶ Substitute the slope and the coordinates of one of the points into the equation $y = mx + b$:

$$11 = \frac{-4}{3}(-5) + b$$

Solve for b: $\frac{33}{3} = \frac{20}{3} + b$

$$\frac{13}{3} = b$$

▶ Therefore, the equation of the line is $y = \frac{-4}{3}x + \frac{13}{3}$.

Wait! That wasn't so bad. We had to do some computation with a couple of fractions, but that really isn't a big deal, is it? We needed to find a process for writing the equation of a line:

- First, determine the value of the slope.
- Second, substitute the coordinates of one of the points and the value of the slope into the equation $y = mx + b$ and solve for b.

We're good to go.

Writing Equations in Point-Slope Form

While determining the equation of the line is $y = \frac{-4}{3}x + \frac{13}{3}$, we don't get the sense of "Oh yeah! I know what that line looks like." It is just easier to have a point whose coordinates are integers. We can deal with fractional slopes by using the components to measure run and rise.

EXAMPLE

▶ Sketch a graph of the line with slope $\frac{-4}{3}$ and which passes through the point (–2, 7).

▶ Plot the point (–2, 7) and then move to the right 3 units and down 4 to reach the point (1, 3). Repeat the movement to reach the point (4, –1). Sketch the line.

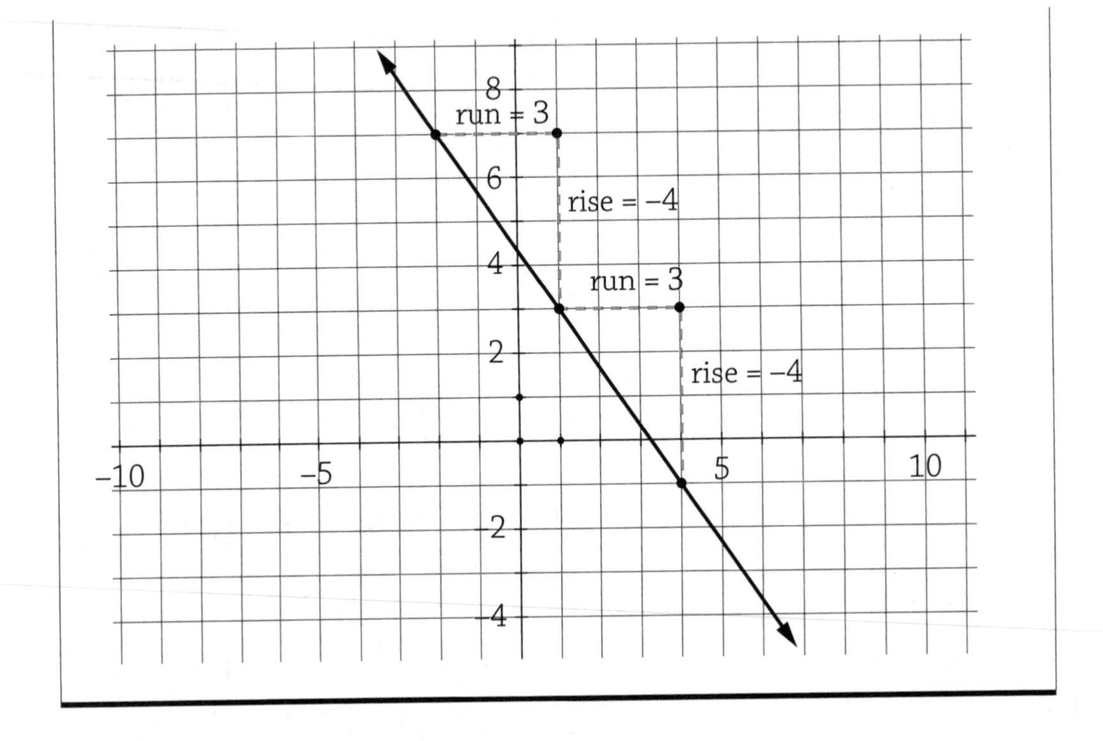

It looks like all we really need to graph a line is the coordinates of one point and directions on how to get to another point (either slope or the coordinates of a second point). The y-intercept has a number of important characteristics. However, as we will find, it can always be determined from an equation by setting x equal to zero.

The **point-slope** form for the equation of a line is based on knowing the slope, m, of the line and the coordinates (x_1, y_1) of a point on the line. The point-slope equation of the line is

$$y - y_1 = m(x - x_1)$$

Be sure that you are clear with subscripted variables such as x_1 and y_1 as opposed to x and y. The coordinates (x_1, y_1) represent the coordinates of a known point, while (x, y) represent the coordinates of any arbitrary point on the line.

EXAMPLE

- Write an equation for the line with slope $\frac{5}{7}$ that passes through the point $(-3, 5)$.
- The point-slope equation for the line is $y-5=\frac{5}{7}(x+3)$.

How easy was that?! It is an easy equation to write and to graph. Need to make a table of values? Use the interpretation of slope as $\frac{\text{rise}}{\text{run}}$ to find other points from the given value $(-3, 5)$. The slope tells us to move to the right 7 and up 5: $(-3 + 7, 5 + 5) = (4, 10)$. We have the coordinate of another point on the graph.

EXAMPLE

- Patty and Max are directed to write an equation for the line that passes through the points $(-3, 5)$ and $(5, 2)$.
- They determine the slope of the line to be $\frac{2-5}{5-(-3)}=\frac{-3}{8}$.
- Patty writes the equation $y-5=\frac{-3}{8}(x+3)$ while Max writes the equation $y-2=\frac{-3}{8}(x-5)$.
- They look at each other and ask, "Which of us is right?"
- They shrug their shoulders and say, "Guess we both are."
- "But how do we know?"
- They look at each other a second time and simultaneously say, "We can apply the distributive property to rewrite the equation in slope intercept form to verify this."

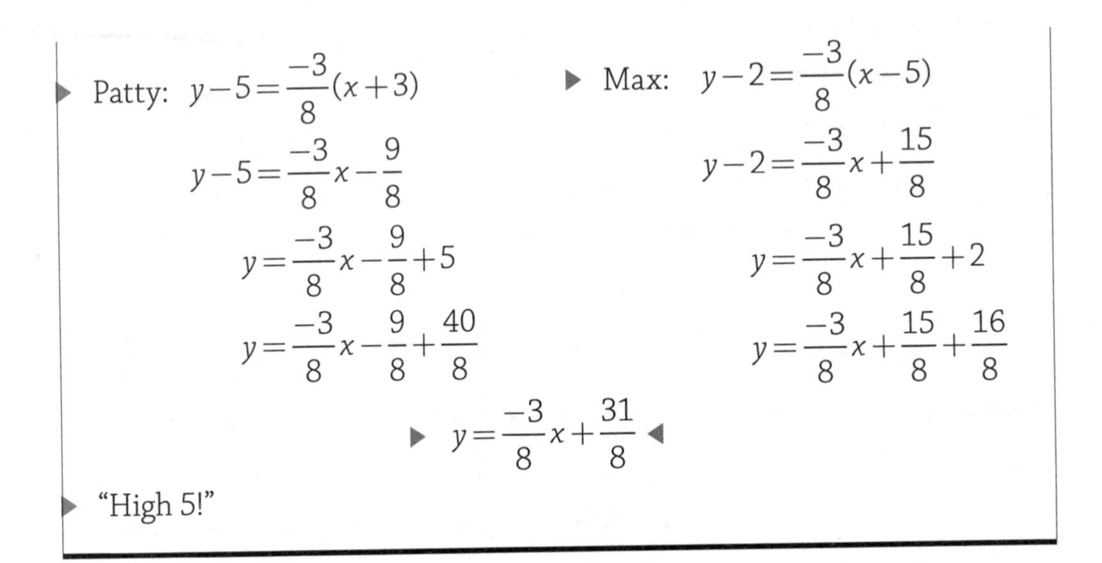

Writing Equations in Standard Form

Equations of the form $Ax + By = C$ (with A, B, and C being integers) are called "in standard form." While it is not often that you will be asked to write the equation of a line in standard, it is not unusual to be asked to manipulate them for some application. Given that, let's take a look at the form of the line and how it can be manipulated to give us useful information.

Solve the standard equation for y. That is, rewrite the standard form of the equation $Ax + By = C$ in the slope intercept form $y = mx + b$.

$Ax + By = C$ becomes $By = -Ax + C$, which in turn becomes $y=\frac{-A}{B}x+\frac{C}{B}$.

We now know that the slope of the line written in standard form is $\frac{-A}{B}$ and the y-intercept is $\frac{C}{B}$. Good to know. But it doesn't really help us write an equation in standard form unless we get very lucky. We are going to take the approach that we will write the equation of the line in either point-slope or slope-intercept form, and then we'll modify the equation to standard form.

EXAMPLE

- Rewrite the equation $y-5=\frac{-3}{7}(x+4)$ in standard form.
- Multiply both sides of the equation by the denominator of the slope:

$$7(y-5)=7\left(\frac{-3}{7}\right)(x+4)$$

- Simplify: $7y - 35 = -3(x + 4)$

$$7y - 35 = -3x - 12$$

- Rewrite the equation with the variable terms on the left:

$$3x + 7y = 23.$$

Let's do another problem and we'll approach it from two ways. You can decide how you prefer to handle future problems involving standard form.

EXAMPLE

- Write the equation of the line joining the points (–4,5) and (7, 11) in standard form.
- The first order of business is to find the slope:

$$m=\frac{11-5}{7-(-4)}=\frac{6}{11}$$

- **Using the point-slope form:**

$$y - 11=\frac{6}{11}(x - 7)$$

- Multiply by the common denominator:

$$11(y-11)=11\left(\frac{6}{11}\right)(x-7)$$

- Simplify: $11y - 121 = 6(x - 7)$

$$11y - 121 = 6x - 42$$

- Gather terms: $-6x + 11y = 79$
- If this had been a multiple-choice question, you would also need to be alert for the choice $6x - 11y = -79$. This is the result of multiplying both sides of the equation by -1.
- **Go directly to the standard form:**
- The slope of the line is $\frac{-A}{B} = \frac{6}{11}$. This gives us $A = -6$ and $B = 11$, so the equation must be of the form

$$-6x + 11y = C.$$

- Substitute the coordinates for one of the points on the line for x and y:

$$\begin{aligned} -6(7) + 11(11) &= C \\ -42 + 121 &= C \\ 79 &= C \end{aligned}$$

- The equation of the line is $-6x + 11y = 79$.

The second approach might look easier (and to some extent it is), but it requires that you remember that the slope is of the form $\frac{-A}{B}$. For some people, that is not an issue. For others, the less they have to memorize the better. Just know that you can get the correct answer using either approach.

Graphing Equations in Standard Form

Theoretically, graphing lines whose equations are written in standard form is pretty easy. All you need to do is determine the value for each intercept, plot them on the graph, and draw a line containing the two points.

EXAMPLE

- Sketch the line determined by the equation $5x - 4y = 20$.
- The x-intercept is found by setting y equal to zero: $5x = 20$, so $x = 4$.
- The y-intercept is found by setting x equal to zero: $-4y = 20$, so $y = -5$.
- Graph (4, 0) and (0, −5) and connect them.

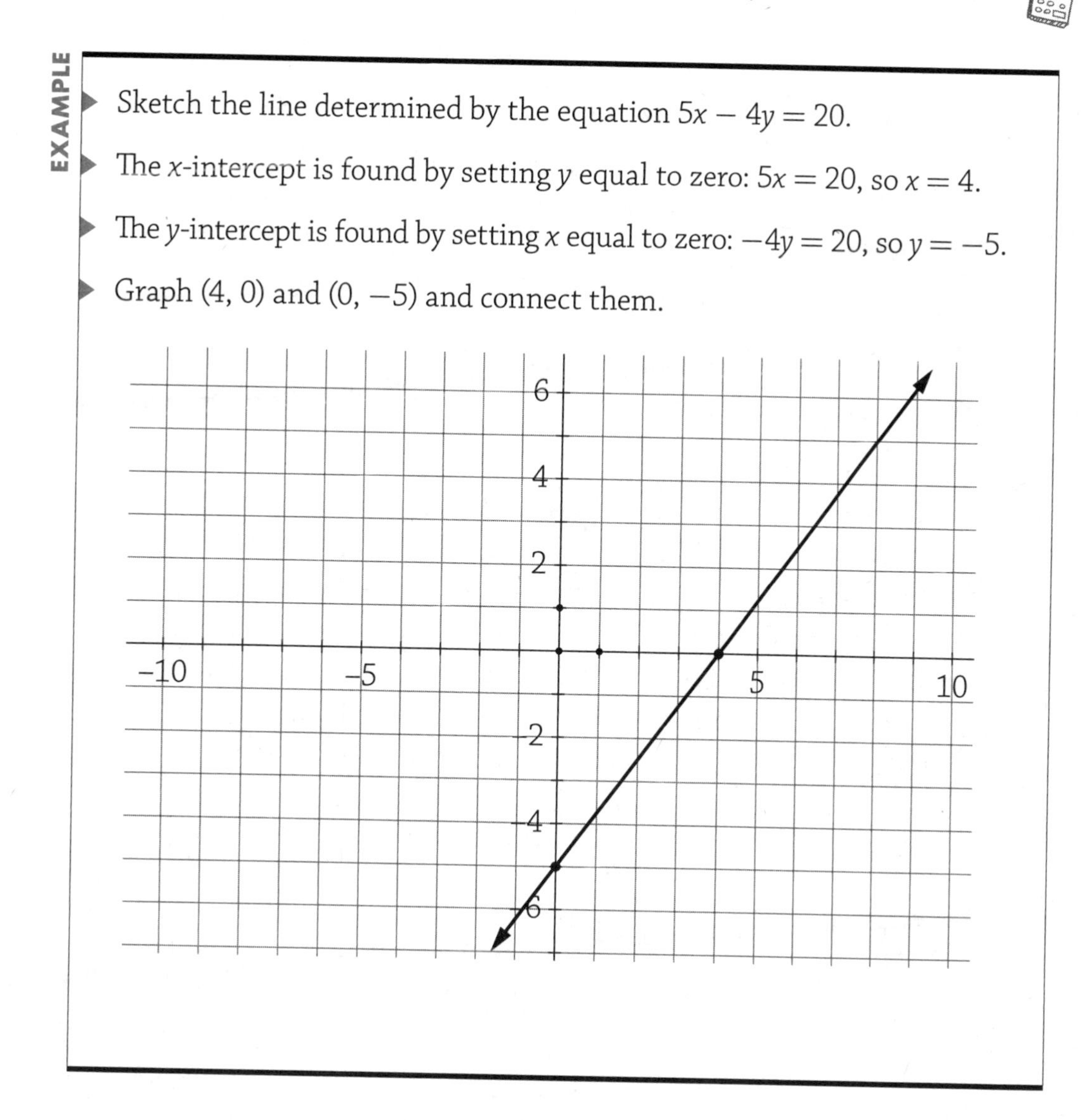

That was easy. However, not all equations are that easy to graph in a practical manner.

EXAMPLE

- Sketch the line determined by the equation $11x - 7y = 18$.
- The x- and y-intercepts are $\left(\frac{18}{11},0\right)$ and $\left(0,\frac{-18}{7}\right)$. These points are easily graphed electronically but very difficult to graph with a pencil on a piece of graph paper. The slope of the line is $\frac{-A}{B}=\frac{-11}{-7}=\frac{11}{7}$. If you can find one point whose coordinates are integers, then you will be able to use the slope to find a second (and third) point whose coordinates are integers.
- We already know that when $x = 0$, y is not an integer (also written as y in a non-integer). Try $x = 1$. $11(1) - 7y = 18$ becomes $-7y = 7$, so $y = -1$. Superb! We have $(1, -1)$ as a point on the line. Use the slope to move right 7 and up 11 to get $(8, 10)$ as a point on the line. (You might want to check that $11(8) - 7(10) = 18$.) A third point is $(15, 21)$. (Has it occurred to you that you could go left 7 and down 11 to get another point on the line? Check that $(1 - 7, -1 - 11) = (-6, -12)$ is also a point on the line.)

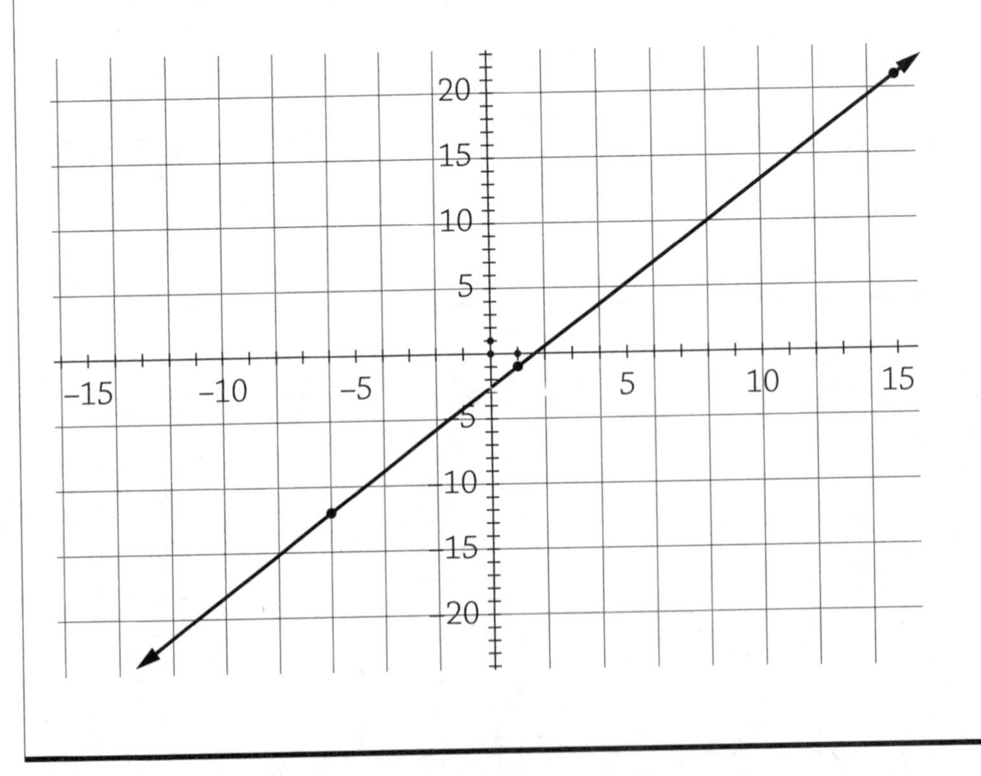

You may need to be patient in finding a point with integer coordinates if the intercepts fail to give you at least one point. Once an acceptable point is found, use the slope to find one or two more.

Parallel and Perpendicular Lines

Two coplanar lines (two lines in the same plane—such as the floor or the ceiling but not one from the floor and one from the ceiling) are parallel if they never intersect. The only way to guarantee that they never intersect is if they move from one point to another point on their respective lines in the same routine. That is, **the two lines have to have the same slope!**

EXAMPLE

- Determine the equation of the line through the point (5, 2) that is parallel to the line with equation $y = 4x + 3$.
- The line with equation $y = 4x + 3$ has a slope of 4 and a y-intercept of 3. The line being sought also has a slope of 4, but we do not know the y-intercept, so its equation is $y = 4x + \text{I}$. Use the ordered pair (5, 2) to determine the value of b:

$$2 = 4(5) + b, \text{ so } 2 = 18 + b \text{ and } b = -16.$$

- Therefore, the equation of the line is $y = 4x - 16$.

If the equation of the line is given to you in standard form, then the equation of the parallel line will look exactly the same on the left side of the equation but will have a different value of C on the right.

EXAMPLE

- Determine the equation of the line through the point (4, −3) that is parallel to the line $5x + 4y = 12$.
- The new line will have the equation $5x + 4y = C$. Substitute the coordinates (4, −3) to determine the value of C.

$$5(4) + 4(-3) = C, \text{ so } 20 - 12 = 8 = C.$$

- The equation of the parallel line is $5x + 4y = 12$.

Of course, it is possible that you will not be given the equation of the first line but the coordinates of two points on the line. "Easy enough," you say. "Simply find the slope of the line through the two points and use that with the third point to find the equation of the line." I say to you, "Nicely done!"

EXAMPLE

- Find an equation of the line through the point (–5,9) that is parallel to the line through the points (–3, 7) and (4, –3).
- The slope of the line through the points (–3, 7) and (4, –3) is $m = \frac{-3-7}{4-(-3)} = \frac{-10}{7}$. You can use the point-slope form for the line to give the equation $y - 9 = \frac{-10}{7}(x+5)$. Done!
- Did you choose the option that since $\frac{-A}{B} = \frac{-10}{7}$, so $A = 10$ and $B = 7$?
- Therefore, the equation must be $10x + 7y = C$. Substituting (–5, 9) for x and y, you get $10(-5) + 7(9) = -50 + 63 = 13$, so the equation of the parallel line is $10x + 7y = 13$. Again, well done!

The next linear relationship that we need to consider is perpendicular lines. We know that horizontal lines (slope = 0) and vertical lines (slope is undefined) are perpendicular. That is a special case of the rule we are about to examine. **Two nonvertical lines are perpendicular if and only if the slopes of the lines are negative reciprocals**. For example, lines with slopes $\frac{3}{4}$ and $\frac{-4}{3}$ are perpendicular, while the lines with slopes $\frac{2}{3}$ and $\frac{3}{2}$ are not.

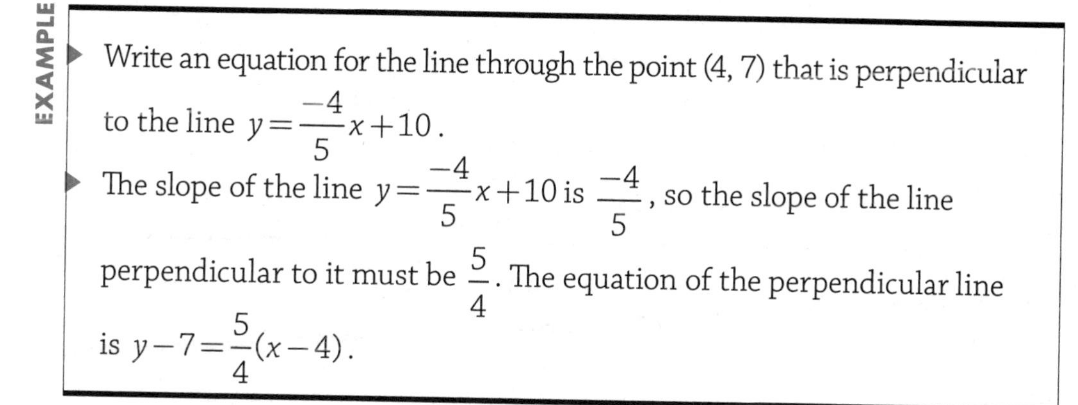

EXAMPLE

▶ Write an equation for the line through the point (4, 7) that is perpendicular to the line $y=\frac{-4}{5}x+10$.

▶ The slope of the line $y=\frac{-4}{5}x+10$ is $\frac{-4}{5}$, so the slope of the line perpendicular to it must be $\frac{5}{4}$. The equation of the perpendicular line is $y-7=\frac{5}{4}(x-4)$.

What happens to the coefficients of a line written in standard form when relating perpendicular lines?

EXAMPLE

▶ Write an equation for the line through the point (7, 11) that is perpendicular to the line with equation $5x + 8y = 10$.

▶ The slope of the given line is $\frac{-A}{B}=\frac{-5}{8}$, so the slope of the perpendicular line is $\frac{8}{5}$. If the line whose equation we are looking to write is $ax + by = c$, then $\frac{-a}{b}=\frac{8}{5}$ so $a = -8$ and $b = 5$. (Note that the coefficients are now written as lowercase letters; this is to prevent getting coefficients of x and y from the two equations mixed up. We could have just as easily written the second equation as $Dx + Ey = F$ if you prefer capital letters. Feel free to write it this way in your work.)

▶ We have the equation for the perpendicular line to be $-8x + 5y = c$. Substitute the coordinates for the given point to get $-8(7) + 5(11) = c$, which gives $-56 + 55 = -1 = c$. Therefore, the equation of the perpendicular line is $-8x + 5y = -1$ or $8x - 5y = 1$ when both sides of the equation are multiplied by -1.

Yes, it is true that the equation of the line perpendicular to the line $Ax + By = C$ will always be $Bx - Ay = D$, where D can be found by substituting the coordinates of the new point for x and y.

EXAMPLE

- Determine the equation for the line through the point (−3, −7) that is perpendicular to the line $4x + 5y = 10$.
- The equation of the line in question is $5x - 4y = D$. Substituting the coordinates, we get $5(-3) - 4(-7) = D$, so $D = -15 + 28 = 13$. The equation of the new line is $5x - 4y = 13$.

Inverse Linear Functions

If we substitute $x = 2$ into the equation $y = 4x + 3$, we would get the answer $y = 11$, and if we substituted $x = -4$ into this equation, we would get the answer $y = -13$. The linear function that would take $x = 11$ and give an answer of 2 while also taking $y = -13$ and giving the answer −4 is called the **inverse** function for $y = 4x + 3$. It is an inverse because it reverses all of the ordered pairs that come from the original function.

EXAMPLE

- Determine the equation of the line that connects the points (−13, −4) and (11, 2).
- The slope of the line is $m = \frac{2-(-4)}{11-(-13)} = \frac{6}{24} = \frac{1}{4}$. The equation of the line is $y = \frac{1}{4}x + b$. Substitute one of the ordered pairs for x and y to find $2 = \frac{1}{4}(11) + b$ so that $b = 2 - \frac{1}{4}(11) = \frac{-3}{4}$. The equation of the inverse function is $y = \frac{1}{4}x - \frac{3}{4}$.

What we did to set up the last problem was to interchange the x and y coordinates and then determine the equation of the new line. We can get the same result by solving the literal equation.

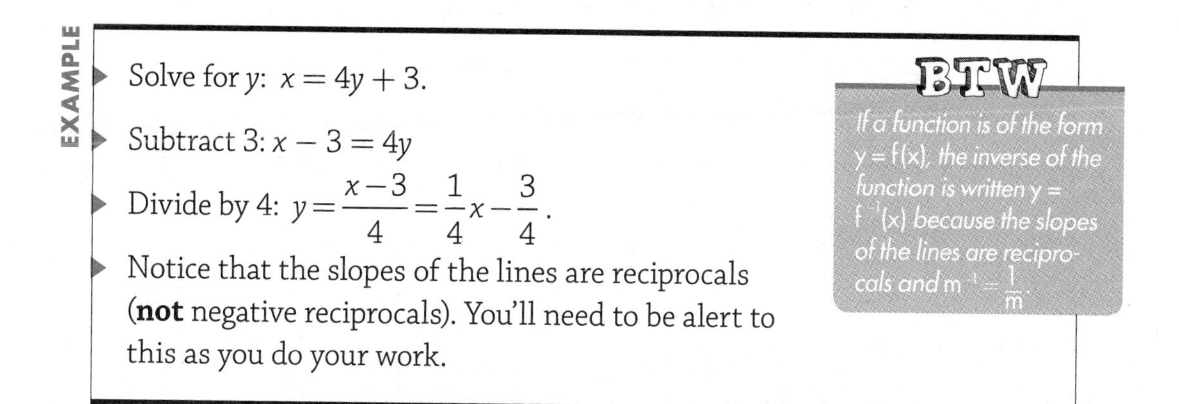

EXAMPLE

- Solve for y: $x = 4y + 3$.
- Subtract 3: $x - 3 = 4y$
- Divide by 4: $y = \frac{x-3}{4} = \frac{1}{4}x - \frac{3}{4}$.
- Notice that the slopes of the lines are reciprocals (**not** negative reciprocals). You'll need to be alert to this as you do your work.

BTW

If a function is of the form $y = f(x)$, the inverse of the function is written $y = f^{-1}(x)$ because the slopes of the lines are reciprocals and $m^{-1} = \frac{1}{m}$.

We pretty much ignore the $f(x)$ notation when we are asked to find the inverse of function written in function form.

EXAMPLE

- Find the equation for the inverse of the function $y = f(x) = \frac{-5}{9}x + 2$.
- We switch the x- and y-coordinates to determine the equation of the inverse: $x = \frac{-5}{9}y + 2$.
- Subtract 2: $x - 2 = \frac{-5}{9}y$
- Multiply by $\frac{-9}{5}$: $\frac{-9}{5}(x-2) = y$
- This is an acceptable form for the equation of the line. We know the slope is $\frac{-9}{5}$ and that the x-intercept has coordinates (2, 0). However,

most people are not comfortable with the equation (especially those who write multiple-choice questions), so let's write the answer in slope-intercept form.

- Apply the distributive property: $y = \frac{-9}{5}(x-2) \Rightarrow y = \frac{-9}{5}x + \frac{18}{5}$
- Do not use the standard form of the answer because the inverse function should always be in the form $y =$.

EXERCISES

EXERCISE 4-1

Determine the slope of the line identified in each problem.

1. The line that passes through the points (3, 0) and (0, 12).
2. The line that passes through the points (−2, −5) and (4, −1).
3. The line that passes through the points (9, −2) and (9, 12).
4. The line that passes through the points (−5, 12) and (13, 12).
5. The line shown in the accompanying diagram.

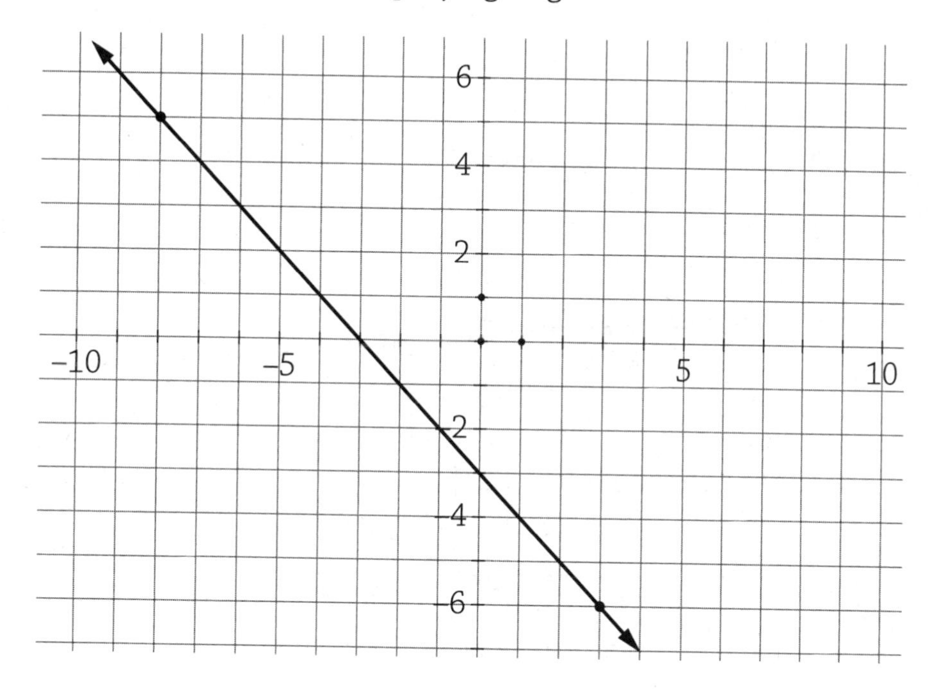

6. The line shown in the accompanying diagram.

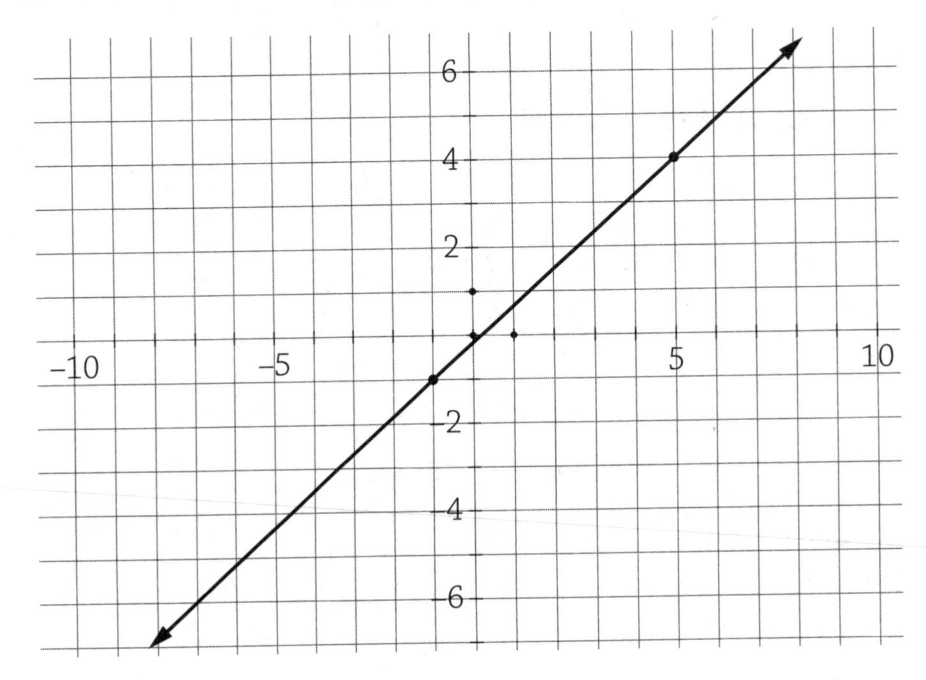

7. The line shown in the accompanying diagram.

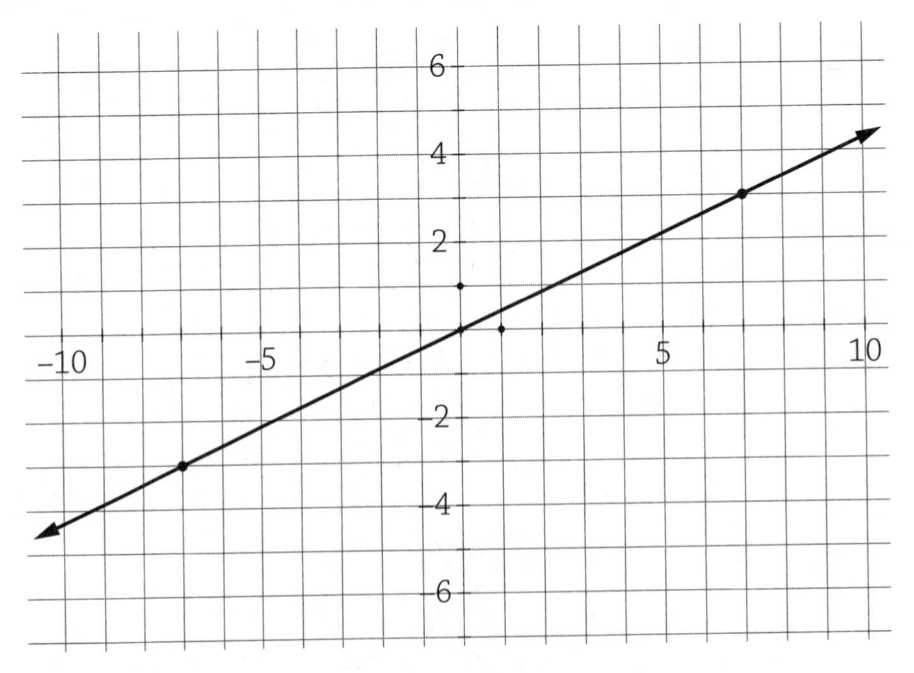

8. The line with equation $y = -3x + 19$.
9. The line with equation $y - 4 = \frac{-2}{3}(x + 5)$.
10. The line with equation $4x + 8y = 18$.
11. The line parallel to the line passing through the points (–3, 4) and (7, 12).
12. The line perpendicular to the line passing through the points (–3, 4) and (7, 12).
13. The line perpendicular to the line $5x - 7y = 35$.

EXERCISE 4-2

Determine the constant rate of change for these questions.

1. Molly is paid an hourly wage and a fixed amount for transportation. She earned \$140 for working 12 hours one week and \$173 for working 15 hours the following week. How much money does Molly earn per hour?
2. Alice is paid a fixed rate for driving packages as a courier and is reimbursed for the mileage driven. She received a check for \$87.00 for driving 80 miles on one trip and another for \$106.50 for a trip of 110 miles. How much money is Alice reimbursed per mile driven?

EXERCISE 4-3

Questions 1 and 2 are based on the following paragraph.

Molly is paid an hourly wage and a fixed amount for transportation. She earned \$140 for working 12 hours one week and \$173 for working 15 hours the following week. How much money does Molly earn per hour?

1. Write an equation for the amount of money Molly is paid based on the number of hours she works.
2. How much money is Molly paid to help with her transportation costs?

EXERCISE 4-4

Questions 1 and 2 are based on the following paragraph.

Alice is paid a fixed rate for driving packages as a courier and is reimbursed for the mileage driven. She received a check for \$87.00 for driving 80 miles on one trip and another for \$106.50 for a trip of 110 miles. How much money is Alice reimbursed per mile driven?

1. Write an equation for the amount of money Alice is paid based on the number of miles she drives.
2. Determine the amount of money Alice is paid if she drives 205 miles.

EXERCISE 4-5

Write the equation of the line described in each question.

1. Write an equation in slope-intercept form for the line that passes through the points (2, 3) and (5, 9).
2. Write an equation in point-slope form for the line that passes through the points (2, 3) and (5, 9).
3. Write an equation in point-slope form for the line that passes through the points (−4, 2) and (3, 5).
4. Write an equation in standard form for the line that passes through the points (−4, 2) and (3, 5).
5. Write an equation for the line that passes though the points (8, 10) and (−2, 12). (You choose the form of the equation.)
6. Write an equation for the line that passes though the points (−7, −4) and (−3, 11). (You choose the form of the equation.)

7. Write an equation of the line that passes through the point (4, –2) and is parallel to the line $y = \frac{-5}{2}x + 3$.
8. Write an equation for the line that passes through the point (5, –3) and is parallel to the line $y = \frac{7}{3}x - 8$.
9. Write an equation for the line that passes through the point (–2, –6) and is parallel $5x + 7y = 13$.
10. Write an equation for the line that passes through the point (4, –7) and is parallel $10x - 11y = 13$.
11. Write an equation for the line that has an x-intercept of 12 and is parallel to the line $7x + 8y = 10$.
12. Write an equation for the line that has a y-intercept at 4 and is perpendicular to the line $y = \frac{3}{5}x + 2$.
13. Write an equation for the line that passes through the point (3, 5) and is perpendicular to the line $y = \frac{-5}{7}x - 2$.
14. Write an equation for the line that passes through the point (–10, 6) and is perpendicular to the line $8x + 3y = 10$.
15. Write an equation for the line that passes through the point (–10, 6) and is perpendicular to the line $8x - 3y = 10$.

EXERCISE 4-6

Answer the following questions.

1. What is the equation of the inverse of the function $y = 5x + 2$?
2. What is the equation of the inverse of the function $y = f(x) = \frac{5}{11}x - 10$?
3. The graph of the function $y = f(x)$ is shown below. Sketch the graph of the inverse function on the same set of axes.

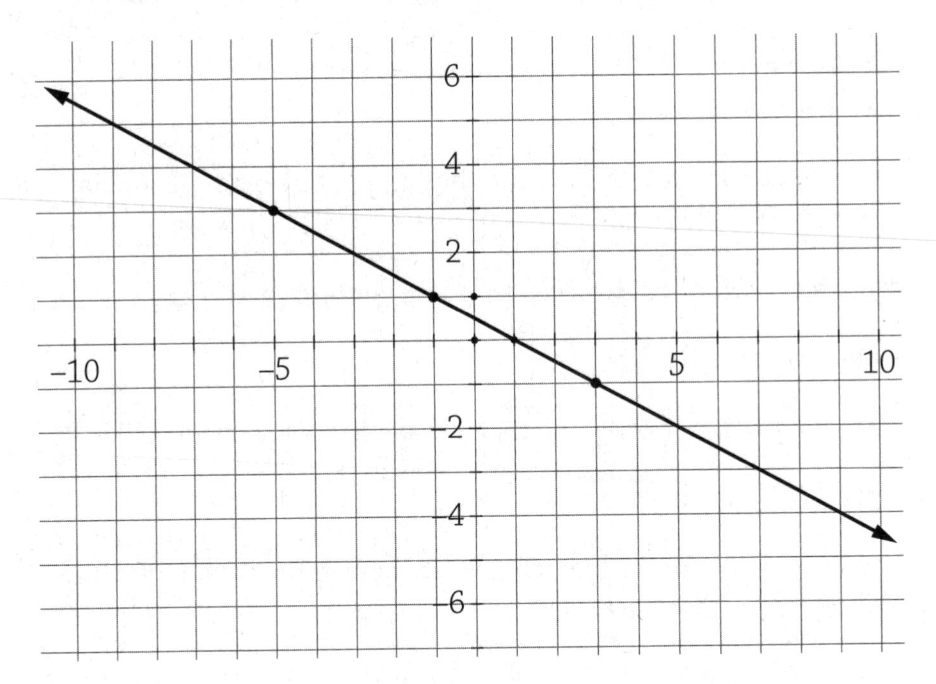

Applications of Linear Functions

Linear functions can be applied in determining the solutions to everyday problems involving money, distance, and motion.

Word Problems: Coins, Geometry, and Motion

Being able to combine algebraic expressions and solve linear equations are the beginning steps to using mathematics in the everyday world. While some of your classmates will pursue careers in the physical sciences, the social sciences, the business world, or in the trades, each will encounter problems that can be solved using algebraic thinking. Even if they do not write down the equation in question, it is the process that matters. We will explore some problems in this section that lend themselves to linear equations. We'll define the variables in question, and set up and solve the equations. Finally, if needed, we'll interpret the answer in the context of the problem. As you learn more skills, the extent of the problems you will be able to solve will grow as well.

Coin Problems

When solving problems involving items of value, you must distinguish between the physical object (a nickel, a quarter, a ten-dollar bill, a stamp) and the value of that object (5 cents, 25 cents, \$10, \$0.49). The equations you write will involve the relationships within each of these ideas.

EXAMPLE

- Colin's piggy bank has 320 coins in it: nickels, dimes, and quarters. He and Diane separate the coins into piles of coins of the same type. They discover that there are twice as many quarters as dimes and 40 more nickels than there are quarters.
- Simple question: How many coins of each type does Colin have in his piggy bank?
- More complicated question: How much money does Colin have in his piggy bank?

- In either case, all we know is the relationship between the numbers of coins of each type. The number of quarters is related to the number of dimes. If we define d as the number of dimes in the piggy bank, then we denote the number of quarters as $2d$. The number of nickels and number of quarters in the bank are defined by their relation to the number of dimes. There are 40 more nickels than quarters, so the number of nickels must be $2d + 40$.
- So we have:

  d: number of dimes

  $2d$: number of quarters

  $2d + 40$: number of nickels
- There are 320 coins in the piggy bank, so the equation for this problem is

  number of nickels + number of dimes + number of quarters = 320

$$2d + 40 + d + 2d = 320$$
$$5d + 40 = 320$$
$$5d = 280$$
$$d = 56$$

- The answer to the simple question is that there are $2(56) + 40 = 152$ nickels, 56 dimes, and $2(56) = 112$ quarters.
- The answer to the more complicated question is that there is $\$.05(152) + \$0.10(56) + \$0.25(112) = \$7.60 + \$5.60 + \$28 = \$41.20$ in the piggy bank.

BTW

The reason for posing both questions is to point out that you must be sure to read the problem and answer the question being asked. Too often students set up the problem correctly and determine that d = 56, and do no more. The more complicated question concerns the value of the money, yet all the information in the problem is about the physical coins, not their value. This indicates that there is more to do once the equation has been solved.

Read this problem carefully so that you can distinguish between what the variable expressions represent and what the difference is between the two questions that are posed:

EXAMPLE

- Cameron and Carson are going to pool their money together so that they can buy a bicycle. Cameron has \$5 more in one-dollar bills than Carson does, while Carson has twice as many five-dollar bills than Cameron has one-dollar bills. Cameron has 2 fewer five-dollar bills than does Carson. Together, they have a total of \$249.
- Simple question: How many bills of each type does Carson have?
- More complicated question: How much money does each boy have?
- The facts of the problem discuss the number of bills of each denomination, but the cumulative statement is about the value of these bills:
  - "Cameron has \$5 more in one-dollar bills (singles) than Carson" also says that Cameron has 5 more singles than Carson. We can let b represent the number of one-dollar bills that Carson has. (I would really like to have s as the variable to represent the number of singles, but a handwritten s is often confused with the number 5, so I choose b for bills. Also, I get tired of using x all the time.)
  - Cameron has 5 more singles than Carson, so he has $b + 5$ singles.
  - "Carson has twice as many five-dollar bills than Cameron has singles" tells us that Carson has $2(b + 5)$ five-dollar bills.
  - "Cameron has 2 fewer five-dollar bills than does Carson" translates to $2(b + 5) - 2$ five-dollar bills for Cameron.

▶ To summarize this, we have:

	Cameron	Carson	Total
Ones	$b+5$	b	$2b+5$
Fives	$2b+8$	$2b+10$	$4b+18$

▶ The total value of these bills is \$249, so the equation is $1(2b+5)+5(4b+18)=249$. (Notice that each term on the left-hand side of the equation is the product of the value of one bill and the number of bills of that denomination.) Solve the equation:

$$2b+5+20b+90=249$$

▶ Combine terms: $22b+95=249$

$$22b=154$$

$$b=7$$

▶ The answer to the simple question is

	Cameron	Carson
Single	12	7
Fives	22	24

▶ The answer to the more complicated question is Cameron has $\$1(12)+\$5(22)=\$122$ and Carson has $\$1(7)+\$5(24)=\$127$.

Geometry Problems

There are basic geometric relationships that can be applied to simple algebraic equations:

- The sum of the measures of the angles of a triangle is 180°.
- The sum of the measures of the angles of a quadrilateral is 360°.

- Two of the sides of an isosceles triangle have the same length.
- The perimeter of any polygonal figure is the sum of the lengths of the sides.

Let's begin with a problem about rectangles.

EXAMPLE

▶ In a rectangle with perimeter 182 inches, the length of one side of the rectangle measures 7 inches longer than twice the length of an adjacent side. Determine the length of each side of the rectangle.

▶ I am a firm believer in drawing pictures is a great way to whenever I can to help me understand a problem. We have a rectangle with a width we'll call w. The adjacent side, the length, has a measure that is 7 inches longer than twice the width.

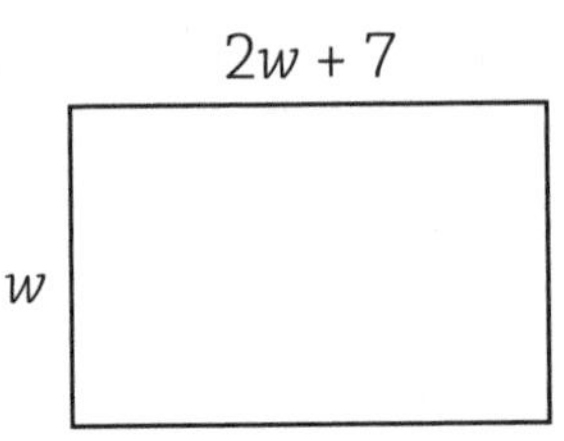

▶ One of the formulas for perimeter is $P = 2(l + w)$. Using this

$$2(w + 2w + 7) = 182$$

$$2(3w + 7) = 182$$

▶ Divide by 2: $3w + 7 = 91$

▶ Subtract: $3w = 84$

▶ Divide: $w = 28$

▶ The sides of the rectangle measure 28 inches and $2(28) + 7 = 63$ inches, respectively.

Recall from your study of geometry in earlier years that the two sides with equal length in an isosceles triangle are called the *legs* and the third side is called the *base*.

EXAMPLE

▶ The base of an isosceles triangle has a length 150 cm less three times the length of either leg. If the perimeter of the isosceles triangle is 400 cm, determine the length of each side of the triangle.

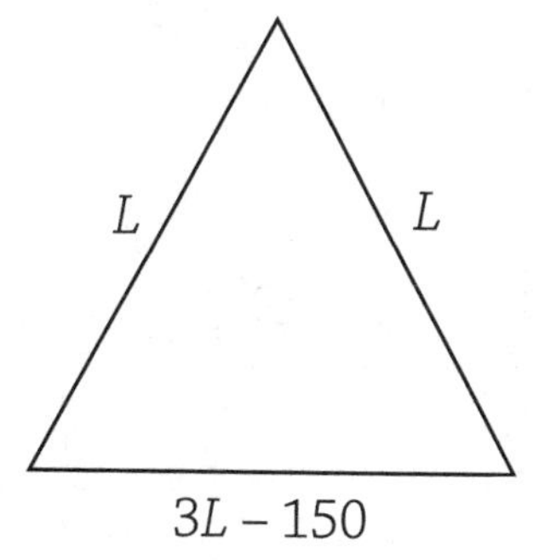

▶ The perimeter equation is

$$\begin{aligned} L + L + 3L - 150 &= 400 \\ 5L - 150 &= 400 \\ 5L &= 550 \\ L &= 110 \end{aligned}$$

▶ The lengths of the sides of the triangle are 110 cm, 110 cm, and 180 cm.

Sometimes an important piece of information will not be stated explicitly but only be implied in the problem.

EXAMPLE

- One of the acute angles of a right triangle measures 20 degrees more than 4 times the other angle. Find the measure of the larger acute angle.

- We know a triangle has three angles and the problem discusses the relationship between two of these angles. Because the triangle is a right triangle, we know that the third angle has a measure of 90°.

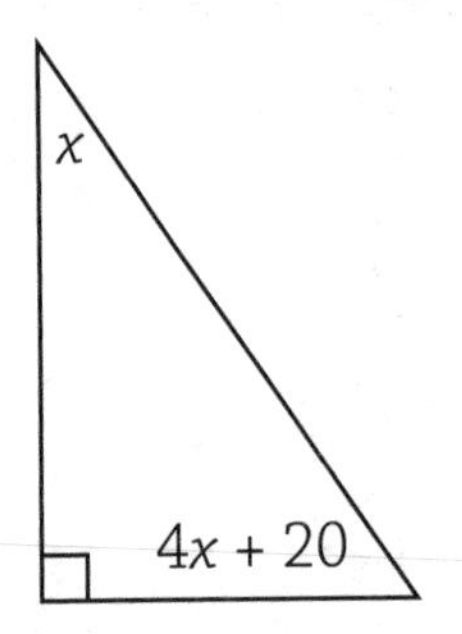

- The equation for this problem can be written using just the two acute angles, since they must add up to 90°.

$$\begin{aligned} x + 4x + 20 &= 90 \\ 5x + 20 &= 90 \\ 5x &= 70 \\ x &= 14 \end{aligned}$$

- The measure of the larger acute angle is 76°.

Motion Problems

Similar to the coin problems, many of the motion problems represent simplistic events but do serve the purpose of providing opportunities to relate rate, time, and distance. Yes, the path described by the objects in these problems is always linear (that's why they are in the chapter of applications of linear functions). We will look at a more complicated motion problems later in the book. Really complicate motion issues cannot be done until you have a more complete background in trigonometric functions.

EXAMPLE

▶ Two trucks leave Des Moines, Iowa, traveling along route I-80. One truck travels west at a rate of 75 mph, and the other travels east at 70 mph. How many hours does it take before the trucks are 1000 miles apart?

▶ Let the time needed be define by the variable t. We know that distance = rate × time.

distance west + distance east = 1000

$$75t + 70t = 1000$$
$$145t = 1000$$
$$t = 6.897$$

▶ Most people will not say "It took 6.897 hours for the two trucks to be a 1000 miles apart." They might say, "It took just less than 7 hours," or they might say, "It took 6 minutes shy of 7 hours" if they want to be more accurate. It seems that an answer like 6.9 hours is reasonable for the math classroom.

Let's take a look at a situation in which the vehicles travel toward each other.

EXAMPLE

▶ A train leaves Dallas, Texas at 9 a.m. heading north to Chicago, Illinois, (970 miles away) at a speed of 60 mph. A second train leaves Chicago at 10 a.m. and heads for Dallas at a speed of 70 mph. If the two trains are travelling on parallel tracks, at what time will they pass each other?

▶ At the time the two trains meet each other, they will have traveled the combined distance from Dallas to Chicago. How far did each train travel? We know the rates for each of the trains. We need to define a variable to represent the time each train has been moving. I choose to define t as the time the northbound train has been in motion making $t - 1$ the time the southbound train was moving. (We could just as easily defined t as the time for the southbound train and $t + 1$ for the northbound train.)

▶ The sum of the distances traveled is 970 miles:

$$60t + 70(t - 1) = 970$$
$$60t + 70t - 70 = 970$$
$$130t = 1040$$
$$t = 8$$

▶ The two trains pass each other at 5 p.m. (8 hours after the Dallas train left the station).

In both examples, we added distances traveled to answer the question. This question is a little different.

EXAMPLE

▶ Frank left his house and got on the local highway, where he began traveling west at 40 miles per hour. After he was gone left, his wife Jen noticed that he left paperwork he needed on the table. By the time she got on the highway, it was an hour after Frank had gotten on the road. If she travels at 65 mph on a parallel route, how much time is needed before she catches up to him?

▶ Similar to the train problem, the two vehicles have a difference in travel time. Let t be the number of hours it takes her to catch up to Frank. Using that definition, Frank has traveled for $t + 1$ hours before Jen catches up to him. What is the relation between the distances?

When Jen catches up to Frank, she will have traveled the same distance as him.

$$40(t+1) = 65t$$
$$40t + 40 = 65t$$
$$40 = 25t$$
$$t = 1.6$$

It takes Jen 1.6 hours to catch up with Frank.

Direct Variation

Direct variations are simply applications of proportions. Because they are applications, we should pay attention to the meaning of the ratio that forms a side of the equation. For example, a person who earns $45 for working 5 hours (and does not receive any extra money for transportation or either reasons) should expect to receive $99 for working 11 hours. The ratio $\frac{\$45}{5 \text{ hours}} = \$9/\text{hr}$, the rate at which she is paid. This constant rate allows us to write the function $p(h) = 9h$, where p is the amount she is paid for the h hours that she works. In this example, we call 9 the **constant of variation**.

EXAMPLE

Under normal driving conditions, Cameron can drive 384 miles using 12 gallons of gasoline. How far will Cameron be able to drive (under similar driving conditions) using 7 gallons of gasoline?

We set up the proportion $\frac{384}{12} = \frac{m}{7}$ and solve to get the answer 224 miles.

We use phrases "varies directly as" or "directly proportional to" to indicate that there is a direct variation. The equation for a direct variation will be of the form $f(x) = kx^n$ (because not all variations are linear). The previous question would be better phrased if the problem contained the information that "the number of miles that Cameron can drive under certain conditions varies directly as the number of gallons of gasoline used." The language used in the preceding problem assumes that the reader understands this concept. So we'll add it here.

EXAMPLE

▶ The number of miles that Cameron can drive under certain conditions varies directly as the number of gallons of gasoline used. If he can drive 384 miles using 12 gallons of gasoline, how far can he drive using 7 gallons of gasoline?

▶ A different approach to solving this problem is: The equation for the mileage, m, driven varies directly with the number of gallons of gasoline, g, used, so $m = kg$, where k is the constant of variation. Solve $384 = 12g$ to determine the consumption rate for the car is 32 miles per gallon (mpg). The equation is now $m = 32g$, and when $g = 7$, $m = 32(7) = 224$ miles.

There a few examples of direct variation other than hourly earnings, gas consumption, and price per gallon worth looking at. Here are a few.

- **Hooke's law** The length a spring is stretched is directly proportional to the weight at the end of the spring.

EXAMPLE

▶ If a 12-kg weight extends the length of a spring by 8 cm, how much weight is needed to extend the length of the same spring 18 cm?

▶ You can set up the proportion $\frac{8}{12} = \frac{18}{w}$ so that $w = 27$ kg. Why did I write the ratio $\frac{8}{12}$ rather than $\frac{12}{8}$? Let's pose that question again with one change. Why write the ratio as $\frac{8\text{ cm}}{12\text{ kg}}$ instead of $\frac{12\text{kg}}{8\text{cm}}$? Does the length of the weight attached to the spring dictate the length the spring is stretched, or does the change in the spring length dictate the weight attached to the spring? The weight attached dictates the change in length, so $\frac{8\text{cm}}{12\text{kg}}$ is more appropriate and also tells us that the length of the spring is stretched $\frac{2}{3}$ cm/kg.

▶ You could have also set up the equation $L = \frac{8}{12}w$, where L is the change in the length of the spring and w is the weight added, and solved the equation $18 = \frac{8}{12}w$.

- **Ideal gas law** The ideal gas law states that under a constant pressure, the volume of a gas is directly proportional to the temperature of the gas (measured in degrees Kelvin, °K).

EXAMPLE

▶ If the volume of a gas is 2.3 liters when the temperature is 300°K, what is the volume of the gas at a temperature of 320°K?

▶ You can set up the proportion $\frac{2.3}{300} = \frac{v}{320}$ to determine that $v = 2.453$ liters (rounded to the nearest thousandth), or you can use the function $v = \frac{2.3}{300}T$ and evaluate this function when $T = 320$.

- **Free fall** The distance an object falls when released from a height varies directly as the square of the time since being released.

EXAMPLE

▶ If an object in free fall falls 64 feet after 2 seconds of being released, how far will the object have fallen 5 seconds after being released?

▶ The distance, d, varies directly as the square of the time, t, so the function is $d = kt^2$, a nonlinear relationship. Use the first piece of information to determine the value of k. $64 = k(2)^2$ to show that $k = 16$ ft/sec^2. You can now determine the distance fallen after 5 seconds: $d = 16(5)^2 = 400$ feet.

The units ft/sec^2 is a measure of acceleration (the rate of change of velocity over a time period) so that $\frac{\text{ft}}{\text{sec}} \div \text{sec} = \frac{\text{ft}}{\text{sec}} \times \frac{1}{\text{sec}} = \frac{\text{ft}}{\text{sec}^2}$.

Sequences

When you were younger, you learned to skip-count. If you are not familiar with that term, let me remind you. Count by 2's, or by 5's, possibly even 10's. As a class, you would recite, "2, 4, 6, 8, 10 . . ., ..." or something similar for 5's and 10's. Invariably, you always began with the number you used as the base. You might have seen a movie in which an army sergeant ordered his troops to "count off by 4." In this case, the soldiers would count "1, 2, 3, 4, 1, 2, 3, 4, . . . ". They would then break into groups so that all the 1's are together, the 2's are together, and the same for the 3's and the 4's. Now consider this: If you were the 73rd person in line, to which group would you be assigned?

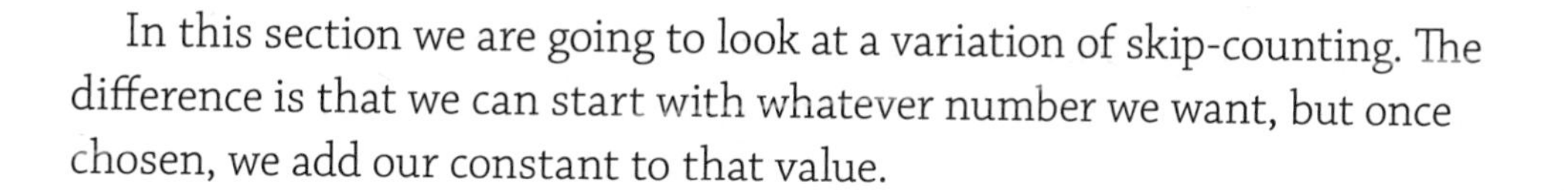

In this section we are going to look at a variation of skip-counting. The difference is that we can start with whatever number we want, but once chosen, we add our constant to that value.

EXAMPLE

▶ Count by 4's beginning with 7.

▶ 7, 11, 15, 19, 23, 27, 31, . . .

We need to get some terminology into the picture. The first term is **sequence**. A sequence is any listing of numbers that follows a pattern.

EXAMPLE

▶ Determine the next five terms in each of the following sequence: 2, 9, 16, 23, . . .

▶ Pattern: Starting with 2, add 7 to get to the next term. The next five terms are 30, 37, 44, 51, 58.

Sequences can decrease as well as increase.

EXAMPLE

▶ Determine the next five terms in each of the following sequence: 87, 79, 71, 63, 55, . . .

▶ Pattern: Starting with 87, subtract 8 to get to the next term. The next five terms are 47, 39, 31, 23, 15.

Not only can sequences be created by adding a constant, but they can also be formed by multiplying by a constant.

EXAMPLE

- Determine the next five terms in each of the following sequence: 3, 6, 12, 24, 48, . . .
- Pattern: Starting with 3, multiply by 2 to get to the next term. The next five terms are 96, 192, 384, 768, 1,536.

Sometimes the constant factor is a fraction.

EXAMPLE

- Determine the next five terms in each of the following sequence: 256, 128, 64, 32, . . .
- Pattern: Starting with 128, multiply by $\frac{1}{2}$ to get the next term.

  The next five terms are 16, 8, 4, 2, 1.

Then there are those sequences in which the pattern is a tad trickier.

EXAMPLE

- Determine the next five terms in each of the following sequence: 1, 2, 4, 7, 11, 16, 22, . . .
- Do you see a pattern here? Adding 1 doesn't work. Neither does multiplying by 2. There's a nice trick for working with sequences such

as this. It is called **successive differences**. See if this diagram gives you a better feeling for the pattern.

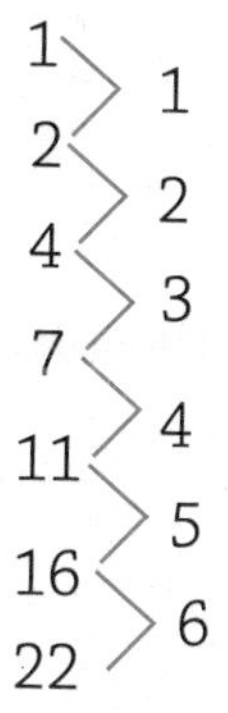

- Pattern: There is not a constant difference or ratio here. The difference between the consecutive terms increases by 1 from term to term. The next five terms are 29, 37, 46, 56, 67.

The last example is a bit more complicated. You'll see that successive differences yield a pattern you won't expect.

EXAMPLE

- Determine the next five terms in each of the following sequence: 1, 1, 2, 3, 5, 8, 13, 21, 34, . . .
- Take a look at the successive differences. With the exception of the zero as the first difference, the pattern repeats itself.
- This is a famous sequence called the Fibonacci sequence. The first two terms are 1. Beginning with the third term, each term is found by adding the two previous terms: $1 + 1 = 2$, $1 + 2 = 3$; $2 + 3 = 5$; $3 + 5 = 8$; $5 + 8 = 13$; and so on. So the next five terms are 55, 89, 144, 233, and 377.

As you can see, the rules for how the sequences can be created can be complicated. We will look at a particular type of sequence in the next section.

Arithmetic Sequences as Linear Functions

The table of values shown below is the output for the function $y = 3x + 2$ using the set of counting numbers as the domain.

x	y
1	5
2	8
3	11
4	14
5	17
6	20
7	23

Observe that the output 5, 8, 11, 14, 17, 20, 23 forms a sequence in which the terms differ by 3. Please note that 3 is the slope of the linear rule that creates the output values. An **arithmetic sequence** is a sequence in which consecutive terms differ by a constant. It is always the case that the terms in an arithmetic sequence can be represented by a linear function.

EXAMPLE

- Determine the rule for the linear function that will generate the arithmetic sequence:
  17, 24, 31, 38, 45, 52, 59, . . .

- The difference between the terms is 7, so the slope of the linear expression must be 7. Therefore, the rule can be written as $y = 7x + b$. The first term of the sequence is 17. Using $x = 1$ and $y = 17$, we can show that $b = 10$ and the rule for generating the arithmetic sequence is $y = 7x + 10$.

We are going to change a couple of things about the result we just got. Rather than use x and y as the input and output, we will go with the convention of using n for the input variable (n is the first letter in "natural" and the set of Natural numbers is the same as the set of counting numbers). We will use a_n for the nth output of the sequence. (It would seem logical to say that we use a because it is the first letter in "arithmetic," but that is not the reason. Some authors choose t to represent the "term" of a sequence, which makes sense, but there is no hard-and-fast reason for using a.)

EXAMPLE

▶ Find the 80th term in the sequence 17, 24, 31, 38, 45, 52, 59, . . .

▶ We just discovered that the rule for creating this sequence is $a_n = 7n + 10$. Therefore, the 80th term is found as $a_{80} = 7(80) + 10 = 560 + 10 = 570$.

Let's put this altogether.

EXAMPLE

▶ Find the 25th term of the arithmetic sequence 19, 30, 41, 52, . . .

▶ The common difference is 11, so $a_n = 11n + b$ and the first term, a_1, is 19, so $a_1 = 11(1) + b$, making $b = 8$. The general for generating this sequence is $a_n = 11n + 8$, and the 25th term is $a_{25} = 11(25) + 8 = 275 + 8 = 283$.

Seating in theaters is often done so that the seats in one row are not directly behind the seats in the row before it so that there is line of vision to the stage and/or screen. This is usually accomplished by adding two more seats to the number of seats in the previous row.

EXAMPLE

- There are 35 seats in the center section of the Hartman Theater, and every row after the first has 2 more seats than the row in front of it. The center section of the Hartman Theater has 32 rows. How many seats are in the last row?

- The rule for the number of seats in a row is of the form $a_n = 2n + b$ (from the 2 extra seats), and we know that $b = 33$ because there are 35 seats in row 1: $a_2 = 2(1) + b = 35$, so $b = 33$. Therefore, the number of seats in row 32 is $a_{32} = 2(32) + 33 = 97$.

How do we deal with a problem in which we know two terms of the arithmetic sequence, neither of which is the first term? As always, we go back to what we know.

EXAMPLE

- If the 9th and 17th terms of an arithmetic sequence are 60 and 116, respectively, what is the value of the first term? What is the value of the 70th term?

- We know that terms of an arithmetic sequence come from the equation of a line. Two of the points on this line must be (9, 60) and (17, 116). We can use this information to determine the equation of the line that defines the sequence and answer the questions using that equation.

- The slope of the line is $\frac{116-60}{17-9} = \frac{56}{8} = 7$, so the equation of the line is of the form $y = 7n + b$. Using the point (9, 60), we can determine the value of b: $60 = 7(9) + b$, so $b = -3$.

- The equation of the line that defines the sequence is $y = 7n - 3$. The first term is $y = 7(1) - 3 = 4$, and the 70th term is $y = 7(70) - 3 = 487$.

Arithmetic Series

At the end of the 20th century (probably before you were born!), there was a major concern about computer systems going haywire. Nicknamed Y2K (year 2000), the scare was caused by the fact that most computers were programmed to hold dates by the six-digit notation of mm/dd/yr (that is, two digits for the month, two for day of the month, and two for the year). For example, April 28, 1954, would be written as 04/28/54, and January 2, 2000, would be tagged as 01/02/00. The problem was that "00" was interpreted by computers to be 1900, not 2000. Panic ensued, with people envisioning major failures in everything from banking systems to weapons launchers. As a result, in the late 1990s, many countries around the world cooperated and shared information to reprogram their computer systems to four-digit years. As it turned out, the century turned with only minor incidents of computer systems failing, and even countries without major preparedness suffered little consequence.

Along with the Y2K panic was an excitement about everything having to do with the start of the new millennium and the ending of the last one. After all, instead of hyping things as "the song of the decade" or "the storm of the century," people could start referring to things as the "[something] of the millennium," and debates arose about who the most influential people of the last 1000 years were. Not wanting to miss this opportunity, professional mathematicians conducted a poll asking who was the most important mathematician of the millennium. The person with the most votes was someone whose name you probably know, Isaac Newton. The person who came in second, with one less vote than Newton, is someone whose name you probably do not know, Carl Friedrich Gauss.

EXAMPLE

Gauss (1777–1855) was 7 years old in a one-classroom school among children of all ages. The story goes that the teacher, wanting to take a break, had the students add the integers from 1 to 100. No sooner had the teacher sat at his desk when Gauss came up to him to tell him the answer was 5050. How did he do this? He thought to himself that the sum, S, or

$$S = 1 + 2 + 3 + \ldots + 98 + 99 + 100$$

could also be written as

$$S = 100 + 99 + 98 + \ldots + 3 + 2 + 1.$$

He then added the two rows together. On the left-hand side of the equation, the sum was $2S$, but on the right-hand side, he had $101 + 101 + 101 + \ldots + 101 + 101 + 101$. That is,

$$2S = 100(101)$$

because he was adding 101 one hundred times. Therefore,

$$S = 50(101) = 5050.$$

(Gauss, clearly a mathematical genius, also found a mistake at age four in his father's accounting books and went on to earn his Ph.D. by age 17!)

What has this got to do with what we are studying? It turns out that Gauss's approach tells us how to add the terms of an arithmetic sequence.

BTW

*A sequence is a listing of numbers. A **series** is the sum of the numbers in a sequence.*

EXAMPLE

Find the sum of the first 60 terms of the sequence defined by the rule $f(n) = 6n + 7$.

The terms of the sequence are 13, 19, 25, 31, . . . , 355, 361, 367. The sum, S, of these terms is

$$S = 13 + 19 + 25 + \ldots + 355 + 361 + 367$$

- Writing the terms in reverse order: $S = 367 + 361 + 355 + \ldots + 25 + 19 + 13$
- Add the equations together: $2S = 380 + 380 + 380 + \ldots + 380 + 380 + 380 = 60(380)$
- Divide by 2: $S = 30(380) = 11{,}400$

What do we really need to know in order to find the sum of the arithmetic series? We need to know the first term, the number of terms, and the last term. If we define a_1 as the first term, n as the number of terms, and a_n as the last term, the sum of the first n terms, S_n, is given by the formula $S_n = \frac{n}{2}(a_1 + a_n)$.

EXAMPLE

- Find the sum of the first 150 terms of the arithmetic series defined by the rule $f(n) = 4n + 9$.
- The first term a_1 is $4(1) + 9 = 13$, and the 150th term is $a_{150} = 4(150) + 9 = 609$. The sum of the 150 terms is $S_{150} = \frac{150}{2}(13+609) = 75(622) = 46{,}650$.

Naturally, if we have an equation, we should be able to solve equations for different terms. For example.

EXAMPLE

- The sum of the first 90 terms of an arithmetic series is 35370. If the first term of the series is 39, what is the value of the 90th term?

▶ Substitute the known values into the equation $S_n = \frac{n}{2}(a_1 + a_n)$ to get $35{,}370 = \frac{90}{2}(37 + a_{90}) = 45(37 + a_{90})$. Divide by 45 to get $786 = 37 + a_{90}$, so that $a_{90} = 749$.

Absolute Value

There are times when all we want to know about a number is its magnitude—that is, how big is it—and we do not care if the number is negative or positive.

How far is 5 from the origin on a number line? → 5 units
How far is −5 from the origin on a number line? → 5 units

Both 5 and −5 have the property that they are the same distance from zero on a number line. The absolute value of a number is often depicted as the number without regard to a positive or negative sign attached to that number. This is a clear-cut, reasonable geometric approach to the subject. The symbol for the absolute value of a number is to use two vertical bars with the number between them. For example, the absolute value of −5 is written $|-5|$.

EXAMPLE

▶ Solve $|x| = 4$.

▶ The problem is asking for those values of x that are 4 units from the origin, so the answer is $x = -4, 4$ (which can also be written as ± 4).

Hopefully, you are not surprised that the number midway between −4 and 4 is 0, as that is the point from which the distance is measured. This point is raised to help you see a geometric solution to a more complicated problem.

EXAMPLE

▶ Solve $|x - 1| = 4$.

▶ As we saw in the last example, if the number inside the absolute value bars is -4 or 4, the absolute value is 4. So if $x - 1 = -4$ or $x - 1 = 4$, then $|x - 1| = 4$. If $x - 1 = -4$, then $x = -3$, and if $x - 1 = 4$, then $x = 5$. Which point is midway between -3 and 5? Why, it is 1! Hold onto that thought.

▶ Solve $|x - 2| = 4$. Do you know the answer? If $x - 2 = -4$, then $x = -2$, and if $x - 2 = 4$, then $x = 6$. Both these points are 4 units from 2.

▶ In fact, the solution to $|x - a| = 4$, will always be $x = a - 4$, and $x = a + 4$, so we can say the solution to the equation $|x - a| = 4$ are those points 4 units from $x = a$ on the number line.

Before we go further into solving equations, there are a few properties of absolute value that we need to learn.

- **Multiplication** $|ab| = |a|\ |b|$

  The absolute value of a product equals the product of the absolute values.

- **Division** $\left|\frac{a}{b}\right| = \frac{|a|}{|b|}$

  The absolute value of a quotient equals the quotient of the absolute values.

The same cannot be said for addition and subtraction. Consider the relationship between $|a + b|$ and $|a| + |b|$ when $a = 5$ and $b = 3$. Both expressions are equal. Now consider the relationship when $a = 5$ and $b = -3$. They are no longer equal. In fact, $|a| + |b|$ is greater than $|a + b|$. When a

and b are both negative, the two expressions are again equal. Consequently, we know $|a| + |b| \geq |a + b|$.

How does $|a - b|$ compare to $|a| - |b|$? If the signs of a and b are the same, we have equality. If the signs are different, then $|a - b| > |a| - |b|$.

EXAMPLE

- The relationship between the terms that are multiplied together are most important to us at this point. We can solve the equation $|3x - 6| = 9$ by:
- Factor 3: $|3(x - 2)| = 9$
- Split the absolute value terms: $|3|\ |x - 2| = 9$
- Divide: $|x - 2| = 3$
- Solve: The points 3 units on the number line from 2 are -1 and 5.
- We could also have solve the problem in a traditional manner:
- Drop the absolute value: $3x - 6 = -9$ or $3x - 6 = 9$
- Add 6: $3x = -3$ or $3x = 15$
- Divide: $x = -1$ or $x = 5$

Here is another example for you to practice.

EXAMPLE

- Solve $|4x - 12| = 24$
- **Geometric approach:** $|4x - 12| = 24$
- Factor 4: $|4|\ |x - 3| = 24$
- Divide: $|x - 3| = 6$
- Solve: The points on the number line 6 units from 3 are -3 and 9.

Algebraic approach: $|4x - 12| = 24$

Drop the absolute value: $4x - 12 = -24$ or $4x - 12 = 24$

Add 12: $4x = -12$ or $4x = 36$

Divide by 4: $x = -3$ or $x = 9$

Notice that all the problems are of the form $|x - a|$. What happens if the form is $|x + a|$? That is pretty simple. We rewrite $|x + a|$ as $|x - (-a)|$ and measure from $x = -a$.

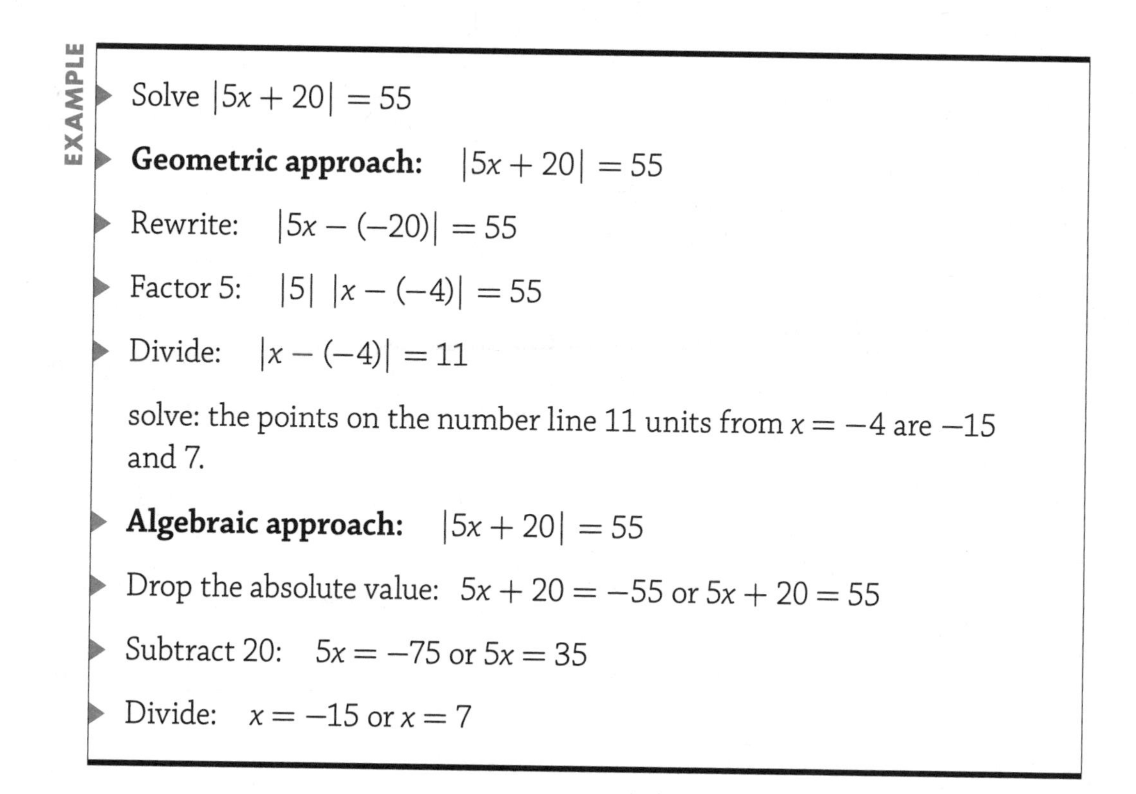

EXAMPLE

Solve $|5x + 20| = 55$

Geometric approach: $|5x + 20| = 55$

Rewrite: $|5x - (-20)| = 55$

Factor 5: $|5|\ |x - (-4)| = 55$

Divide: $|x - (-4)| = 11$

solve: the points on the number line 11 units from $x = -4$ are -15 and 7.

Algebraic approach: $|5x + 20| = 55$

Drop the absolute value: $5x + 20 = -55$ or $5x + 20 = 55$

Subtract 20: $5x = -75$ or $5x = 35$

Divide: $x = -15$ or $x = 7$

Not all problems have nice integer responses, so you will have to choose how you prefer to deal with such issues.

EXAMPLE

- Solve $|-3x+16|=23$
- **Geometric approach:** $|-3x+16|=23$
- Factor 3: $|-3|\left|x-\frac{16}{3}\right|=23$
- Divide by 3: $\left|x-\frac{16}{3}\right|=\frac{23}{3}$
- Solve: The points on the number line $\frac{23}{3}$ units from $x=\frac{16}{3}$ are $\frac{-7}{3}$ and $\frac{39}{3}=3$.
- **Algebraic solution:** $|-3x+16|=23$
- Drop the absolute value: $-3x+16=-23$ or $-3x+16=23$
- Subtract 16: $-3x=-39$ or $-3x=7$
- Divide: $x=3$ or $x=\frac{-7}{3}$

EXERCISES

EXERCISE 5-1

Use the following problem to answer questions 1 and 2.

Cameron opened his piggy bank and determined that he had 165 coins in it—a combination of nickels, dimes, and quarters. The number of dimes exceeded the number of nickels by 15, and the number of quarters exceeded three times the number of nickels by 5.

1. How many coins of each type did Cameron have in his piggy bank?
2. How much money did Cameron have in his piggy bank?

EXERCISE 5-2

Use the following information to answer questions 1 and 2.

Charlie is a cashier at a convenience store, and each day when his shift ends, he needs to count the amount of money is his register. One day he notices that the number of five-dollar bills was 20 more than twice the number of singles and the number of ten-dollar bills was 50 more than the number of five-dollar bills. All told, he had 240 bills in his drawer.

1. How many bills of each type did Charlie have in his drawer?
2. If the amount of change in his drawer was $12.35, how much money was in his drawer at the end of the shift?

EXERCISE 5-3

Answer the following questions.

1. The length of a rectangle is 1 foot less than twice the width of the rectangle. If the perimeter of the rectangle is 52 feet, determine the dimensions of the rectangle.

2. The width of a rectangle measures 2 cm less than one-half the length of the rectangle. If the perimeter of the rectangle is 734 cm, determine the dimensions of the rectangle.

3. The base of an isosceles triangle is 3 inches shorter than twice the length of a leg of the triangle. If the perimeter of the triangle is 53 inches, determine the dimensions of the triangle.

4. The vertex angle of an isosceles triangle measures 5 degrees more than one-half the measure of either base angle. Find the measure of each angle of the triangle.

5. One of the angles of a quadrilateral is a right angle. The largest angle of the quadrilateral measures 3 times the measure of the smallest angle. The fourth angle of the quadrilateral measures 10 degrees less than the largest angle. Find the measure of each angle of the quadrilateral.

6. Two vehicles leave Indianapolis at 1 p.m. One vehicle travels north at a steady rate of 45 mph, while the other travels south at a steady rate of 55 mph. At what time are the two vehicles 300 miles apart?

7. John leaves his house at 9 a.m. and drives along local roads that have a speed limit of 25 mph. His sister Joanne leaves at 9:30 a.m. and travels along a parallel road with a speed limit of 40 mph. At what time does Joanne pass her brother?

8. Two trucks leave at 8 a.m. from depots that are 1000 miles apart and travel toward each other on parallel roads. If the trucks travel at speeds of 65 and 60 mph, respectively, at what time will the trucks pass each other?

9. If a weight of 20 pounds will extend a spring 7 inches, how far will a weight of 35 pounds extend the spring?

10. If the volume of a gas in a balloon has a volume of 1.4 cubic centimeters (cc) when the temperature is 280°K, what will be the volume of the gas when the temperature rises 5°K (assuming pressure remains constant)?

11. If a parachutist falls 19.8 meters 2 seconds after jumping out of a plane, how far will the parachutist fall after 3 seconds?

EXERCISE 5-4

Determine the linear rule for each of the arithmetic sequences.

1. 81, 104, 127, 150, . . .

2. 99, 93, 87, 81, . . .

3. 123, 150, 177, 204, . . .

4. 42, 29, 16, 3, . . .

EXERCISE 5-5

Answer the following questions.

1. Find the 30th term in the sequence in problem 1 from Exercise 5-4.

2. Find the 50th term in the sequence in problem 2 from Exercise 5-4.

3. Find the 90th term in the sequence in problem 3 from Exercise 5-4.

4. Find the 15th term in the sequence in problem 4 from Exercise 5-4.

5. The 18th term in an arithmetic sequence is 44 and the 33rd term in the sequence is 164. Determine the 200th term in the sequence.

6. The 7th term in an arithmetic sequence is 74, and the 12th term in the sequence is 44. Find the 25th term in the sequence.

7. Find the sum of the first 250 terms in the sequence described in problem 12.

8. Find the sum of the first 120 terms in the sequence described in problem 13.

9. There are 35 seats in the center section of the Hartman Theater, and every row after the first has 2 more seats than in the row in front of it. The center section of the Hartman Theater has 32 rows. How many seats are in the center section?

EXERCISE 5-6

Solve each of the following equations.

1. $|x-7|=4$
2. $|x+3|=2$
3. $|2x-6|=10$
4. $|3x+9|=15$
5. $|2x-7|=11$
6. $|4x+5|=13$
7. $|9-3x|=12$
8. $|8-5x|=10$
9. $|3x+2|+10=15$

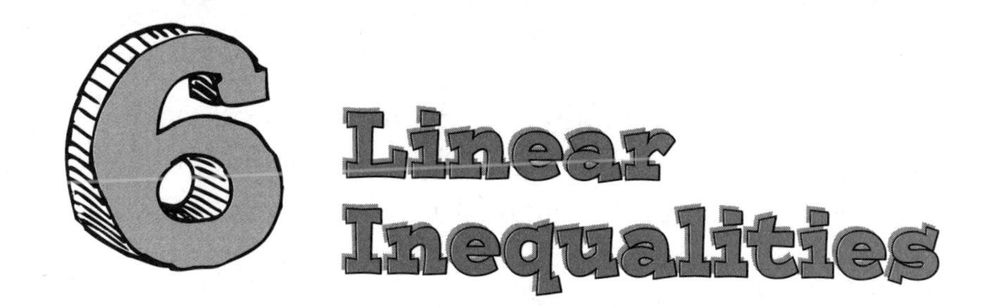

6 Linear Inequalities

MUST KNOW

- Linear inequalities represent scenarios that have more than one solution.
- Some inequalities include the boundary as part of the solution while others do not.

ot all problems are about finding a single value that will make the statement of the problem true. There are many times when the solution can be a range of values. We will examine the notion of linear inequalities in this chapter finding both algebraic and graphical solutions.

Solving Inequalities with the Four Operations

You know that the inequality $5 > 3$ is true. You also know that for any value of x, $x + 5 > x + 3$, because the left-hand side of the equation will always be two larger than the right-hand. It is also true that for any value of x, $5 - x > 3 - x$ for the same reason. Frankly, if $a > b$, then it does not matter what the value of x might be; the orientation of $x + a$ and $x + b$ will not change. That is, the magnitude of $x + a$ will always be larger than $x + b$. Consequently, we know we can solve linear inequalities that involve only addition and subtraction in exactly the same way we would solve linear equations.

EXAMPLE

▶ Solve $x + 5 > 21$.

▶ Subtract 5 from each side of the inequality to get that $x > 16$.

The issue of multiplication and division is a bit trickier. Certainly if x is a positive value and $a > b$, then $ax > bx$ and $\frac{a}{x} > \frac{b}{x}$. Things get more interesting when x is a negative number. Again, we all know that $5 > 3$, but when both sides of the inequality are multiplied by -1, it certainly not true that $-5 > -3$, but rather $-5 < -3$. The orientation is reversed! The same is true if you divide by a negative value.

EXAMPLE

- Solve $-5x > 35$.
- Divide by -5 to get $x < -7$.

Graphing representation of solutions is an important skill. We'll begin with linear inequalities on a number line and will then work our way to linear equations and inequalities in two dimensions.

EXAMPLE

- Sketch the graph of all real numbers that are greater than or equal to 5. That is, show the graph for all values of x for which $x \geq 5$.
- We will not take the time to draw a number line and denote a large number of integers from -5 to 5. Rather, we will show where 0 is on the number line and then the value(s) of importance. In this case, that value is 5.

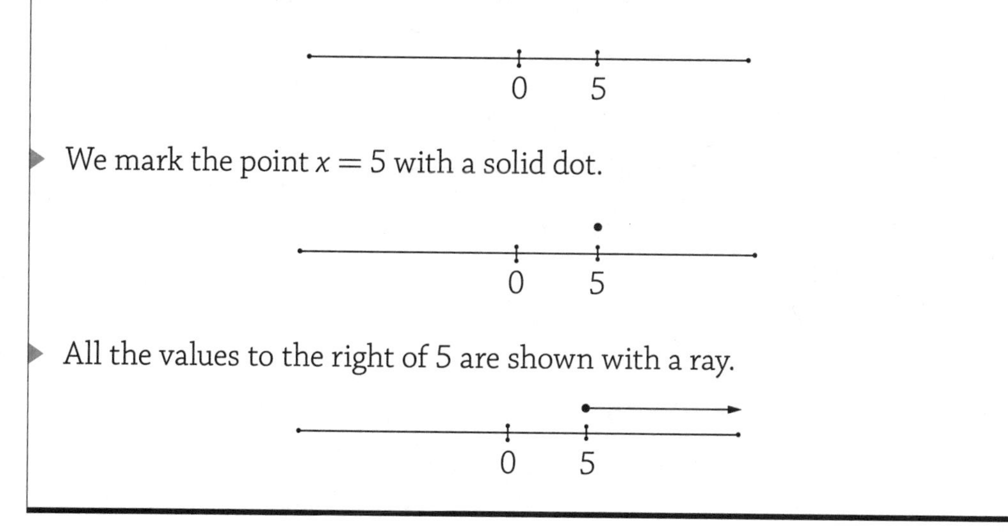

- We mark the point $x = 5$ with a solid dot.
- All the values to the right of 5 are shown with a ray.

The difference between the graph of $x \geq 5$ and $x > 5$ is that we use an open dot at $x = 5$ to show that it is a boundary for the inequality but is not part of the inequality. That is,

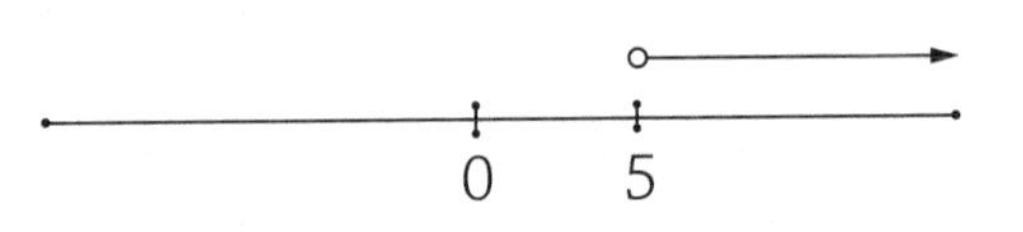

Solving Multistep Inequalities

Solving multistep inequalities is no different than solving multistep equations **except** for the step in which you might multiply or divide both sides of the inequality by a negative value. In that case, you will be to change the orientation of the inequality.

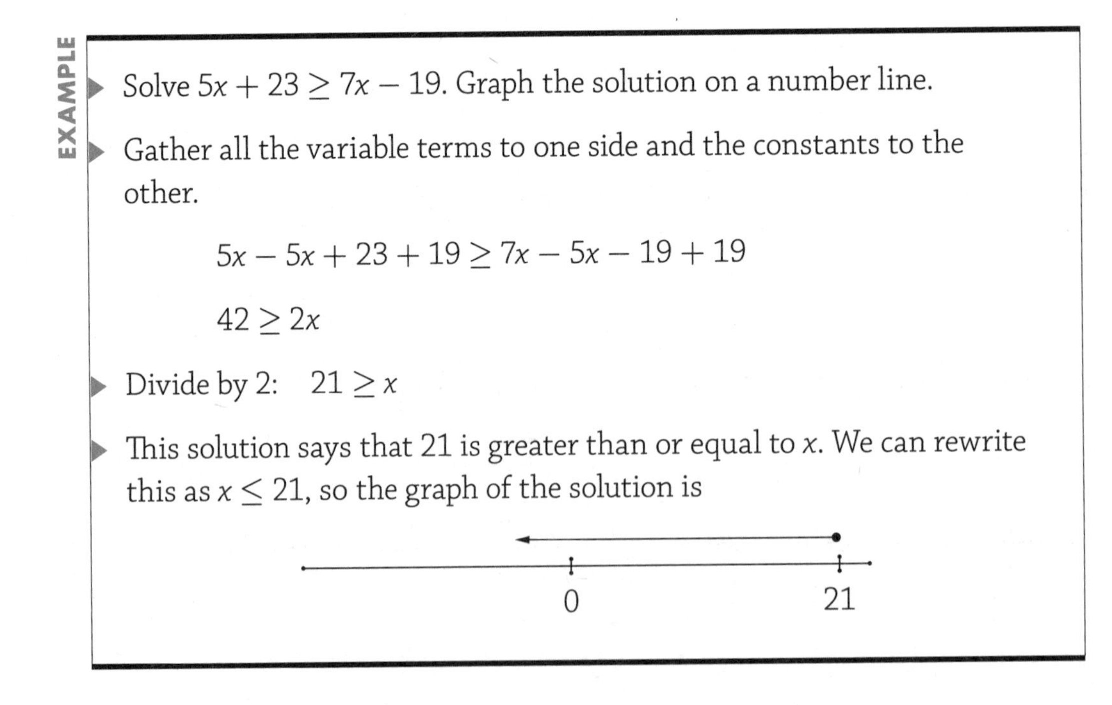

EXAMPLE

▶ Solve $5x + 23 \geq 7x - 19$. Graph the solution on a number line.

▶ Gather all the variable terms to one side and the constants to the other.

$$5x - 5x + 23 + 19 \geq 7x - 5x - 19 + 19$$

$$42 \geq 2x$$

▶ Divide by 2: $21 \geq x$

▶ This solution says that 21 is greater than or equal to x. We can rewrite this as $x \leq 21$, so the graph of the solution is

It is not critical that the variable be on the left side of the inequality so long as you make sure to read the inequality correctly. "This solution says that 21 is greater than or equal to x" tells us that the solution contains those values that are smaller than 21 and your graph should reflect that (as opposed to "I graph to the right when I see $>$ and to the left when I see $<$ no matter what the rest of the sentence says").

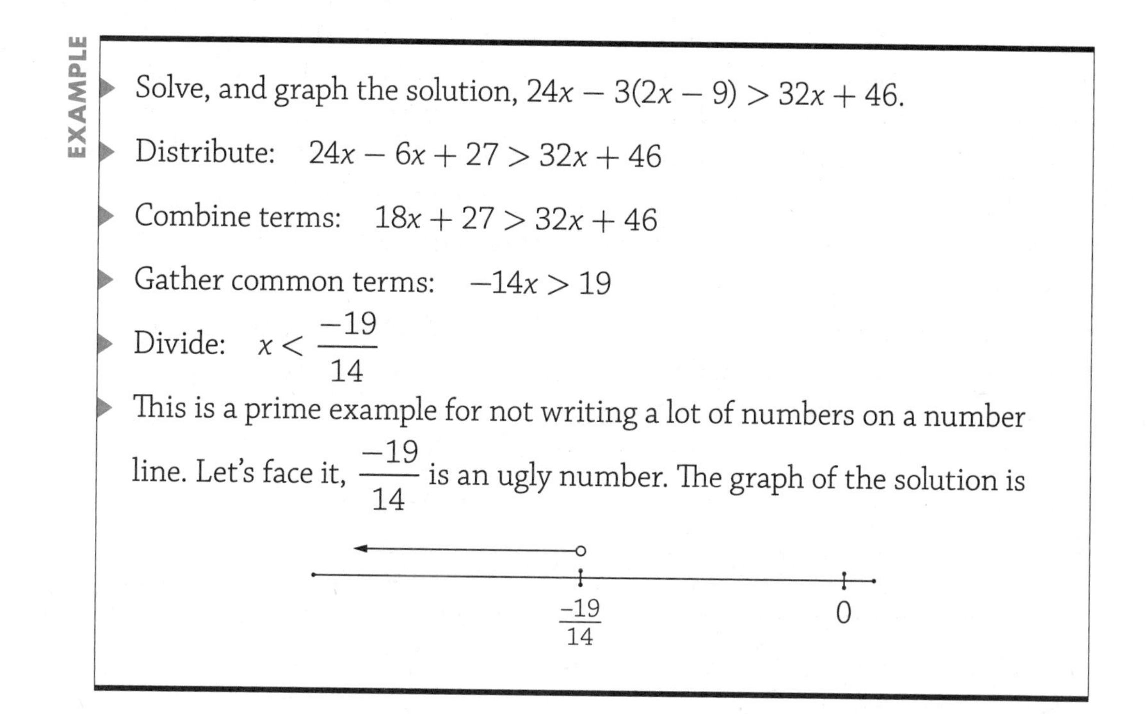

EXAMPLE

- Solve, and graph the solution, $24x - 3(2x - 9) > 32x + 46$.
- Distribute: $24x - 6x + 27 > 32x + 46$
- Combine terms: $18x + 27 > 32x + 46$
- Gather common terms: $-14x > 19$
- Divide: $x < \dfrac{-19}{14}$
- This is a prime example for not writing a lot of numbers on a number line. Let's face it, $\dfrac{-19}{14}$ is an ugly number. The graph of the solution is

Solving Compound Inequalities

You hear the statement, "The attendance at the concert is between 18,000 and 20,000," and being the good algebra student that you are (you're reading this book, right?), you think to yourself "if p represents the number of people at the concert, then $p \geq 18{,}000$ and $p \leq 20{,}000$." Ok, so maybe you didn't think that but I just put it into your head, didn't I? A more compact representation for these inequalities is $18{,}000 \le p \le 20{,}000$. Notice that the inequality reads "18,000 is less than or equal to p, and p is less than or equal to 20,000." This is an example of a compound inequality and, of course, the graphical representation is

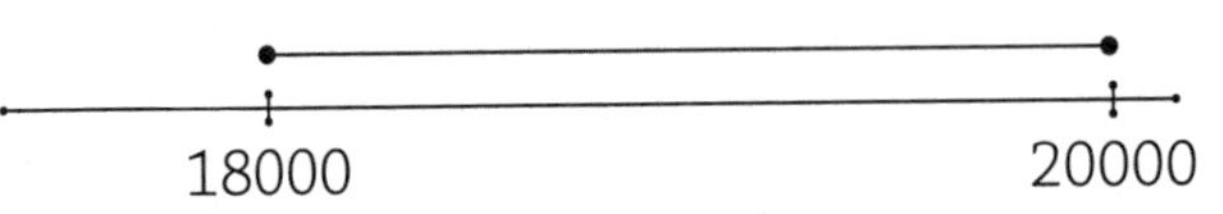

Now consider this statement: "Only seniors and minors are eligible for this government program." If a minor is defined to be someone who is under the age of 18 and a senior is defined to be someone who is aged 60 or older, then the age groups who meet the eligibility requirements stated are $g < 18$ or $g \ge 60$, where g represents the age of the person. We **cannot** write this compound inequality in a more compact manner. The logic would not make any sense: $18 < g \ge 60$ would read as "18 is less than g and g is greater than or equal to 60." Do you know a number that is simultaneously smaller than 18 and bigger than 60? Neither do I. The graphical representation for the compound inequality $g < 18$ or $g \ge 60$ is

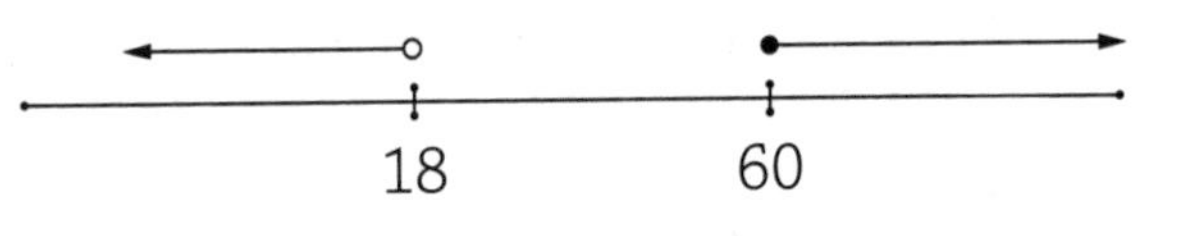

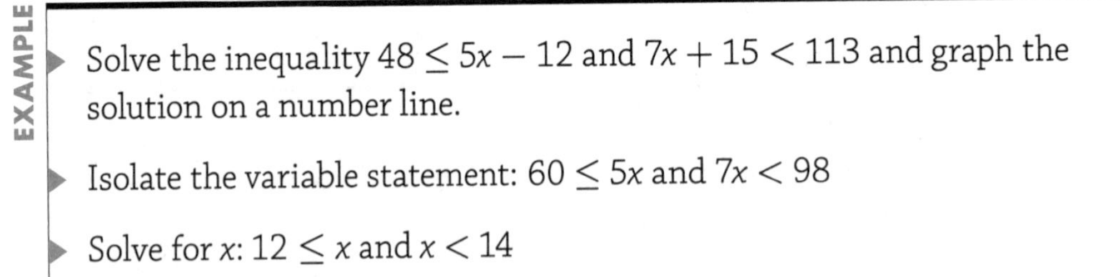

EXAMPLE

- Solve the inequality $48 \le 5x - 12$ and $7x + 15 < 113$ and graph the solution on a number line.
- Isolate the variable statement: $60 \le 5x$ and $7x < 98$
- Solve for x: $12 \le x$ and $x < 14$
- Write as a compound inequality: $12 \le x < 14$
- Graph the solution:

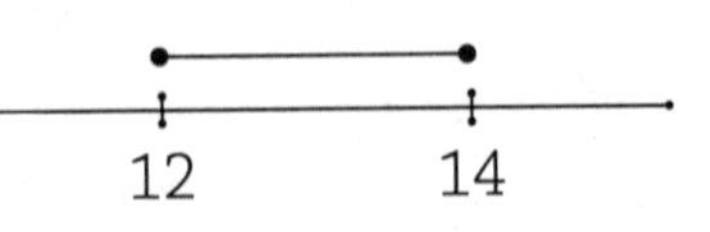

Very often the variable expression of the compound inequality will be the same.

EXAMPLE

- Solve the inequality $28 < 5x - 7 \leq 43$ and graph the solution on a number line.
- We can write the problem as two inequalities: $28 < 5x - 7$ and $5x - 7 \leq 43$
- Add 7 to both sides: $35 < 5x$ and $5x \leq 50$
- Divide by 5: $7 < x$ and $x \leq 10$
- Write as one inequality: $7 < x \leq 10$
- While this works just fine, it is easier to just work with the compound inequality as a whole.

$$28 < 5x - 7 \leq 43$$

- Add 7: $35 < 5x \leq 50$
- Divide: $7 < x \leq 10$
- Graph:

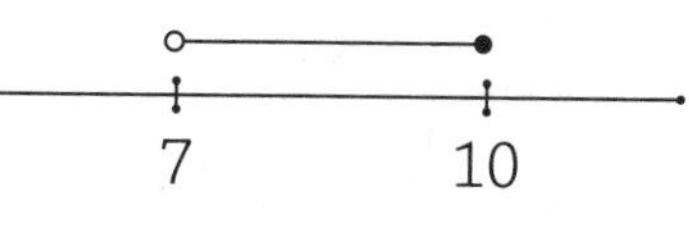

Be careful with the orientation of the inequalities in this example.

EXAMPLE

▶ Solve the inequality $28x - 47 < 37$ or $5x + 19 > 38$ and graph the solution on a number line.

▶ These can't be treated as a single unit because the variable expressions are different.

$$28x - 47 < 37 \text{ or } 5x + 19 > 38$$

▶ Isolate the variable: $28x < 84$ or $5x > 19$

▶ Divide: $x < 3$ or $x > 3.8$

▶ Graph:

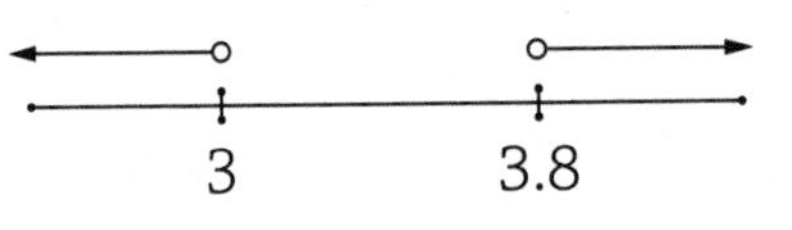

There will be times when the solution yields something completely unexpected.

EXAMPLE

▶ Solve the inequality and sketch the solution on a number line.

$$14 - 3x > -13 \text{ or } 5x + 7 > 12$$

▶ Isolate the variable: $-3x > -27$ or $5x > 5$

▶ Divide: $x < 9$ or $x > 1$

▶ Graph:

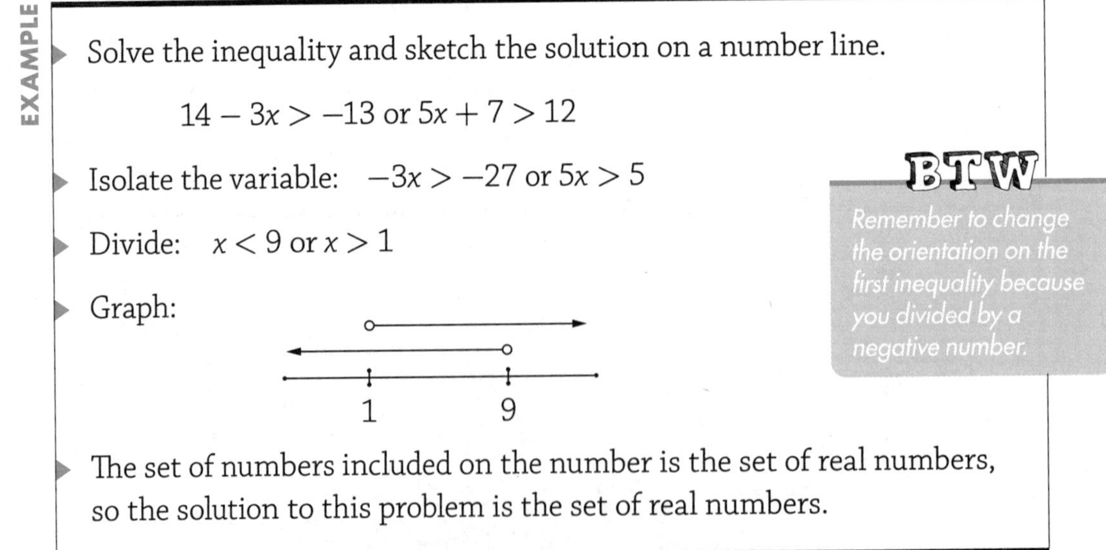

BTW

Remember to change the orientation on the first inequality because you divided by a negative number.

▶ The set of numbers included on the number is the set of real numbers, so the solution to this problem is the set of real numbers.

Inequalities Involving Absolute Value

As we saw in Chapter 5, equations involving absolute value can be interpreted from a geometric perspective. This can be particularly useful when working with absolute value inequalities. If the equation $|x-4|=3$ represents the points that are 3 units from 4 on the number line so that $x=1$ or 7, then it is not too difficult to understand that $|x-4|>3$ must represent those points that are more than 3 units from 4 on the number line and consequently, $x<1$ or $x>4$. The algebraic approach requires that you memorize a couple of rules. Whichever approach you take, you can still solve absolute value inequalities.

$$\text{If } |x|>a, \text{ then } x<-a \text{ OR } x>a.$$
$$\text{If } |x|<a, \text{ then } -a<x<a.$$

Solve $|2x-7|>11$.

Geometric approach: Factor the 2 from the absolute value
$|2x-7|>11 \Rightarrow |2|\,|x-3.5|>11 \Rightarrow |x-3.5|>5.5$

The set of points greater than 5.5 units from 3.5 on the number line are those less than −2 or greater than 9, so $x<-2$ or $x>9$.

Algebraic approach: $|2x-7|>11 \Rightarrow 2x-7<-11$ OR $2x-7>11$
$\Rightarrow$ Add 11: $2x<-4$ or $2x>18 \Rightarrow x<-2$ or $x>9$.

There are plenty of applications for absolute value inequalities.

EXAMPLE

▶ Solve $|r - 2| < 0.02$.

▶ **Geometric approach:** The set of points no more than 0.02 units from 2 are in the interval $1.98 \leq r \leq 2.02$.

▶ **Algebraic approach:** $|r - 2| \leq 0.02 \Rightarrow -0.02 \leq r \leq 0.02$. Add 0.02 to get $1.98 \leq r \leq 2.02$.

A manufacturing application that would use this example might be that the radius of a ball bearing should be 2 mm with a maximum error of 0.02 mm. So long as the radius of the ball bearing is in the interval $1.98 < r < 2.02$ mm, it is considered acceptable for distribution.

You are more than likely already familiar with certain phrases that translate into mathematical inequalities:

at most	$\geq$
no more than	$\leq$
at least	$\geq$
a maximum of	$\leq$
a minimum of	$\geq$
must surpass	$>$
must not surpass	$\leq$

Did you notice that the last two phrases are negations of each other and that the negation of greater than is not less than but is less than or equal to? The example of the radius of the ball bearing could have been written as "the difference between the radius of the ball bearing and the ideal radius of 2 mm must not exceed 0.02 mm." This is why you should read the specifications carefully for things that you are having built or are putting together yourself (that includes models as well as houses).

EXAMPLE

▶ A data value is considered to be an outlier if it is more than 1.5 times the inter-quartile range (IQR) from the mean of the data. If a set of data has a mean of 58 and an IQR of 14, what is the set of values that are considered to be outliers?

▶ The inequality reads "if it is more than," not "if it is more than or equal to," so this inequality is strictly greater than. If x is a data value, then the difference (positive or negative) cannot be more than 1.5 times the IQR from the mean. The inequality for this problem is $|x - 58| > 1.5(14)$ or $|x - 58| > 21$. Solve the inequality:

$$x - 58 < -21 \text{ or } x - 58 > 21$$

$$x < 37 \quad \text{or} \quad x > 79$$

▶ Data values smaller than 37 or greater than 79 are considered to be outliers.

Graphing Inequalities in Two Variables

The last thing to look at in this chapter is the graphing of inequalities in two variables. We know that the graph of the line $y = 3x + 5$ contains all the values for the ordered pair (x, y) for which the inequality is true. That is, the ordered pair (2, 11) is a point on this line, but the ordered pair (2, 12) is not. So how does one go about graphing the inequality $y \geq 3x + 5$?

Just as we graphed $x \geq 5$ on the number line, we begin with the boundary. For the number line, it was the point $x = 5$ and it is a solid dot because the point is included in the solution. For $y \geq 3x + 5$, the boundary is the line $y = 3x + 5$, and it is solid because it is included in the inequality.

While we can easily see which points satisfy $x > 5$ on the number line, we need to be a bit more careful when dealing with inequalities in two variables. We can apply a little bit of logic to help us out here. We already have determined that the points on the line satisfy the equation $y = 3x + 5$. That means, the ordered pairs for all the other points in the plane will not satisfy the equation. These ordered pairs will satisfy the inequality $y > 3x + 5$ or the inequality $y < 3x + 5$. How do we know which is which? Easy. Test one. We just wrote that the ordered pair (2, 12) will not satisfy the equation $y = 3x + 5$. Does the ordered pair satisfy the inequality $y > 3x + 5$? Substitute the values for x and y to see: $12 > 3(2) + 5 \Rightarrow 12 > 11$. This is a true statement, so the ordered pair (2, 12) must be part of the solution.

Here is the most important part of this process: If an ordered pair that is not on the line satisfies the inequality, then all the points on that side of the line have ordered pairs that will solve the inequality and NONE of the points on the other side of the line have points whose ordered pairs that will satisfy the inequality. This is why you only need to check one point. So long as the line that serves as the boundary does pass through the origin, it's a good idea to always use the ordered pair (0, 0) as the test point, and if the boundary does pass through the origin, use the point (1, 0). But you can choose any point that you like so long as the point is not on the boundary line.

What does the graph of $y \geq 3x + 5$ look like? We have the solid line that serves as the boundary and then we shade the side of the line that have the points whose ordered pairs make the inequality a true statement.

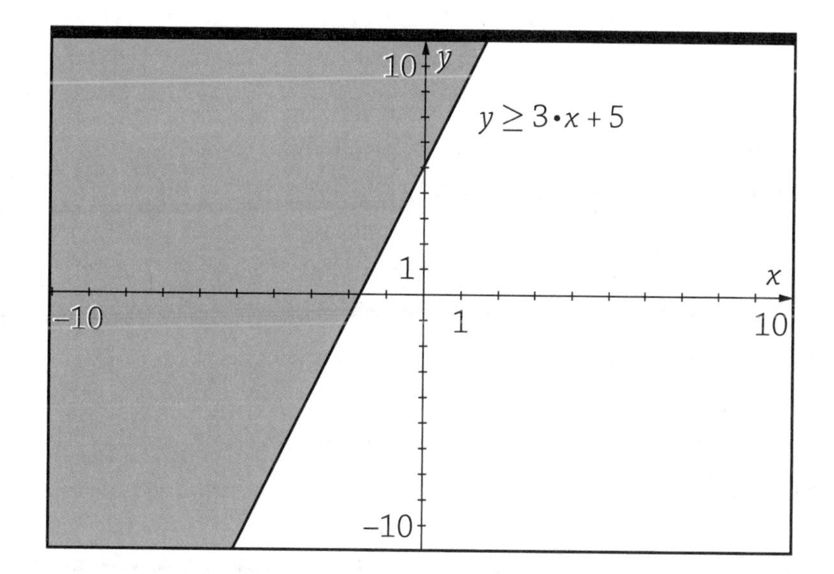

EXAMPLE

Sketch the solution to the inequality $y \leq -2x + 3$.

We know that the y-intercept for this line is (0, 3), so we can use (0, 0) to test the inequality. Is $0 \leq -2(0) + 3$? Yes, $0 \leq 3$, so the side of the line containing the origin is shaded.

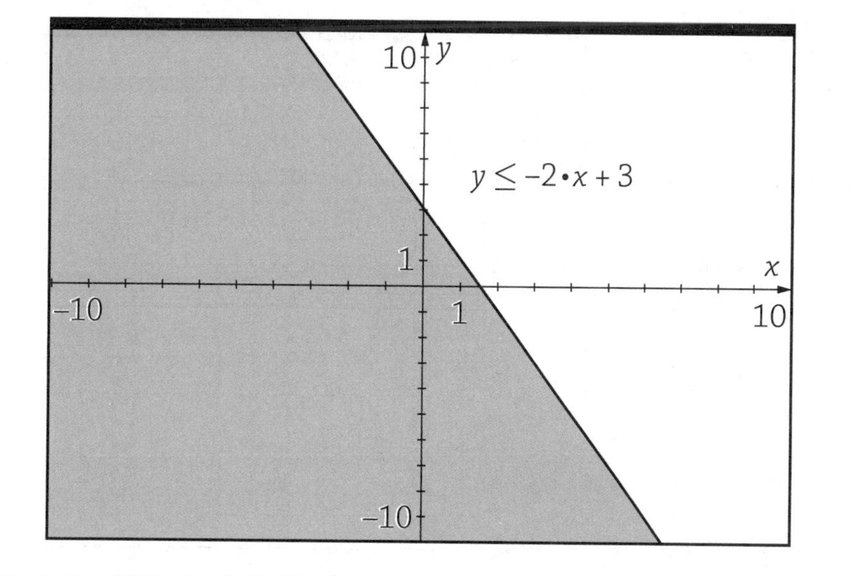

The next question is how do we graph the inequality if there is no sense of equality as part of the statement. That is, what do we do about the boundary if the statement reads $y >$ or $y <$? On the number line, we simply replaced the solid dot that formed the boundary with an open dot that indicated the boundary is not part of the solution. In two dimensions, the equivalent of this process is to use a dotted line rather than a solid line to show where the boundary is located. Everything else remains the same for graphing the solution.

Inequalities in two (or more) variables are important in the study of linear programming problems. In linear programming, decisions are made on such issues as maximizing profit, minimum cost, and minimum calorie intake. We can look at simple problems with two variables but need more sophisticated mathematical techniques when the problems have more than two variables.

EXAMPLE

▶ Sketch the solution to the inequality $y > 2x + 5$.

▶ The boundary will be the line $y = 2x + 5$, but the line will be graphed as a dotted line as opposed to a solid line. The origin is not a point on the line, so we test if $0 > 2(0) + 5$ and determine that, no, $0 > 5$. That means that the origin is not part of the solution, so we shade the other side of the line.

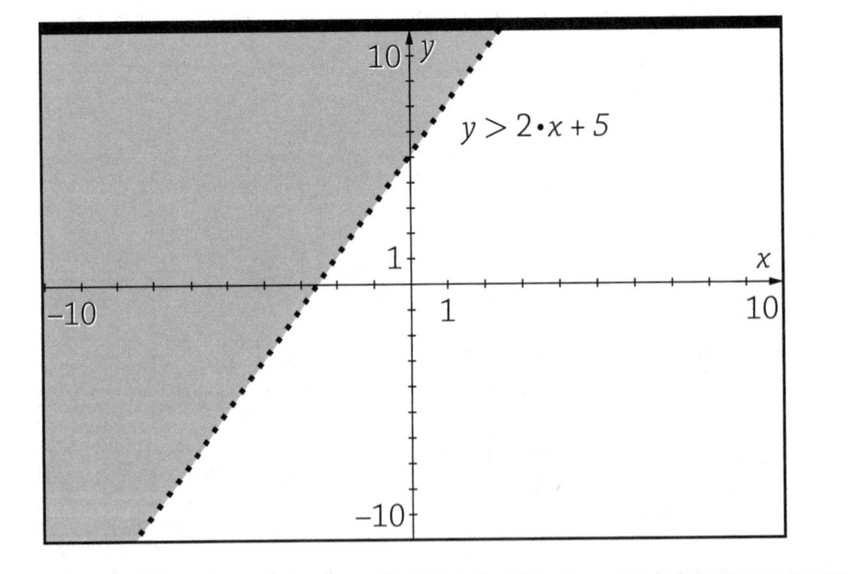

If you see directions such as, "Sketch the set of ordered pairs so that $x > -2$," you'll need to be careful "Sketch the set of ordered pairs" tells you that you are graphing the solution on the coordinate plane and not on the number line.

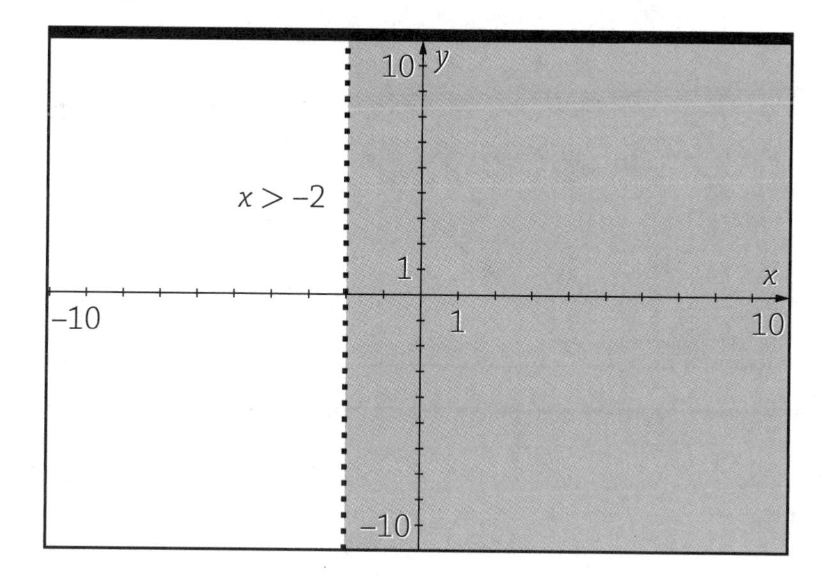

EXAMPLE

▶ Sketch the solution to the inequality $4x - 3y \leq 12$.

▶ Remember that you can determine the coordinates of the intercepts quickly when the equation is written in standard form. The intercepts for $4x - 3y = 12$ are $(0, -4)$ and $(3, 0)$. Test the origin (since this point is not on the boundary line): $4(0) - 3(0) \leq 12$ is a true statement, so the origin is part of the graph. Note that the boundary is part of the solution in this case.

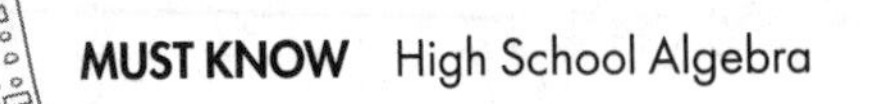

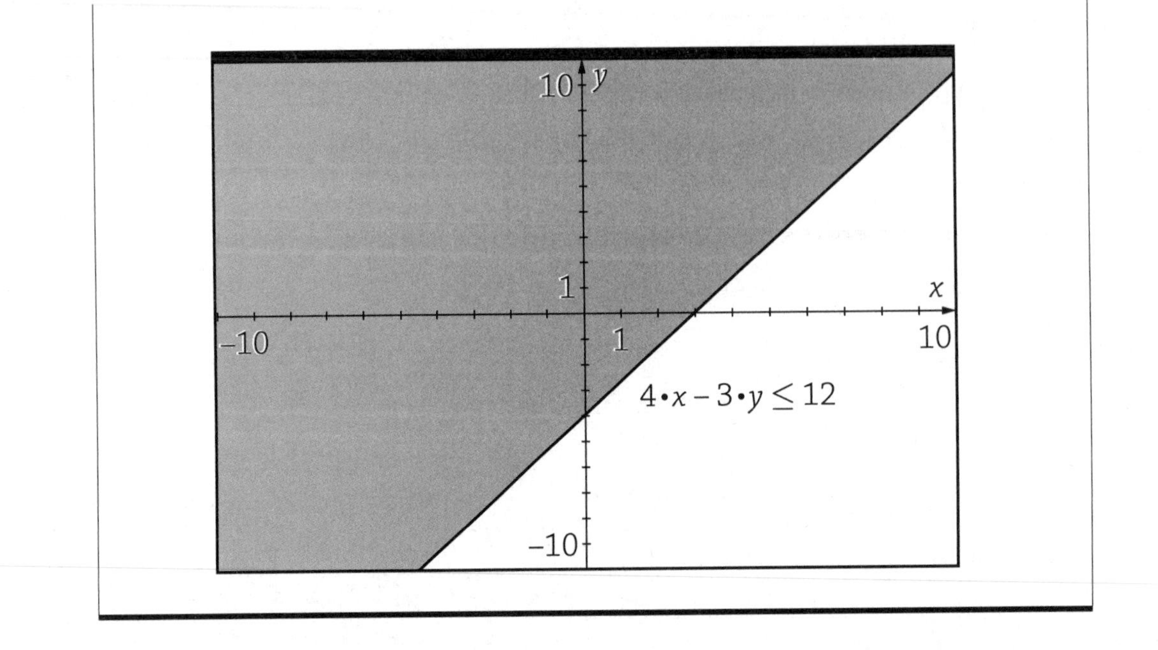
y
x
10
1
-10
1
10
-10
$4 \cdot x - 3 \cdot y \leq 12$

EXERCISES

EXERCISE 6-1

Solve each of the following inequalities and graph the solution on a number line.

1. $5x - 9 > 16$
2. $14 - 8x \geq 42$
3. $3(2x - 5) - 4(7 - x) < 17$
4. $8(4 - 3x) + 3(2x + 7) \leq 20 - 13x$
5. $11 < 2x - 7 \leq 29$
6. $32 < 4 - 7x < 53$
7. $7x + 19 \leq 33$ or $12x - 35 > 49$
8. $17 - 3x < 26$ or $14 + 8x > 5x - 7$
9. $3\frac{2}{3} \leq \frac{5}{9}x - 4\frac{1}{2} < 4\frac{5}{12}$
10. $810{,}000 > 20{,}000x + 197{,}000$ or $15{,}000x - 210{,}000 > 60{,}000$

EXERCISE 6-2

Solve each of these absolute value inequalities.

1. $|x - 9| \leq 5$
2. $|x + 3| > 7$
3. $|2x - 5| \geq 13$
4. $|3x + 10| < 14$

EXERCISE 6-3

Answer the following questions.

1. The number of ounces dispensed by a soft drink dispenser should be 12 ounces with an error of no more than 0.25 ounces. Express this statement as an absolute inequality with z representing the number of ounces of the soft drink that is dispensed.

2. Write an absolute value inequality to represent the set of numbers shown in the graph.

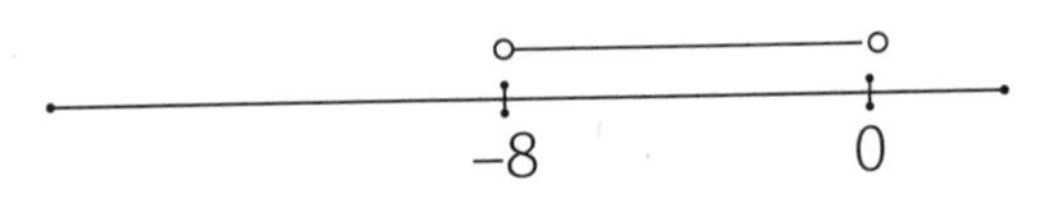

EXERCISE 6-4

Sketch the graph of each inequality.

1. $y > -4x - 2$
2. $y \leq 3x - 4$
3. $2x + 5y \geq 20$
4. $3x - 2y < 12$

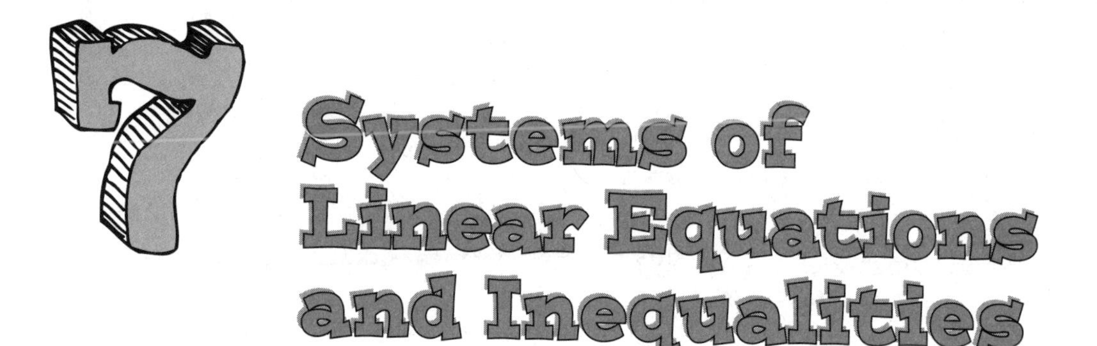

7 Systems of Linear Equations and Inequalities

MUST KNOW

- Algebraic solutions to systems of equations are accomplished by reducing the system to one equation in one variable.

- Graphical solutions to systems of linear equations determines the values of *x* and *y* at the same time. Until we develop tools such as holography (!) to graph in three dimensions, graphical solutions are limited to two equations with two variables.

- Matrix solutions to systems of algebraic equations can be used to solve systems of equations. Although it can be done with paper and pencil, it is more commonly used with graphing calculators and computer systems.

- Systems of inequalities are not solved algebraically but graphically because the solutions are infinite sets of points rather than a finite number of ordered pairs.

There are some interesting problems that involve just two variables. More often, the problems have more than two variables. The techniques we will look at in this chapter provide us with the guidelines to solve systems of linear equations with as many variables as might be needed.

Graphing Systems of Equations

If we graph the lines defined by two linear equations on the same set of axes, one of three things can happen:

- The lines intersect at a point.
- The lines never intersect.
- The equations represent the same line, so the "lines" lie on top of one another.

How can we tell which scenario will occur when we look at the equations? If the equations represent lines with different slopes, then the lines will intersect at one point. If the equations represent lines with the same slope, we can easily find the y-intercept for each line (set $x = 0$ and find the value of y). If the y-intercepts are different, the lines are parallel, and if the y-intercepts are the same, then the equations represent the same line. Most of the work with systems of linear equations involves the first scenario as we try to find the coordinates of the point of intersection. We'll run into the second and third cases on occasion (and will most likely not remember to check the slopes ahead of time, but we can use the slope statements to verify our work).

EXAMPLE

Sketch the graphs of the lines represented by the following two equations on the same set of axes, and determine the coordinates of the point of intersection.

$$y = 3x - 4$$
$$y = -2x + 11$$

BTW

A heads up for you—the directions "Sketch the graphs of the lines represented by the following two equations on the same set of axes, and determine the coordinates of the point of intersection" are usually just written as "Solve graphically."

You checked the slopes first, didn't you? Good job! If you are using graph paper, you can plot the points (0, −4), (1, −1), and (2, 2) (for example) for $y = 3x - 4$ and draw the line through these points. You can plot the points (0, 11), (1, 9), and (2, 7) (for example) for $y = -2x + 11$ and draw the line through these points. The two lines will pass through the point (3, 5).

Using technology, enter the equations into equation editor, graph each of the lines, and use the intersect feature to find the points of intersection. (Screen shots from two different Texas Instruments calculators are shown: TI-Nspire and TI-84.)

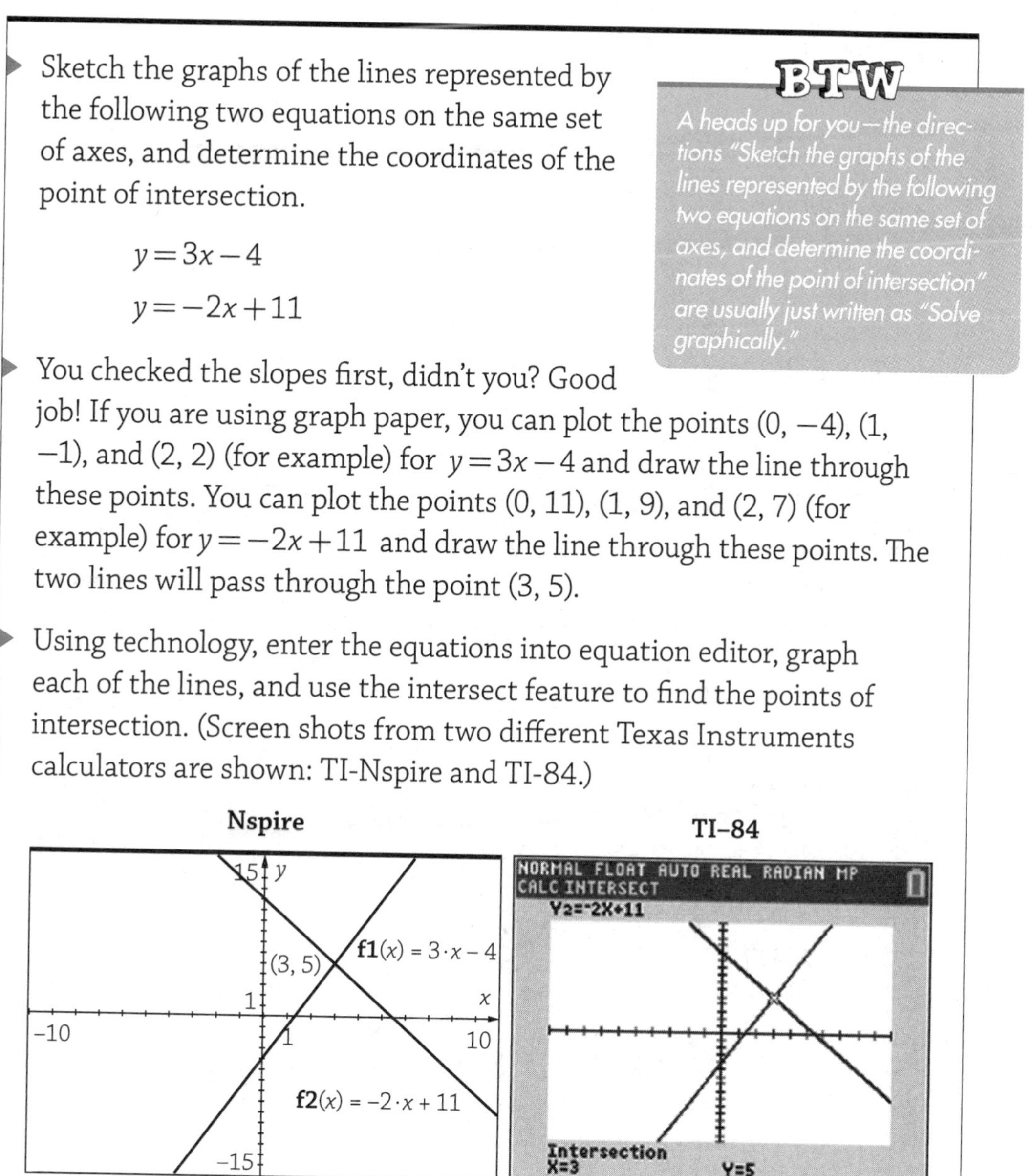

You can do this problem on graph paper with a pencil if you wish, but you should also practice with a graphing utility.

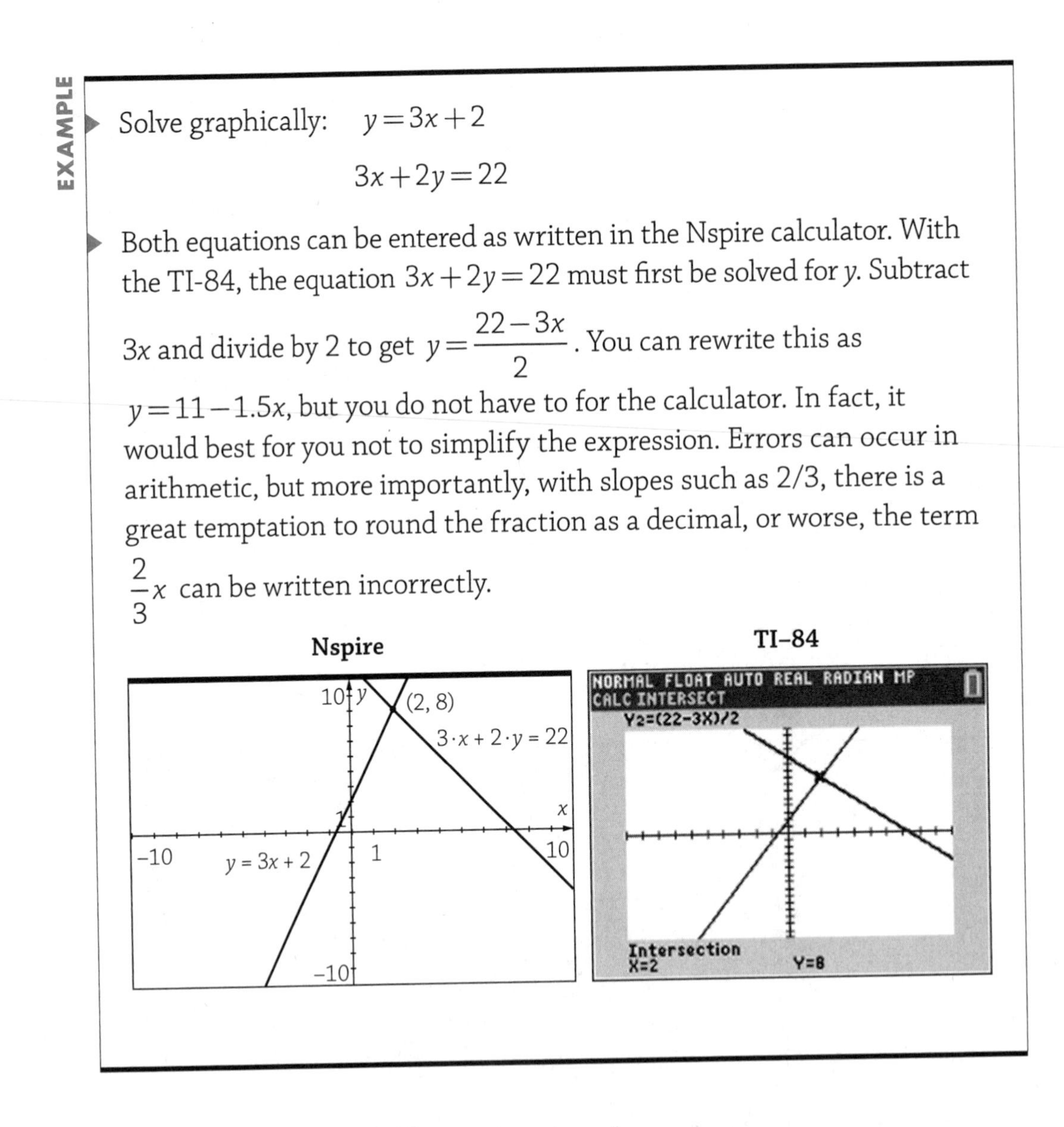

EXAMPLE

▶ Solve graphically: $y = 3x + 2$

$$3x + 2y = 22$$

▶ Both equations can be entered as written in the Nspire calculator. With the TI-84, the equation $3x + 2y = 22$ must first be solved for y. Subtract $3x$ and divide by 2 to get $y = \dfrac{22 - 3x}{2}$. You can rewrite this as $y = 11 - 1.5x$, but you do not have to for the calculator. In fact, it would best for you not to simplify the expression. Errors can occur in arithmetic, but more importantly, with slopes such as 2/3, there is a great temptation to round the fraction as a decimal, or worse, the term $\dfrac{2}{3}x$ can be written incorrectly.

Let's do one more before looking at some algebraic methods for solving systems of equations.

EXAMPLE

Solve graphically:

$$3x-5y=12$$
$$-6x+10y=25$$

We can enter the equations as they are with the Nspire. The entries for the TI-84 would be $y=(12-3x)/-5$ for the first equation and $y=(25+6x)/10$ for the second equation.

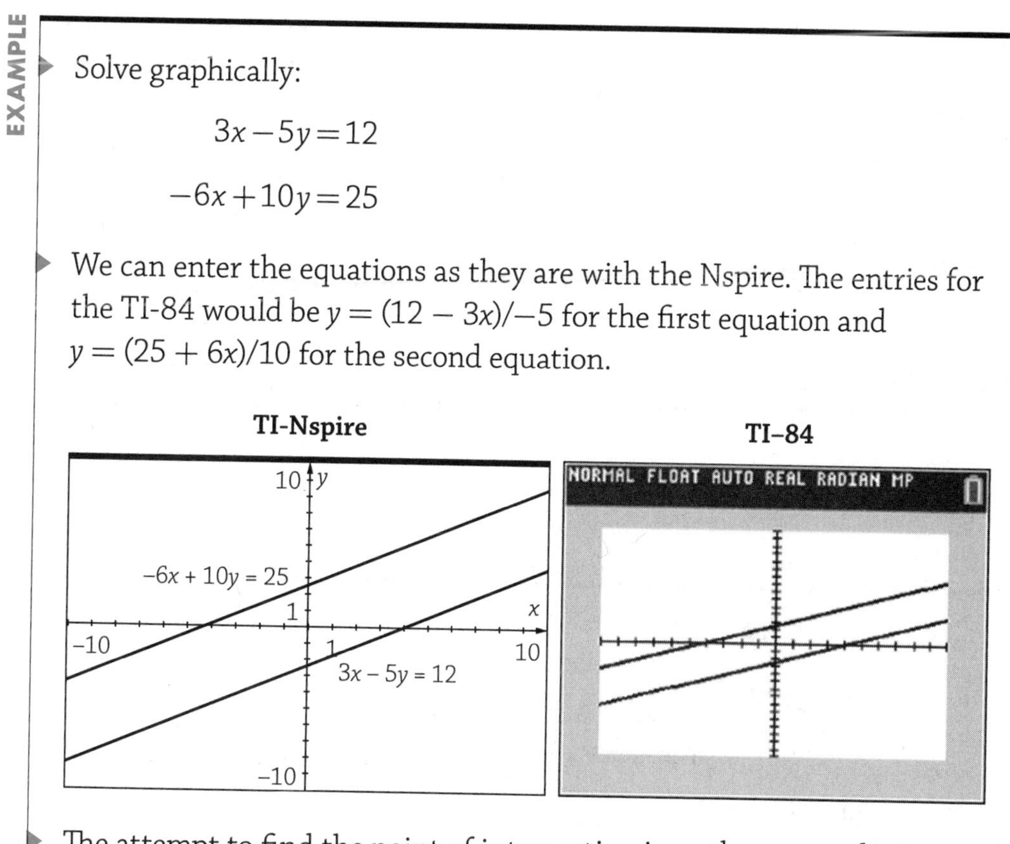

The attempt to find the point of intersection in each case results in an error message. Why? You know why; the lines are parallel. Each line has a slope of 0.6 and they have different y-intercepts. As I said earlier, we might not stop to look at the slopes initially, but we can verify the result after we take a look at the graphs.

Algebraic Solutions

Graphical solutions have become much easier with the development of affordable technology for students. However, while the graphical approach works with two equations in two variables, it does not prove to be so useful with more variables and more equations. We are only responsible to solve systems of equations in two variables in Algebra I, but we are also responsible to set the groundwork for what is to come in Algebra II. The three techniques we will look at are substitution (which sounds exactly like it is), elimination (of a variable), and the technology oriented matrix solution.

Substitution

The substitution approach works best when one of the equations is of the form $y =$ or $x =$ (although all equations can be transformed into that form). We are able to substitute an algebraic expression into the second equation in the system to form one equation in one variable.

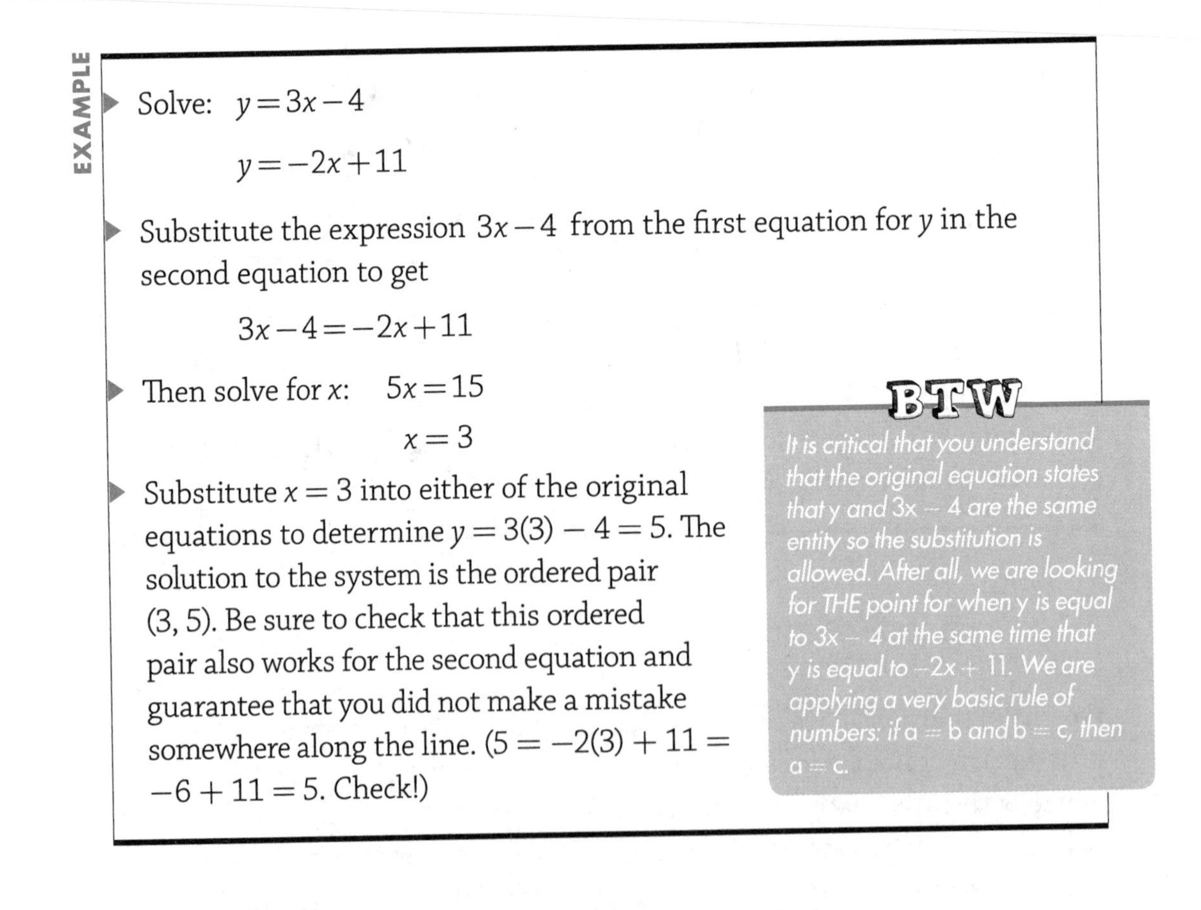

EXAMPLE

- Solve: $y = 3x - 4$

$$y = -2x + 11$$

- Substitute the expression $3x - 4$ from the first equation for y in the second equation to get

$$3x - 4 = -2x + 11$$

- Then solve for x: $5x = 15$

$$x = 3$$

- Substitute $x = 3$ into either of the original equations to determine $y = 3(3) - 4 = 5$. The solution to the system is the ordered pair (3, 5). Be sure to check that this ordered pair also works for the second equation and guarantee that you did not make a mistake somewhere along the line. ($5 = -2(3) + 11 = -6 + 11 = 5$. Check!)

BTW

It is critical that you understand that the original equation states that y and $3x - 4$ are the same entity so the substitution is allowed. After all, we are looking for THE point for when y is equal to $3x - 4$ at the same time that y is equal to $-2x + 11$. We are applying a very basic rule of numbers: if $a = b$ and $b = c$, then $a = c$.

The substitution method is also a good approach when one of the equations is in slope-intercept while the other equation is in standard form.

EXAMPLE

- Solve: $y = 3x + 2$

$$3x + 2y = 22$$

- Substitute $3x + 2$ for y in the second equation to get

$$3x + 2(3x + 2) = 22$$

- Then solve for x: $3x + 6x + 4 = 22$

$$9x = 18$$

$$x = 2$$

- Substitute $x = 2$ into $y = 3x + 2$ to determine $y = 3(2) + 2 = 8$. Check $(2, 8)$ into the equation $3x + 2y = 22$.

$$3(2) + 2(8) = 22 \Rightarrow 6 + 16 = 22 \Rightarrow 22 = 22.$$

- The solution is the ordered pair $(2, 8)$.

Substitution doesn't usually work well when both equations are written in standard form. Let's take a look at one example in case you ever need to apply the technique.

EXAMPLE

- Solve: $3x + 2y = -2$

$$4x + 3y = 3$$

- Rewrite both equations in the form $y =$.

$$3x + 2y = -2 \Rightarrow 2y = -2 - 3x \Rightarrow y = \frac{-2 - 3x}{2}.$$

$$4x + 3y = 3 \Rightarrow 3y = 3 - 4x \Rightarrow y = \frac{3 - 4x}{3}.$$

- Set both fractional expressions equal to each other

$$\frac{-2-3x}{2} = \frac{3-4x}{3}$$

- Solve for x: $3(-2-3x)=2(3-4x)$

$$-6-9x=6-8x$$

- Add $9x$ and -6 to both sides of the equation to get $-12=x$.
- Substitute -12 into one of the original equations to determine the value for y.

$$3(-12)+2y=-2 \Rightarrow -36+2y=-2$$
$$\Rightarrow 2y=34 \Rightarrow y=17.$$

- Use the other original equation to check your result.

$$4(-12)+3(17)=3 \Rightarrow -48+51=3$$
$$\Rightarrow 3=3.$$

- The solution is $(-12, 17)$.

BTW

It is my experience that students often want to use one of the transformed equations to determine the value of y and/or check their work. The problem is that if a mistake is made in transforming the equation, it might be possible they will have the wrong answer and not know it. When in doubt, always go back to the original source.

Elimination Using Addition and Subtraction

The elimination technique takes advantage of two basic principles:

- Multiplication of both sides of an equation by the same number does change the truth of the equation.
- Addition of both sides of an equation by the same number does change the truth of the equation.

The process is to arrange the equation so that one of the variables has coefficients that equal in magnitude and opposite in sign. When the two equations are added together, that variable will be eliminated from the problem. We can then solve for the value of the remaining variable. As before, we then use one of the original equations to determine the value of the second variable.

EXAMPLE

- Solve: $3x+2y=-2$

  $4x+3y=3$

- We can multiply the first equation by -3 and the second equation by 2 so that the coefficients of y will be -6 and 6.

$$-3(3x+2y=-2) \Rightarrow -9x-6y=6$$

$$2(4x+3y=3) \Rightarrow 8x+6y=6$$

- Add the two equations: $-x=12$

  $x=-12$

- Solve for y: $3(-12)+2y=-2 \Rightarrow -36+2y=-2 \Rightarrow 2y=34 \Rightarrow y=17$
- Check the answer: $4(-12)+3(17)=3 \Rightarrow -48+51=3 \Rightarrow 3=3$
- The answer is $(-12, 17)$. (As we saw in the last section.)

You can see how much simpler the elimination method is than is substitution when the equations are written in standard form.

EXAMPLE

- Solve: $3x-2y=30$

  $2x+5y=1$

- We can multiply the first equation by 5 and the second equation by 2 so that the coefficients of y are -10 and 10.

$$5(3x-2y=30) \Rightarrow 15x-10y=150$$

$$2(2x+5y=1) \Rightarrow 4x+10y=2$$

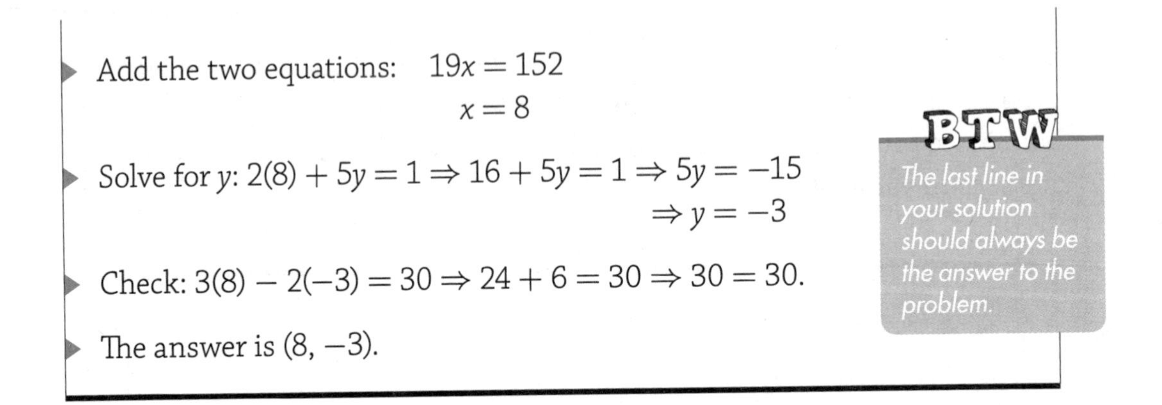

- Add the two equations: $19x = 152$

$$x = 8$$

- Solve for y: $2(8) + 5y = 1 \Rightarrow 16 + 5y = 1 \Rightarrow 5y = -15 \Rightarrow y = -3$
- Check: $3(8) - 2(-3) = 30 \Rightarrow 24 + 6 = 30 \Rightarrow 30 = 30.$
- The answer is $(8, -3)$.

BTW

The last line in your solution should always be the answer to the problem.

There are times when the solution to the system will not require as many actions as do others.

EXAMPLE

- Solve:

$$\begin{aligned} 2x + 3y &= -5 \\ -4x + 5y &= -111 \end{aligned}$$

- Two things to note with this problem: it will be easier to eliminate the variable x and we'll only have to multiply one equation by a constant.

$$2(2x + 3y = -5) \Rightarrow 4x + 6y = -10$$

$$-4x + 5y = -111 \Rightarrow -4x + 5y = -111$$

- Add the two equations: $11y = -121$

$$y = -11$$

- Solve for x: $2x + 3(-11) = -5 \Rightarrow 2x - 33 = -5 \Rightarrow 2x = 28 \Rightarrow x = 14$
- Check: $2(14) + 3(-11) = -5 \Rightarrow 28 - 33 = -5 \Rightarrow -5 = -5$
- The answer is $(14, -11)$. (Even though we solved for y first, the ordered pair is still (x, y).)

Here are two examples whose solutions you'll need to be able to recognize.

EXAMPLE

- Solve: $2x + 5y = 19$

  $4x + 10y = 23$

- Multiply the first equation by -2:

$$-2(2x + 5y = 19) \Rightarrow \quad -4x - 10y = -38$$

$$4x + 10y = 23$$

- Add the two equations: $0 = -15$

- We'll be back to finish this in a moment.

- Solve: $3x - 4y = 17$

  $12x - 16y = 68$

- Multiply the first equation by -4.

$$-4(3x - 4y = 17) \Rightarrow \quad -12x + 16y = -68$$

$$12x - 16y = 68$$

- Add the two equations: $0 = 0$

- In each case both variables were eliminated from the problem. This means that there is no single point that solves the system. The difference between the problems is that in one case, $0 = -15$, the statement is false while in the other, $0 = 0$, the statement is true. The false statement tells us that there is no point that solves the system of equations, while the true statement tells us that there are many points that solve the problem.

- Go back and look at the equations. Notice that the slopes are equal for the equations in each system.

BTW

If both variables in a system of linear equation are eliminated, the equations represent parallel lines if the resulting equation is a false statement. If the resulting equation is a true statement, the equations in the system represent the same line.

In the first system, the lines have different y-intercepts, so the lines are parallel. In the second system, the y-intercepts are the same, so the equations represent the same line and there are infinite number of points that satisfy the system.

Solution Using Matrices

The third algebraic method for solving systems of linear equations is to use an algebraic process called *matrices* (the singular form of matrices is matrix). Look at the following systems of equations:

$$\begin{array}{llll} x + 4y = 17 & m + 4n = 17 & p + 4q = 17 & g + 4h = 17 \\ 5x - 2y = 8 & 5m - 2n = 8 & 5p - 2q = 8 & 5g - 2h = 8 \end{array}$$

They are all the same! The variables used in each problem different but the coefficients are the same. Matrix solutions concern themselves with just the coefficients from the left side of the equation and the constants on the right side of the equation. However, there are some special aspects of matrices that you will learn in time. The most important items at this time are the following:

- You cannot divide matrices.
- Multiplication of matrices is not commutative.

We will also need to pull out a mathematical term that you do not use much. You know that the reciprocal of $\frac{3}{5}$ is $\frac{5}{3}$ and that their product is 1. It is because the product is 1 that we also call these numbers *multiplicative inverses*. In general, the mathematical inverse of x is $\frac{1}{x}$ and that this is also written as x^{-1}.

A matrix is essentially a rectangle of numbers. We need to learn to set up the matrices. The coefficient matrix, A, has two rows (one for each equation) and two columns (one for each variable). The variable matrix, X, has two rows but only one column, and the constant matrix, B, has two rows and one column. The matrix equation is

$$AX = B$$

and the solution to the matrix equation is

$$X = A^{-1}B$$

Open your calculator and activate the MATRIX menu on the TI-84 by pressing 2nd x^{-1}.

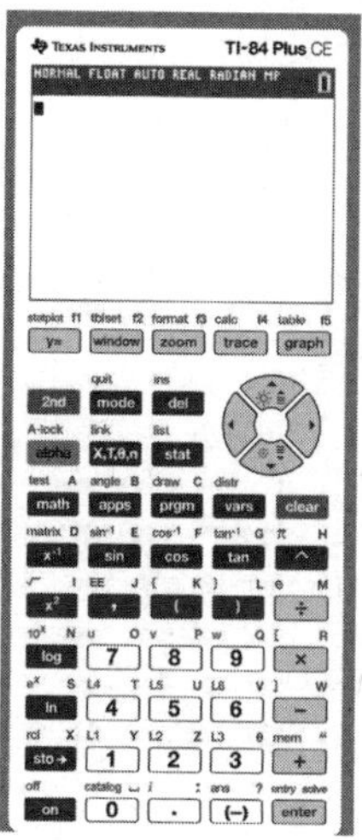

The screen will display the names of the available matrices. Press the left arrow once to get to the Edit menu and press Enter to select Matrix A. By default, matrix A has one row and one column (1 x 1).

Press 2, Enter, 2, Enter to change the dimensions to 2 x 2 (two rows and two columns). It is time to make the entries into the matrix. Press 1, Enter, 4, Enter, 5, Enter, −2 Enter to fill in matrix A with the coefficients. Press 2nd x^{-1} to go back to the matrix menu, left arrow to Edit, and down arrow to matrix B. Make the dimensions of matrix B 2 x 1 (two rows and one column). Enter the constants 17, Enter, 8, Enter. Press 2nd mode to quit the matrix menu. If there is text on the screen, press the Clear button.

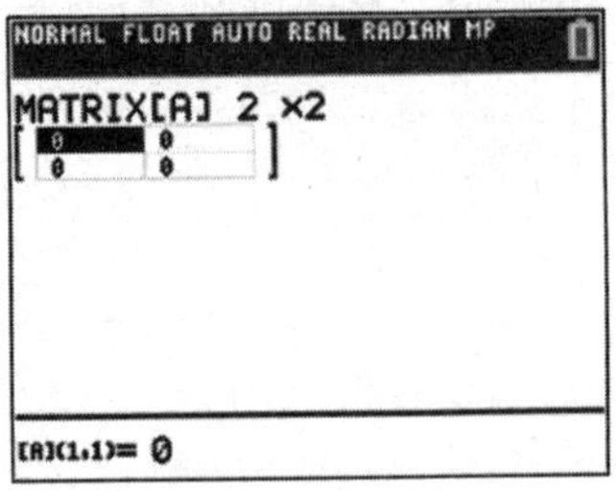

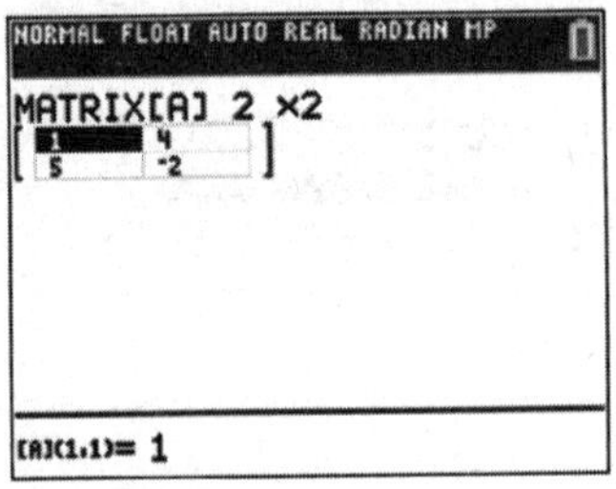

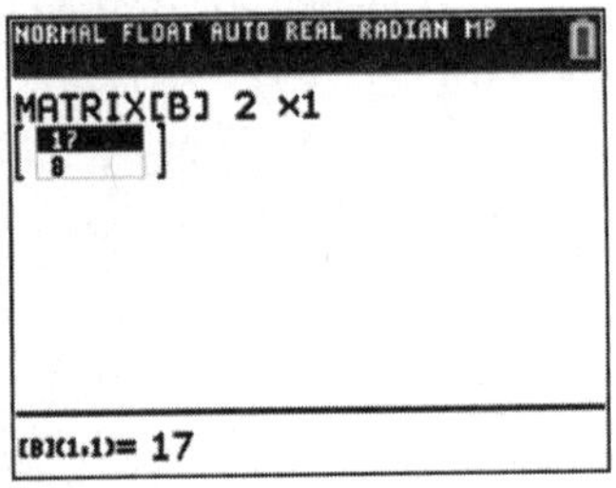

It's time to enter the matrix multiplication to get the solution. Press 2nd x^{-1} to enter the matrix menu. Press Enter to access matrix [A], press the x^{-1} button to get the inverse of [A]. Press 2nd x^{-1} to enter the matrix menu. Press down arrow to access matrix [B]. Press Enter. Finally, press Enter one more time to get the solution to the system.

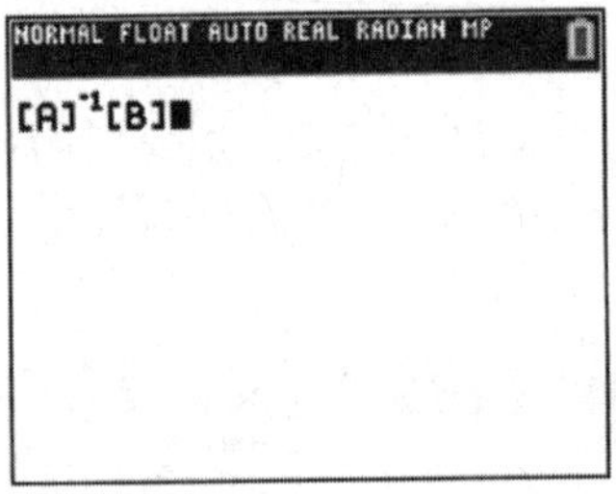

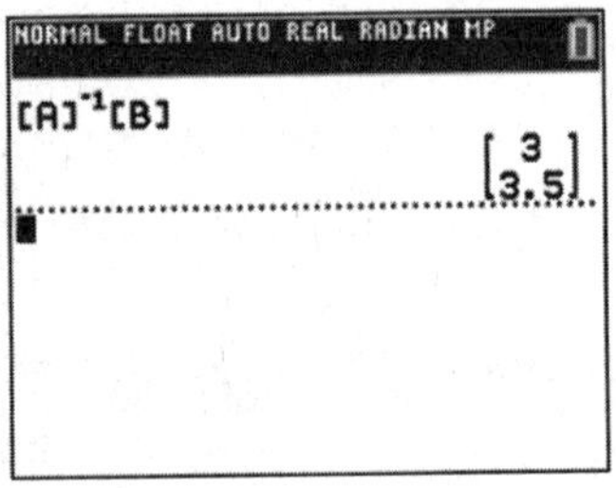

Technically, matrix [X], using the variables from the first system, equals $\begin{bmatrix} x \\ y \end{bmatrix}$, so this solution is $\begin{bmatrix} x \\ y \end{bmatrix} = \begin{bmatrix} 3 \\ 3.5 \end{bmatrix}$ and the solution is (3, 3.5).

Let's practice with a few of the systems we solved in the elimination section. We can reuse matrices A and B. Go to the Edit menu within the Matrix menu. You do not need to change the dimensions, so press the down arrow until you are able to change the entries within the matrix.

EXAMPLE

▶ Solve: $3x + 2y = -2$

$4x + 3y = 3$

▶ Matrix $[A] = \begin{bmatrix} 3 & 2 \\ 4 & 3 \end{bmatrix}$ and matrix $[B] = \begin{bmatrix} -2 \\ 3 \end{bmatrix}$ and $[A]^{-1}[B] = \begin{bmatrix} -12 \\ 17 \end{bmatrix}$ so the answer is (–12, 17).

Let's do one more of the systems with a unique solution before we examine the special cases of parallel lines and a duplicate line.

EXAMPLE

▶ Solve: $2x + 3y = -5$

$-4x + 5y = -111$

▶ Matrix $[A] = \begin{bmatrix} 2 & 3 \\ -4 & 5 \end{bmatrix}$ and matrix $[B] = \begin{bmatrix} -5 \\ -111 \end{bmatrix}$ and $[A]^{-1}[B] = \begin{bmatrix} 14 \\ -11 \end{bmatrix}$ so the answer is (14, –11).

What does the matrix solution look like when we have the special cases?

EXAMPLE

▶ Solve: $2x + 5y = 19$

$4x + 10y = 23$

▶ Matrix $[A] = \begin{bmatrix} 2 & 5 \\ 4 & 10 \end{bmatrix}$ and matrix $[B] = \begin{bmatrix} 19 \\ 23 \end{bmatrix}$ and $[A]^{-1}[B]$ produces the screen

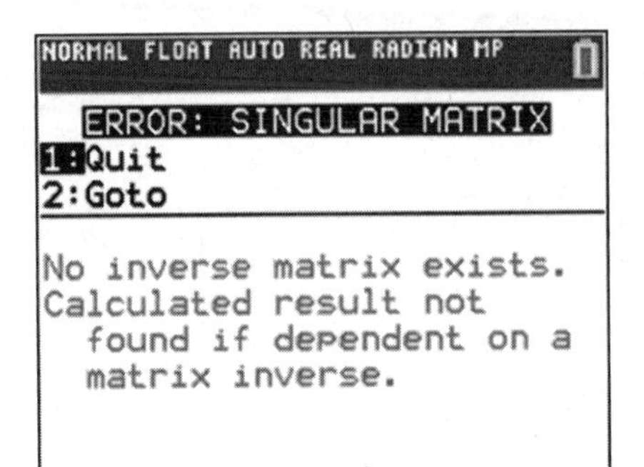

▶ This tells us we are dealing with one of the special cases. After we've made sure that we entered the information into the calculator correctly, we need to go through the process of recognizing equal slopes and determining whether or not the y-intercepts are equal.

Applying Systems of Linear Equations

There are many situations that involve two characteristics in which we do not know the value of either characteristic but do know something about the combination of them. We'll take a look at three classic situations in this chapter: currency, motion, and mixture.

Currency

We've already looked at some of these questions in Chapter 5. Here are a couple of variations for those problems.

EXAMPLE

▶ When Colin opened his piggy bank, he found that he had 156 coins in nickel and dimes that had a total value of $11.40. How many of each type of coin were in his piggy bank?

- Letting n represent the number of nickels and d the number of dimes, we have the two equations:

  Coins: $n + d = 156$

  Value: $5n + 10d = 1140$

- Multiply the first equation by -5:

$$-5(n + d = 156) \Rightarrow -5n - 5d = -780$$
$$5n + 10d = 1140$$

- Add: $5d = 360 \Rightarrow d = 72$
- Find the number of nickels: $n + 72 = 156 \Rightarrow n = 84$
- Check the total value: $5(84) + 10(72) = 420 + 720 = 1140$.
- Colin had 84 nickels and 72 dimes in his piggy bank.

Here's a similar type of problem using paper currency.

EXAMPLE

- Sadie is ending her shift at the register when she decided to have some fun with her shift supervisor, Anna Grace.

  Anna: "How much money did you have in your register?"

  Sadie: "Not counting the change, I had $404 from a total of 60 bills, and there were only singles, five-dollar bills, and ten-dollar bills in the cash drawer."

  Anna: "How many singles did you have?"

  Sadie: "14."

  Anna: "OK. Thanks."

- Sadie seems a bit perplexed that the only question the supervisor asked was for the number of singles. Anna reports the story to the store manager, Maura. Maura chuckles at the story and simply says, "She'll learn."

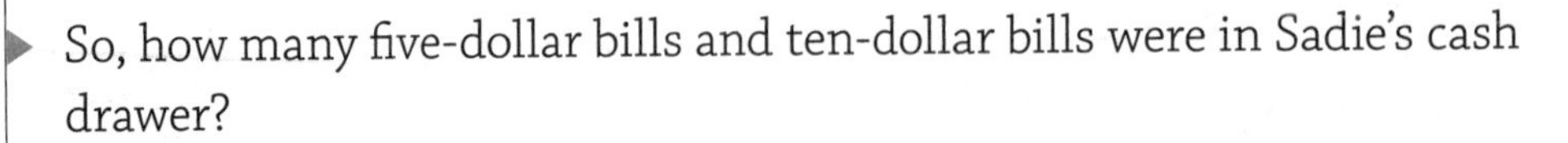

- So, how many five-dollar bills and ten-dollar bills were in Sadie's cash drawer?
- Letting f represent the number of five-dollar bills and n the number of ten-dollar bills, (if your t's look significantly different than your plus signs when you write, then you can use t to represent the number of ten dollar bills) we have two equations:

  Bills: $14 + f + n = 60$

  Value $1(14) + 5f + 10n = 404$

- Simplifying these equations: $f + n = 46$

  $$5f + 10n = 390$$

- Multiply the first equation by -5: $-5(f + n = 46) \Rightarrow \quad -5f - 5n = -230$

  $$5f + 10n = 390$$

- Add: $5n = 160$

  $$n = 32$$

- Find f: $f + n = 46 \Rightarrow f = 14$.
- Check the value: $14 + 5(14) + 10(32) = 14 + 70 + 320 = 404$.
- Sadie had 14 five-dollar bills and 32 ten-dollar bills in her register.

Motion

The basic formula rate $\times$ time = distance still applies, but now we get to consider such things as the current in a river and the wind speed when flying in an airplane. Understand that we are still looking at simple examples in that the boat and plane will either move with the current/wind at their back or directly into their face. We are not ready to consider the problems in which there is a 30° angle, for example, between the river current and the boat.

What happens when the direction of the current and the direction of the boat are the same? The boat goes as fast as the sum of the two speeds. When the current and boat travel against each other, the net speed is the difference of the speed of the current and the speed of the boat in still water (such as the speed of the boat on a lake).

EXAMPLE

- Charlie and Max make a 16-mile trip downstream in 2 hours. The return trip upstream takes 8 hours. How fast can their boat travel in still water, and what is the speed of the river's current?
- Letting r be the speed of the boat in still water and c be the speed of the current, we get:

$$\text{Downstream:} \quad 2(r + c) = 16 \Rightarrow r + c = 8$$

$$\text{Upstream:} \quad 8(r - c) = 16 \Rightarrow r - c = 2$$

- Add:

$$2r = 10$$

$$r = 5$$

- Find the current: $5 + c = 8 \Rightarrow c = 3$.
- The rate of the boat in still water is 5 mph and the speed of the current is 3 mph.

I am sure you noticed that rather than multiply through the parentheses, I divided both sides of the equation by the time factor. It is another way in which we can arrange for the coefficients of one of the variables to be the same magnitude and opposite in sign.

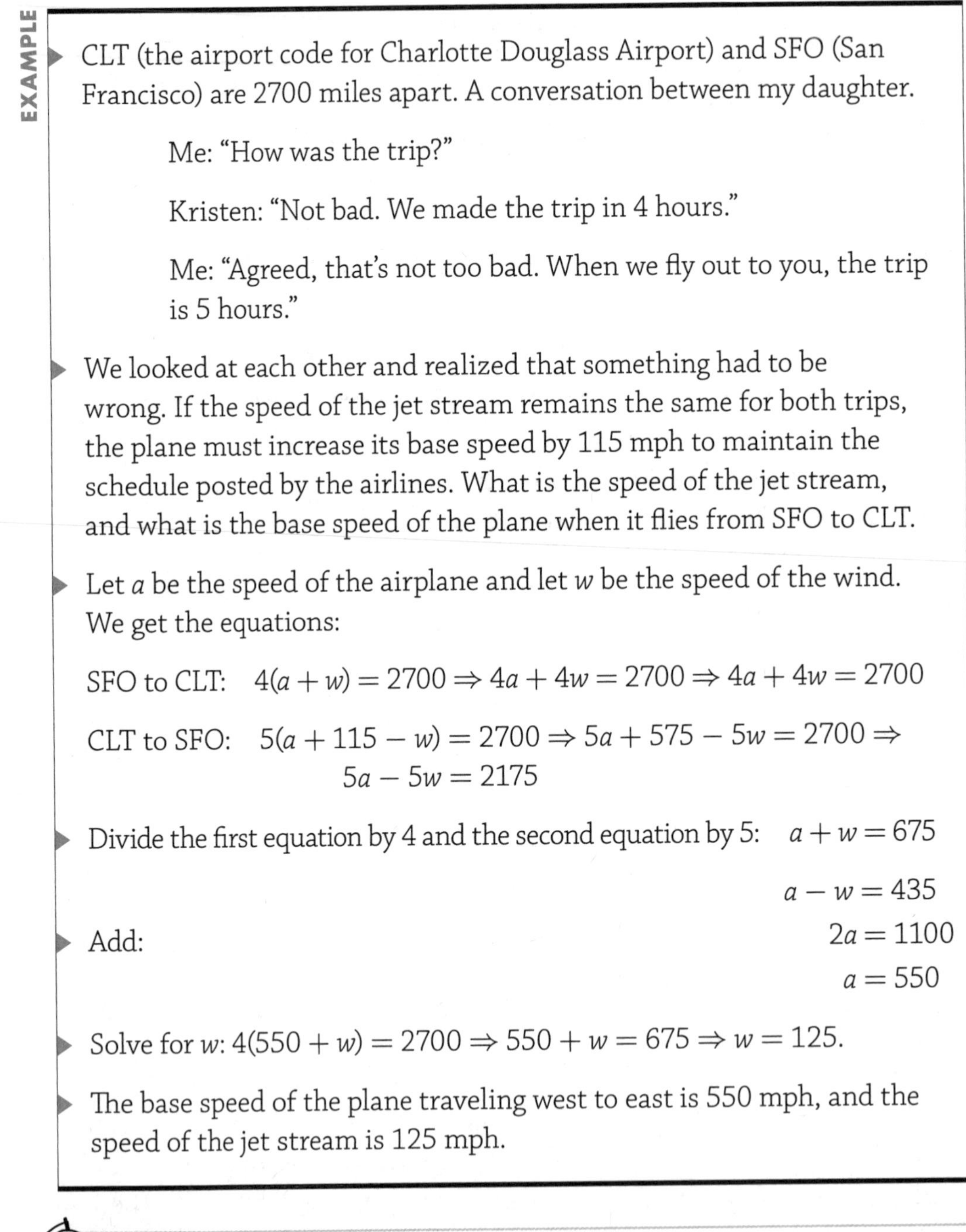

EXAMPLE

▶ CLT (the airport code for Charlotte Douglass Airport) and SFO (San Francisco) are 2700 miles apart. A conversation between my daughter.

> Me: "How was the trip?"
>
> Kristen: "Not bad. We made the trip in 4 hours."
>
> Me: "Agreed, that's not too bad. When we fly out to you, the trip is 5 hours."

▶ We looked at each other and realized that something had to be wrong. If the speed of the jet stream remains the same for both trips, the plane must increase its base speed by 115 mph to maintain the schedule posted by the airlines. What is the speed of the jet stream, and what is the base speed of the plane when it flies from SFO to CLT.

▶ Let a be the speed of the airplane and let w be the speed of the wind. We get the equations:

SFO to CLT: $4(a + w) = 2700 \Rightarrow 4a + 4w = 2700 \Rightarrow 4a + 4w = 2700$

CLT to SFO: $5(a + 115 - w) = 2700 \Rightarrow 5a + 575 - 5w = 2700 \Rightarrow$
$5a - 5w = 2175$

▶ Divide the first equation by 4 and the second equation by 5: $a + w = 675$

$$a - w = 435$$

▶ Add:

$$2a = 1100$$
$$a = 550$$

▶ Solve for w: $4(550 + w) = 2700 \Rightarrow 550 + w = 675 \Rightarrow w = 125$.

▶ The base speed of the plane traveling west to east is 550 mph, and the speed of the jet stream is 125 mph.

IRL **The speed of the jet stream varies from 110 mph to 250 mph. Airlines take this into account when determining the speed their planes will fly in order to maintain their schedule.**

Mixture

Mixture problems come in all shapes and sizes. Chemists will add water to an acid solution to dilute it or add pure acid to the solution to increase the concentration. Retailers that sell mixed nuts will balance the number of pounds of each type of nut with the cost of those nuts to determine the price at which the mixture will be sold. The same is true for a mixed fruit basket or a cheese tray. As with currency, we need to separate the equation that describes the weight or volume of the mixture from the price or concentration of the mixture.

EXAMPLE

- A merchant has a 45 pound display of mixed nuts that she is selling at $5 per pound. The mixture is made of almonds that are worth $4 per pound and cashews, worth $7 per pound. If the merchant does not lose any money in this process, how many pounds of each type of nut are used to make the mixture?
- Let a represent the number of pounds of almonds in the mixture and c represent the number of pounds of cashews in the mixture. The equations are:

  Weight: $a + c = 45$

  Value: $4a + 7c = 5(45)$

- Multiply the first equation by -4: $-4(a + c = 45) \Rightarrow -4a - 4c = -180$

  $4a + 7c = 225$

- Add: $3c = 45$

  $c = 15$

- Find a: $a + 15 = 45 \Rightarrow a = 30$
- Check: $4(30) + 7(15) = 120 + 105 = 225$
- The merchant mixes 30 pounds of almonds with 15 pounds of cashews.

A statement such as "use a 40% alcohol solution" indicates that 40% of the volume of the solution is alcohol and the remainder is some other liquid (usually water). Here are two problems involving percent solution.

EXAMPLE

- How many liters of a 20% ethyl alcohol solution must be added to a 40% ethyl alcohol solution to make 150 liters of a 25% ethyl alcohol solution?
- Let's begin by defining the variables. Let e represent the number of liters of the 25% solution and v represent the number of liters of the 40% solution that will be used.
- What do we know? We know the final solution will contain 150 liters:

  $$e + v = 150$$

- We know that all the alcohol in the final solution came from the two solutions that were mixed together.

  $$0.2e + 0.4v = 0.25\ (150)$$

- We have our system of equations. Multiply the first equation by -0.2

  $$-0.2(e + v = 150) \Rightarrow -0.2e - 0.2v = -30$$
  $$0.2e + 0.4v = 37.5$$

- Add the two equations together: $0.2v = 7.5$

  $$v = 37.5$$

- Determine the value of e: $e + 37.5 = 150 \Rightarrow e = 112.5$
- Does this give the right concentration: $0.2(112.5) + 0.4(37.5) = 22.5 + 15 = 37.5$ liters of alcohol
- Mix 112.5 liters of a 20% ethyl alcohol solution with 37.5 liters of a 40% alcohol mixture to create 150 liters of a solution that is 25% ethyl alcohol.

You can also add water to dilute a solution:

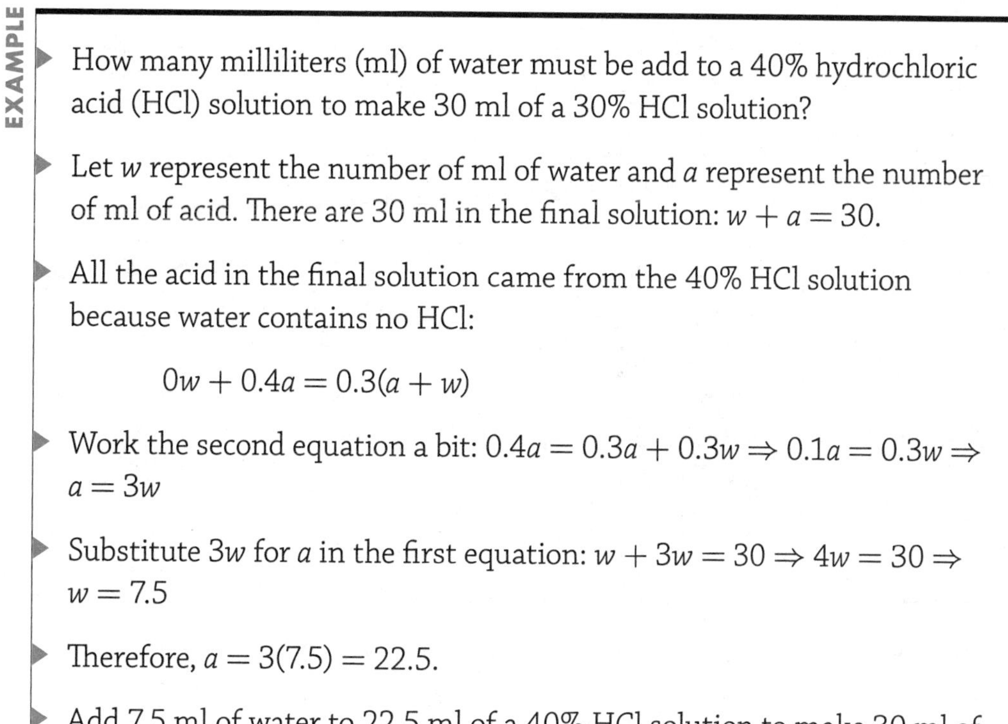

EXAMPLE

- How many milliliters (ml) of water must be add to a 40% hydrochloric acid (HCl) solution to make 30 ml of a 30% HCl solution?
- Let w represent the number of ml of water and a represent the number of ml of acid. There are 30 ml in the final solution: $w + a = 30$.
- All the acid in the final solution came from the 40% HCl solution because water contains no HCl:

$$0w + 0.4a = 0.3(a + w)$$

- Work the second equation a bit: $0.4a = 0.3a + 0.3w \Rightarrow 0.1a = 0.3w \Rightarrow a = 3w$
- Substitute $3w$ for a in the first equation: $w + 3w = 30 \Rightarrow 4w = 30 \Rightarrow w = 7.5$
- Therefore, $a = 3(7.5) = 22.5$.
- Add 7.5 ml of water to 22.5 ml of a 40% HCl solution to make 30 ml of a 30% HCl solution.

Systems of Inequalities

The solution to a linear inequality is a region of the plane. The solution to a system of linear inequalities will be those points that satisfy both inequalities. In writing our solution, we need to be clear where the boundaries are located and which region represents the common solution.

EXAMPLE

- Solve: $x > 2$

  $x + y > 3$

- Both inequalities lack the notion of equality, so both boundaries will be dotted lines.

- The graph of $x > 2$ is shown on the left below, while the graph of $x + y > 3$ is shown on the right.

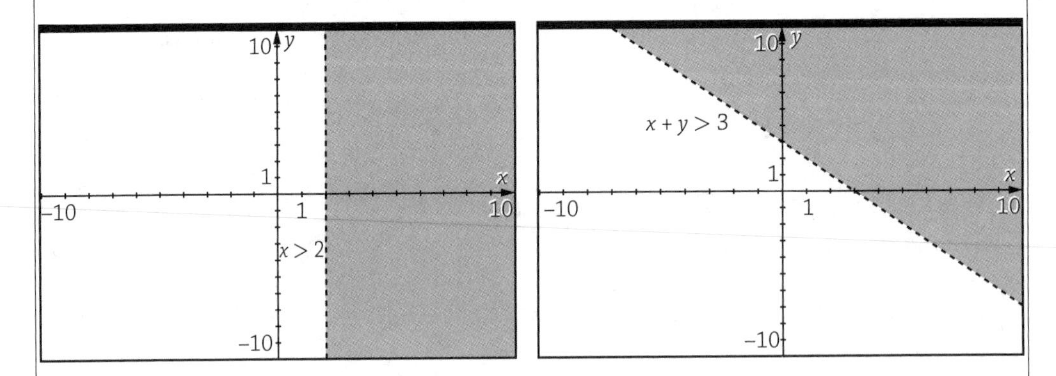

- The common solution then is represented by the dark region in the graph.

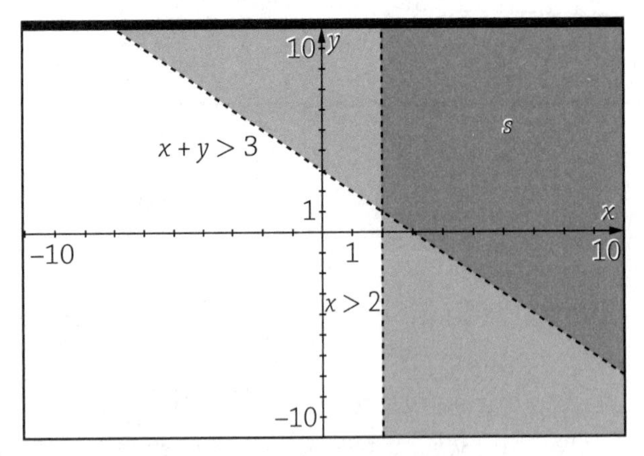

- It is not uncommon to write the letter S to represent the common solution.

Another approach to showing the solution to a system of inequalities is to only show the common solution. To do this, you will sketch the boundary as usual. Rather than shade the region that solves the individual inequality, you'll place an arrow along the boundary indicating which side of the boundary is part of the solution. Once all the boundaries have been formed and the solution determined, that part of the plane is shaded (with or without the S to indicate the solution). The picture would look like this:

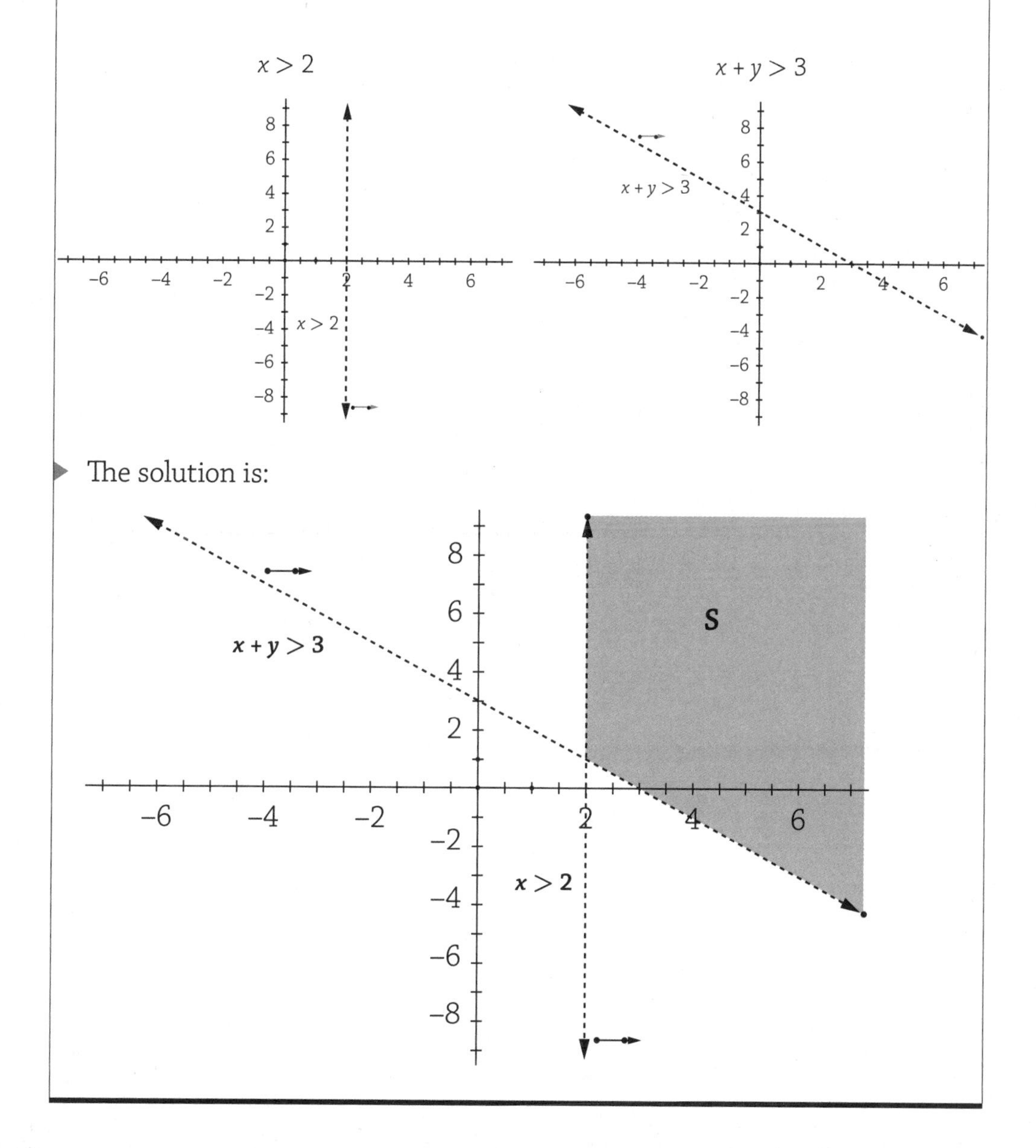

The solution is:

Shading can be done with pencil and paper using slanted lines in different directions for each individual solution and the crosshatch of these slanted lines will indicate the common solution. This can be a bit cumbersome when working with more than two boundaries.

EXAMPLE

- Solve the system of inequalities: $x \geq 4 \quad y \geq 2 \quad 2x + 3y \leq 24$
- The boundaries in this inequality are all included in the solution. The common solution is

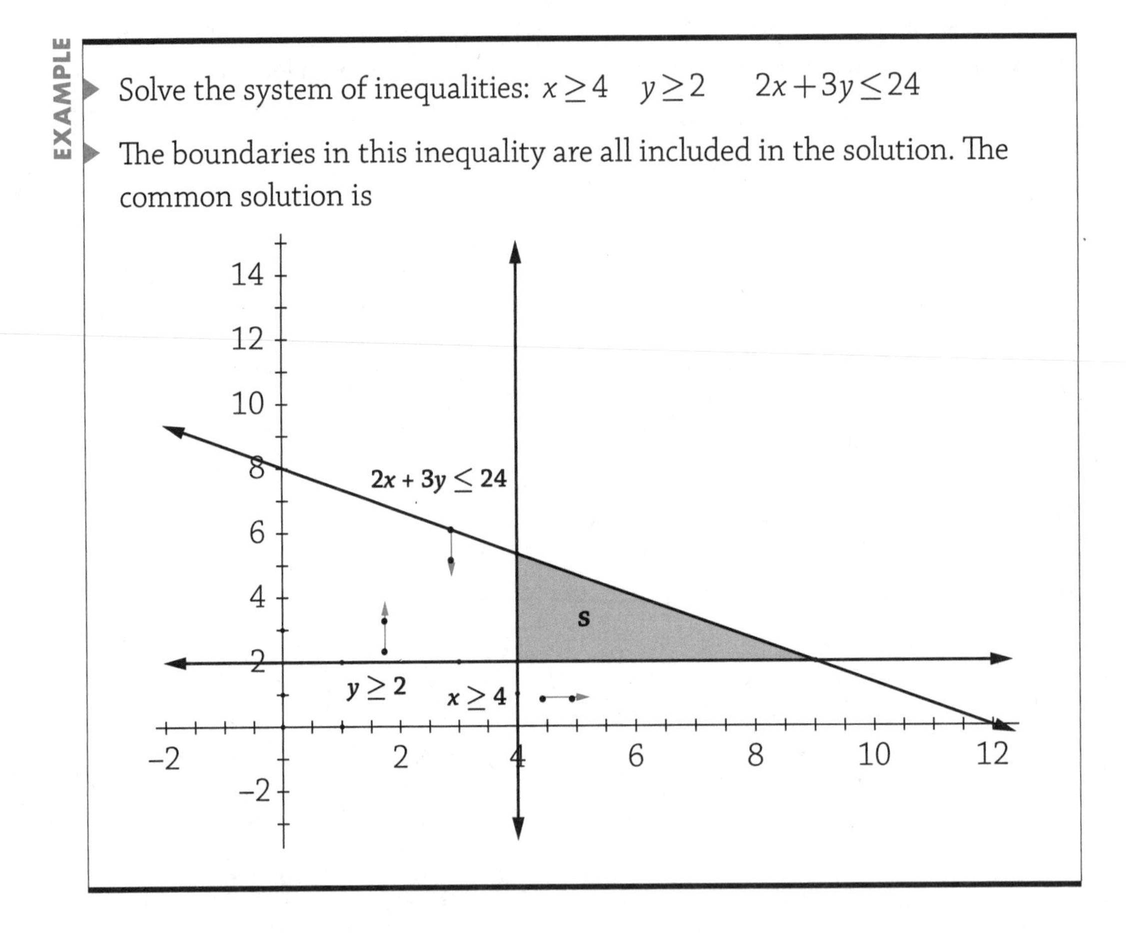

Let's take a look at a more complicated system of linear inequalities.

EXAMPLE

- $x \geq 0 \quad y \geq 0$
- $3x + 4y \geq 24$
- $x + 3y \geq 12$
- $2x + y \geq 10$
- Again, the boundaries are included in the solution. It is easiest to use the intercepts for each lot to plot the boundaries:
- $3x + 4y \geq 24$; (8, 0); (0, 6)
- $x + 3y \geq 12$; (12, 0); (0, 4)
- $2x + y \geq 10$; (5, 0); (0, 10)

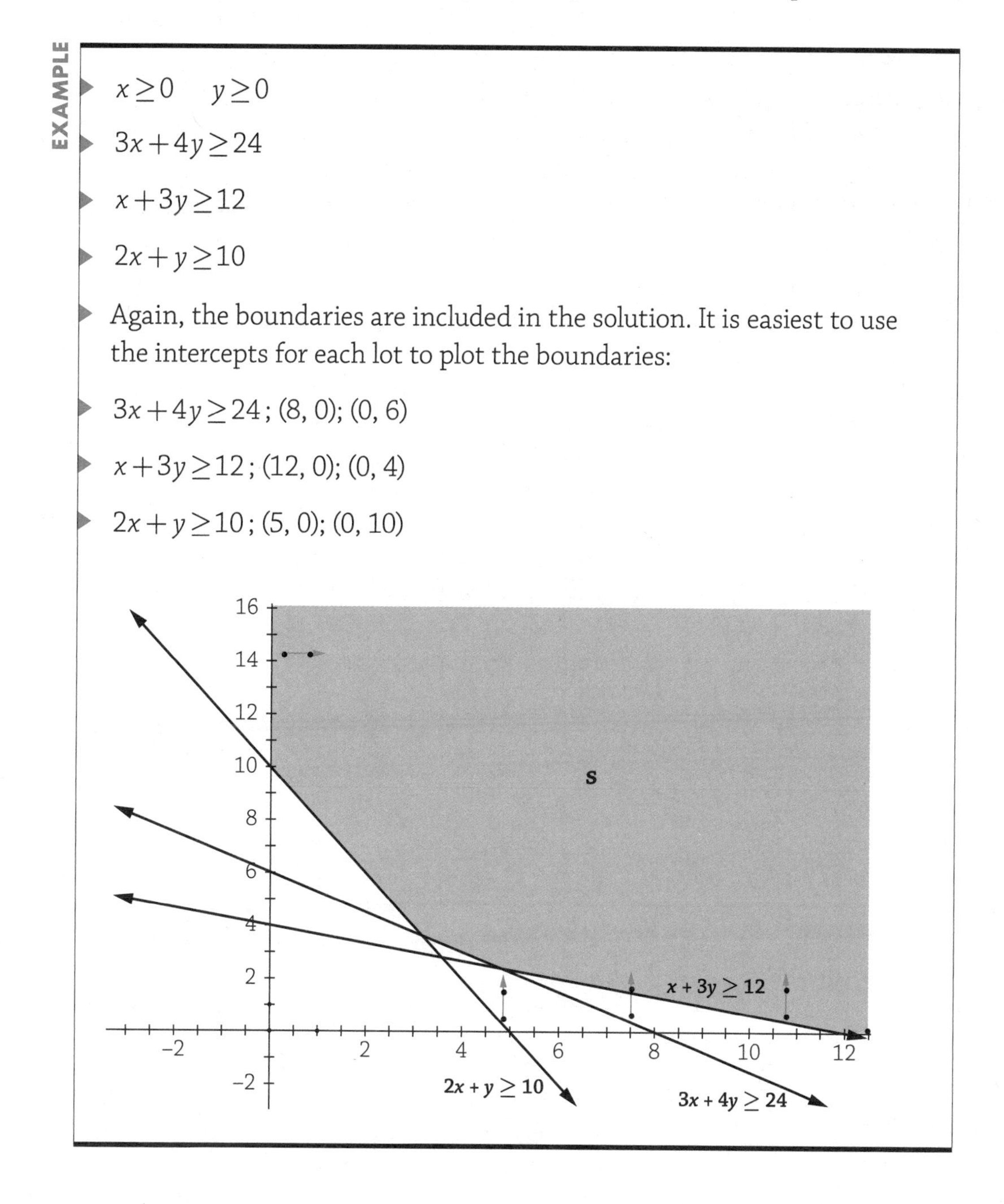

I made reference to a topic called Linear Programming in Chapter 6. In a linear programming problem, the inequalities in the last example might represent the minimum daily requirements for vitamins, zinc, and iron and the variables x and y are two supplements that provide these nutrients. A function of the form $C(x, y)$, called the objective function could represent the cost of taking these supplements. The goal would be to be sure that the minimum daily requirements be met but do so at the least possible cost. Linear Programming theory claims that the optimal (best) solution will occur at one of the turning points that form the boundary of the common solution (which in Linear Programming terminology is called the feasibility region).

EXERCISES

EXERCISE 7-1

Solve the system of equations by graphing.

1. $y = x - 5$
   $x + y = 1$

2. $y = -2x + 1$
   $y = -x + 3$

3. $y = 3x - 5$
   $2x - y = 1$

EXERCISE 7-2

Solve the system of equations by using the substitution method.

1. $y = -x - 1$
   $y = 3x + 15$

2. $y = -2x + 5$
   $3x + 2y = 12$

3. $y = 2x + 1$
   $7x - 3y = 1$

EXERCISE 7-3

Solve the system of equations by using the elimination method.

1. $4x + 3y = 17$
   $5x - y = 7$

2. $5x + 3y = 71$
   $2x + 5y = 74$

3. $5x - 8y = -1$
   $3x + 5y = 68$

4. $6x + 9y = 27$
   $2x + 3y = 10$

EXERCISE 7-4

Solve the system of equations by using matrices.

1. $9x + 18y = 21$
   $12x - 24y = -12$

2. $3x - 2y = 206$
   $5x + 7y = 931$

3. $4x + 7y = -211$
   $5x - 2y = -349$

EXERCISE 7-5

Solve the system of equations by using the method of your choice.

1. $x + y = 15$
   $y = x - 1$

2. $3x - 2y = 17$
   $y = x - 5$

3. $2x + 3y = -9$
   $3x + 2y = 14$

4. $y = 2x - 1$
   $10x - 3y = 21$

5. $3x + 5y = 11.76$
   $-7x + 3y = 16.12$

6. $7x + 4y = 20$
   $y = 1.75x + 5$

7. $12x + 20y = -268$
   $42x - 34y = 1454$

EXERCISE 7-6

Solve these applied problems.

1. Cameron and Carson have been saving their dimes and quarters when they receive them as change. (The pennies and nickels they throw into the "Someday Piggy Bank" shaped like a large crayon.) The day finally came to count the money, and they found they had a total of 200 coins worth $37.40. How many dimes and how many quarters had they saved?

2. Diane and Chris returned from their trip to Australia and discovered that they had $96 in $1 and $2 Australian coins. If they had a total of 59 coins, how many of each type of coin did they have?

3. A motorboat makes an 18-mile trip downstream in 2 hours while the return trip takes 6 hours. Determine the speed of the current and the speed of the boat in still water.

4. The flight from Brisbane, Queensland, Australia to Los Angeles, California takes 15 hours to travel the near 7200 miles. The trip from Los Angeles back to Brisbane also takes 15 hours. To make this happen, the pilot must fly the jet at 240 mph faster than the initial trip to allow for the impact of the jet stream. What is the speed of the jet stream and at what speeds does the pilot set his speed during each leg of the round trip?

5. How much water must be added to a 70% acid solution to make 105 ml of a 40% acid solution?

6. How many liters of a 10% acid solution and how many liters of a 40% acid solution are needed to make 270 liters of a 20% acid solution?

EXERCISE 7-7

Solve the system of linear inequalities.

1. $x > -2$
   $y < 3$

2. $y \geq -3x + 2$
   $x + y \leq 10$

3. $2x + 3y \leq 12$
   $3x - 2y \geq 12$

4. $x + 2y < 10$
   $2x - 5y > 20$

Exponents and Exponential Functions

MUST KNOW

- The rules for combining terms in exponential form reduces multiplication and division to addition and subtraction.

- Knowing and applying the rules for the properties of exponents enables us to answer questions about phenomena such as compound interest, population growth, and radioactive decay that can be modeled with exponential functions.

lbert Einstein was once quoted as saying, "Compound interest is the most powerful force in the universe." As we watch the impact of the borrowing and lending habits across the globe, it is hard to argue with him. Compound interest is just one example for the use of exponential functions. In this chapter, we'll take a look at the rules governing exponents and also look at a few other applications of exponential functions.

Multiplication Properties of Exponents

We know that x^n, where n is a positive integer, means to use x as a factor n times. For example, $x^3 = x \cdot x \cdot x$. Consequently, we get some basic rules for multiplication when using exponents.

- **Exponent Rule 1** $x^m \cdot x^n = x^{m+n}$

  When multiplying two terms with a common base, keep the base and add the exponents.

$x^5 \cdot x^4 = x^{5+4} = x^9$

There are nine factors of x on the left-hand side of the equation, so there need to be nine factors on the right-hand side as well.

- **Exponent Rule 2** $(xy)^n = x^n y^n$

  When a product is raised to a power, the answer is the same as taking the product of each of the factors raised to that power.

EXAMPLE

$(3x)^4 = (3x)(3x)(3x)(3x) = 3^4 x^4 = 81x^4$

■ **Exponent Rule 3** $(x^n)^m = x^{nm}$

When a term raised to a power is in turn raised to a power, the answer is the same as taking the term and raising it to the product of the powers.

EXAMPLE

▶ $(4x^5)^3 = (4x^5)(4x^5)(4x^5) = 4^3x^5 = 4^3(x^5)^3$

Let's practice with a few problems.

EXAMPLE

▶ Simplify: $(12x^4)(5x^3)$

▶ We can regroup these terms so that the problem reads $(12 \times 5)(x^4 \cdot x^3) = 60x^7$.

Here's a problem that is a bit more involved in that it uses all three rules.

EXAMPLE

▶ Simplify: $(4m^5)^3(2m^4)^5$.

▶ Let's deal with the exponents that are outside of each set of parentheses and work our way in.

$(4m^5)^3(2m^4)^5 = (4^3(m^5)^3)(2^5(m^4)^5) = (64m^{15})(32m^{20}) = 2048m^{35}$

You may have thought to break 4 to be 2^2 so that the constants were all written as a power of 2. If so, good for you! Just be sure you understand that the coefficient 2 in the second set of parentheses might have been a

3 instead. In this case, there would not have been a common base for you to use.

EXAMPLE

▸ Simplify: $(5x^6)^2(3xy^2)^2(y^3z^4)^5$

▸ While it may look complicated because there are three different variables in the problems, the process is fairly straightforward. We'll deal with each set of parentheses as a separate problem and then deal with the final product.

$$(5x^6)^2(3xy^2)^2(y^3z^4)^5 = (5^2(x^6)^2)(3^2x^2(y^2)^2)((y^3)^5(z^4)^5)$$

$$= (25x^{12})(9x^2y^4)(y^{15}z^{20}) = (25 \times 9)(x^{12} \cdot x^2)(y^4 \cdot y^{15})z^{20} = 225x^{14}y^{19}z^{20}$$

Division Properties of Exponents

Division of exponents behaves in a way that is very predictable, especially after you have looked at the rules for multiplication.

- **Exponent Rule 4** $\frac{x^m}{x^n} = x^{m-n}$ when $m > n$.

  When there are more factors of x in the numerator than there are in the denominator, the number of factors of x remaining in the numerator is equal to the difference in the number of factors.

There are two interesting ramifications of this statement:

- What if there is the same number of factors of x in the numerator and the denominator?
- What if there are more factors of x in the numerator than the denominator?

What if there is the same number of factors? So long as x does not equal zero, $\frac{x^5}{x^5} = \frac{x^{17}}{x^{17}} = \frac{x^n}{x^n} = 1$ because any number (except zero) divided by itself is 1. (Remember, we cannot divide by zero.) If we ignore the caveat from rule 1 that $m > n$, then $\frac{x^n}{x^n} = x^0$. We now have a rule for the zero exponent.

- **Exponent Rule 5** $x^0 = 1$ provided $x \neq 0$.

  What if the number of factors in the denominator is greater than the number of factors in the numerator? Well, that really isn't a big deal, is it? The answer to $\frac{x^3}{x^5}$ has to be $\frac{1}{x^2}$ because there are two more factors of x in the denominator than there are in the numerator. BUT, if we look back at rule 1 again (and ignoring the caveat), we would get $\frac{x^3}{x^5} = x^{3-5} = x^{-2} = \frac{1}{x^2}$. You can apply this logic to any pair of exponents that you like. We now have a meaning for a negative exponent—it is simply the reciprocal.

- **Exponent Rule 6** $x^{-n} = \frac{1}{x^n}$

  Let's practice with these rules.

EXAMPLE

▶ Simplify: $\frac{12x^4y^3}{6x^2y^5}$

▶ We can separate the terms into three problems: $\frac{12}{6}\frac{x^4}{x^2}\frac{y^3}{y^5} = \frac{2x^2}{y^2}$ using positive exponents (usually the preferred form of the answer), but $2x^2y^{-2}$ is a correct answer as well.

You will find that it is almost always the case that answers to problems like these will have the terms with positive exponents. The few times that the answers may contain a negative exponent are in some applications (which we will see later in the chapter) and in multiple-choice questions.

EXAMPLE

▶ Simplify: $\dfrac{50x^{-4}y^{6}}{100x^{3}y^{-2}}$

▶ I will usually approach a problem like this by reducing the constants and then subtracting the exponents as with rule 4. I'll deal with any negative exponents in the answer as a last step.

▶ $\dfrac{50x^{-4}y^{6}}{100x^{3}y^{-2}} = \dfrac{x^{-7}y^{8}}{2} = \dfrac{y^{8}}{2x^{7}}$

Now let's put all the rules to work.

EXAMPLE

▶ Simplify: $\dfrac{(6x^{4}y^{-2})^{3}(2x^{2}y^{3})^{4}}{108x^{-5}y^{6}}$

▶ Begin with computing the terms with the exponents outside the parentheses.

$$\frac{(6x^{4}y^{-2})^{3}(2x^{2}y^{3})^{4}}{108x^{-5}y^{6}} = \frac{(216x^{12}y^{-6})(16x^{8}y^{12})}{108x^{-5}y^{6}}$$

▶ Simplify the variable terms in the numerator while reducing the constant coefficients.

$$\frac{(216x^{12}y^{-6})(16x^{8}y^{12})}{108x^{-5}y^{6}} = \frac{32x^{20}y^{6}}{x^{-5}y^{6}}$$

▶ Subtract exponents:

$$\frac{32x^{20}y^{6}}{x^{-5}y^{6}} = 32x^{25}y^{0}$$

▶ Since $y^{0} = 1$, the answer is $32x^{25}$

Rational Exponents

We now have an understanding of what x^n for any integer value of n. Now for a relationship that is a bit more complicated. We take advantage of rule 3: $(x^m)^n = x^{mn}$.

EXAMPLE

- Solve for n: $(x^2)^n = x$
- Apply rule 3 so the equation reads $x^{2n} = x^1$. (I wrote the 1 for the exponent for the right-hand side of the equation for emphasis.) The only way these two terms can be equal (without having x equal 0 or 1) is if the two exponential statements are equal. Consequently, $2n = 1$ so that $n = \frac{1}{2}$.

Wow! We can have exponents be fractions. But what do we interpret what they mean? After all, x^3 tells us to use x as a factor 3 times. We can't use x as a factor for one-half a time. Let's look at the problem with a specific value for x.

$(5^2)^{\frac{1}{2}} = 5$ becomes $(25)^{\frac{1}{2}} = 5$ while $(7^2)^{\frac{1}{2}} = 7$ becomes $(49)^{\frac{1}{2}} = 7$. Do you see the relationship? We know $\sqrt{25} = 5$ and that $\sqrt{49} = 7$, so it stands to reason that the one-half exponent corresponds to a square root. That is, $x^{\frac{1}{2}} = \sqrt{x}$, provided $x \geq 0$. We could continue this discussion to discuss problems like $(5^3)^n = 5$, $(7^4)^n = 7$, or $(13^9)^n = 13$. In each case, the value of n will equal the reciprocal of the exponent within the parentheses and would have to be interpreted as the root (cube root, fourth root, ninth root) of the expression in the parentheses. This given us a new rule:

- **Exponent Rule 7** $x^{n} = \sqrt[n]{x}$.

  Care must be taken with this rule. If n is an even number, then $x \geq 0$, whereas if n is an odd number, then x can be any real number. We know that the square of any nonzero will be positive. The same is true for any even exponent. However, the cube of a negative number is a negative number, and we extend this to note that whenever a negative number is raised to an odd exponent, the product must be negative. Stop and give that some thought before you move on.

EXAMPLE

▶ Evaluate: $(-27)^{\frac{1}{3}}$

▶ $(-27)^{\frac{1}{3}} = ((-3)^3)^{\frac{1}{3}} = -3$

What do we do if there is a fractional exponent but the numerator of the fraction is not 1? We simply use rule 3 and rule 7.

EXAMPLE

▶ Simplify: $(64)^{\frac{2}{3}}$.

▶ Apply rule 3: $(64)^{\frac{2}{3}} = (64^2)^{\frac{1}{3}}$. Hang on! How is that easier? Yes, I could break out a calculator and figure out that $64^2 = 4096$, but there is no way I can be expected to know the cube root of this number. Let's try this again.

▶ Apply exponent rule 3: $(64)^{\frac{2}{3}} = (64^{\frac{1}{3}})^2$. I know the cube root of 64 is 4 so $(4)^2$ and I know that $4^2 = 16$. (And now I know that the cube root of 4096 is 16—for what that's worth!)

While we really don't need a new rule, it might help if we write this down in our list.

- **Exponent Rule 8** $x^{\frac{m}{n}} = (x^{\frac{1}{n}})^m$

There may be an occasion in which we would apply the rule $x^{\frac{m}{n}} = (x^m)^{\frac{1}{n}}$, but this does not happen as often.

EXAMPLE

▶ Simplify: $\left(\dfrac{32x^{11}y^{-9}}{162x^{-7}y^3}\right)^{\frac{-3}{2}}$

▶ This is a whopper! Rather than be startled, take a look at the problem and realize that the expression within the parentheses can be simplified. We know that 32 and 162 have a common factor of 2, so they can be reduced and the terms in x and y can be simplified. Let's do this.

$$\left(\frac{32x^{11}y^{-9}}{162x^{-7}y^3}\right)^{\frac{-3}{2}} = \left(\frac{16x^{18}}{81y^{12}}\right)^{\frac{-3}{2}}.$$

▶ That wasn't too bad. The exponent outside the parentheses is negative. Rule 6 tells us to take the reciprocal of what is inside the parentheses and make the exponent positive. That's easy to do.

$$\left(\frac{16x^{18}}{81y^{12}}\right)^{\frac{-3}{2}} = \left(\frac{81y^{12}}{16x^{18}}\right)^{\frac{3}{2}}$$

▶ Exponent rule 8 tells us to break up the exponent.

$$\left(\frac{81y^{12}}{16x^{18}}\right)^{\frac{3}{2}} = \left(\left(\frac{81y^{12}}{16x^{18}}\right)^{\frac{1}{2}}\right)^3$$

- Exponent rule 2 tells us how to deal with multiple factors within the parentheses.

$$\left(\left(\frac{81y^{12}}{16x^{18}}\right)^{\frac{1}{2}}\right)^3 = \left(\frac{\left(81^{\frac{1}{2}}\right)(y^{12})^{\frac{1}{2}}}{\left(16^{\frac{1}{2}}\right)(x^{18})^{\frac{1}{2}}}\right)^3 = \left(\frac{9y^6}{4x^9}\right)^3$$

- That's a lot of writing, but the concept is easy enough (you can probably just do the arithmetic without all that writing). Apply the exponent to get your final answer.

$$\left(\frac{9y^6}{4x^9}\right)^3 = \frac{729y^{18}}{64x^{27}}$$

- Congratulations! You can work through problems like these, and this is as about as tough as they get.

Scientific Notation

The sun is approximately 93,000,000 miles from Earth, while Neptune is approximately 2,793,000,000 miles from the sun. The speed of light is 670,616,629 miles per hour. The diameter of a hydrogen atom is approximately 0.000000000106 meters. There comes a point in which writing all of the digits in a number gets to be exhausting without really being informative. In such cases, we tend to write large (and small) numbers in **scientific notation**.

Numbers in scientific notation take the form $m.bbb \times 10^n$, where n is an integer and m is a digit from 1 to 9. (The value of b can be any digit.) The distance from the sun to the Earth is 9.3×10^7 miles. The distance from the sun to Neptune is 2.793×10^9 miles, and the speed of light is 6.706×10^8. The diameter of the hydrogen atom is 1.06×10^{-10} meters.

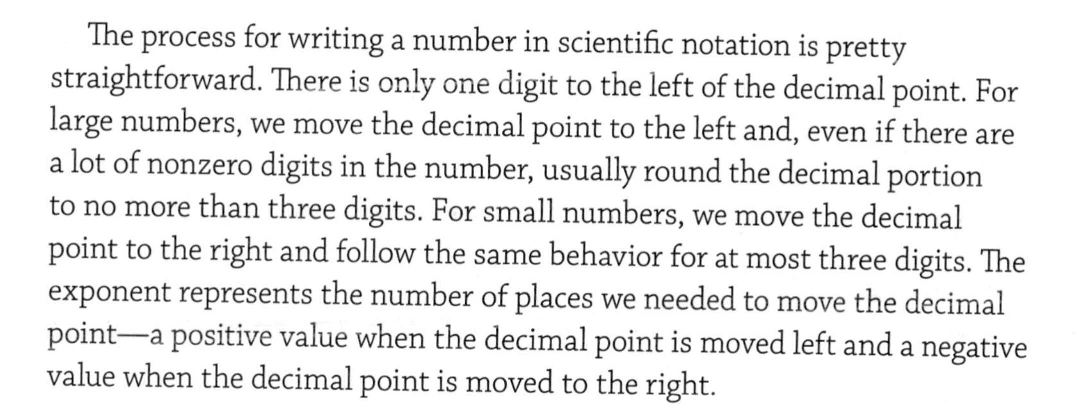

The process for writing a number in scientific notation is pretty straightforward. There is only one digit to the left of the decimal point. For large numbers, we move the decimal point to the left and, even if there are a lot of nonzero digits in the number, usually round the decimal portion to no more than three digits. For small numbers, we move the decimal point to the right and follow the same behavior for at most three digits. The exponent represents the number of places we needed to move the decimal point—a positive value when the decimal point is moved left and a negative value when the decimal point is moved to the right.

EXAMPLE

▶ Write 27,280,000,000 in scientific notation:

▶ $27{,}280{,}000{,}000 = 2.728 \times 10^{10}$

Let's try a problem from the other end of the spectrum.

EXAMPLE

▶ Write 0.00000000000235 in scientific notation:

▶ $0.00000000000235 = 2.35 \times 10^{-13}$

When doing addition or subtraction with scientific notation, the exponents need be the same so you can add or subtract the constants. Rewrite your answer in scientific notation.

EXAMPLE

▶ Compute: $(4.35\times10^4)+(6.21\times10^5)$

$$(4.35\times10^4)+(6.21\times10^5) = (4.35\times10^4)+(62.1\times10^4)$$

$$=(4.35+62.1)\times10^4 = 66.45\times10^4 = 6.645\times10^5$$

Multiplication and division are done by combining the constants and then the exponential terms. That is, $(a \times 10^m)(b \times 10^n) = ab \times 10^{m+n}$, while $\frac{a \times 10^m}{b \times 10^n} = \frac{a}{b} \times 10^{m-n}$, and answers are written in scientific notation. The number of decimal places in the answer should be the same as the smaller number of decimal places in *a* or *b*.

EXAMPLE

- Compute: $(4.35 \times 10^5)(6.123 \times 10^3)$

$$(4.35 \times 10^5)(6.123 \times 10^3) = (4.35 \times 6.123) \times 10^{5+3}$$
$$= 26.6351 \times 10^8 = 2.66 \times 10^9$$

- There are only two decimal places in 4.35, so there are two decimal places in the result.

Let's take a stab at a division problem.

EXAMPLE

- Compute: $\frac{3.45 \times 10^6}{2.1 \times 10^4}$

- $$\frac{3.45 \times 10^6}{2.1 \times 10^4} = \frac{3.45}{2.1} \times 10^{6-4} = 1.64286 \times 10^2 = 1.6 \times 10^2$$

This is similar to the last problem, but there is a slight difference.

EXAMPLE

- Compute: $\frac{1.75 \times 10^2}{2.0 \times 10^5}$

- $$\frac{1.75 \times 10^2}{2.0 \times 10^5} = \frac{1.75}{2.0} \times 10^{2-5} = 0.875 \times 10^{-3} = 8.8 \times 10^{-4}$$

Exponential Functions

Functions of the form $f(x) = b^x$, where b is a constant, are called exponential functions. The restrictions on b are that it must be a positive number but cannot equal 1. Why not 1? Because it is boring. One raised to any power is always 1, and that would make the function a linear function. If $b > 1$, the graph will increase as the values of x get larger, while if $0 < b < 1$, the graph will decrease as the values of x get larger.

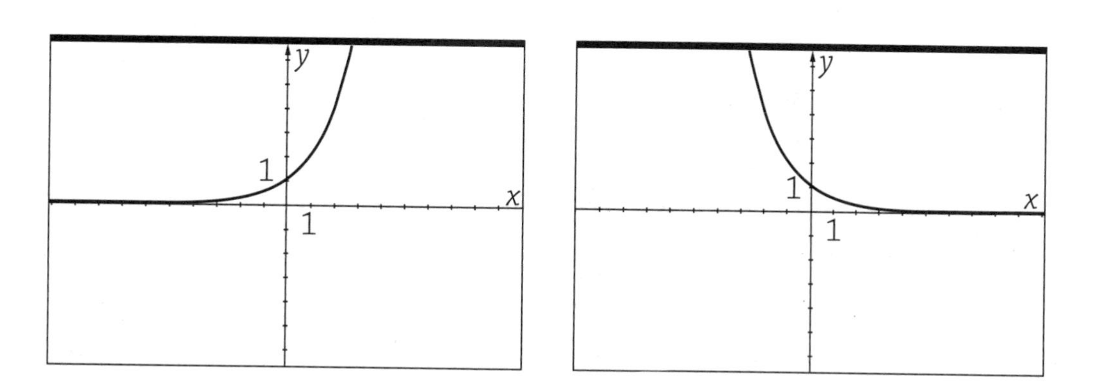

The domain for all exponential functions is the set of real numbers. The range of the exponential function is $y > 0$ because there is no value of x for which $b^x = 0$.

Why do we study exponential functions? These functions have many applications. Some of the applications are compound interest (including mortgage and car payments, as well as individual retirement accounts), population growth, nuclear decay, and forensic science. We'll take a look at some of these applications in the next section. We must first take some time to learn how to solve exponential equations.

The basic rule for solving exponential functions is this: if $b^m = b^n$, then $m = n$. For example, if $2^{x+1} = 2^{3x-5}$, then $x + 1 = 3x - 5$ and $x = 3$. Some of the problems we will have to solve are fairly straightforward, while others are not. First, let's look at the more straightforward problems.

EXAMPLE

▶ Solve: $2^x = 32$.

▶ How is this a straightforward problem? Once you realize that $32 = 2^5$, the problem becomes $2^x = 2^5$ so $x = 5$.

That wasn't too bad. We can jazz these questions up to have a little more algebra included.

▶ Solve: $3^{3x+2} = 9^{x-4}$.

▶ We know that $9 = 3^2$, so the equation can be rewritten as $3^{3x+2} = (3^2)^{x-4}$, which in turn becomes $3^{3x+2} = 3^{2x-8}$. We now know that $3x + 2 = 2x - 8$ so that $x = 10$.

Here is a problem that is just a bit trickier.

EXAMPLE

▶ Solve: $2^{5x+4} = 4(8^x)$.

▶ We know that if we are expected to be able to solve this problem, we should be able to write both sides with a common base. Both 4 and 8 are powers of 2, so we can rewrite the equation as $2^{5x+4} = 2^2(2^3)^x = 2^{3x+2}$. We can solve the equation $5x + 4 = 3x + 2$ to determine that $x = -1$.

The bottom line is that if we can rewrite both sides of the equation with a common base, then the problem is reasonably straightforward. But what if we cannot write both sides of the equation with the same base? In the not too distant past, we would have been out of luck in trying to get the answer. (That is, until we studied Algebra II and found this wonderful thing called logarithms. But this is not Algebra II.) What we have available to us is a graphing device that can give us the result. Take a look at this example.

EXAMPLE

▸ Solve: $2^x = 7$.

▸ We know the answer is between 2 and 3 because $2^2 = 4$ and $2^3 = 8$, so the power of 2 needed to yield 7 has to fall between 2 and 3. We'll graph the two functions $f(x) = 2^x$ and $y = 7$ and determine the point of intersection, just as we did with the two linear equations.

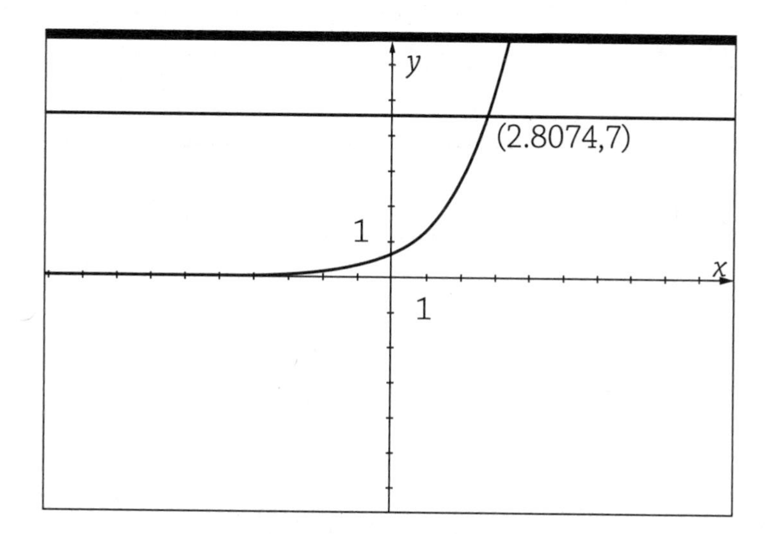

▸ The solution to the problem is that x is approximately equal to 2.8074. The reality is that when we cannot write the equation with common bases, we will always get approximate answers. Nonetheless, a four-decimal digit approximate is pretty good. Enter $2^{2.8074}$ into your calculator. That's a good approximation for 7.

Growth and Decay

Recall that the formula for simple interest, I, is $I = Prt$, with P standing for the principal (the amount of money in the account), r is the rate of interest, and t is the amount of time. It's important to remember that the time unit for the rate of interest and the time the money is earning interest agree. For instance, if the rate of interest is 1.4% per year and the time is 6 months, the 6 months needs to be converted to 0.5 years.

So if $\$P$ are invested, then the amount of money, A, in the account after the first interest period is over is $A = P + I = P + Prt = P(1 + rt)$. This equation is going to be the basis for creating the formula for compound interest. We'll work through a problem with numbers so that you can see what is happening.

EXAMPLE

- Suppose that \$1000 is deposited into an account that pays 1.4% annual interest that is compounded quarterly. How much money will be in the account after 2 years?

- It is the phrase "compounded quarterly" that distinguishes this problem from a simple interest question. What does it mean? As you know, the word quarterly implies four units. In this case, it means that interest is credited to the account four times a year. The annual rate of interest is 1.4%, so the quarterly rate of interest will be one-fourth of this amount, 0.35%. So, at the end of the first quarter (making $t = 1$), the amount of money in the account will be $\$1000(1 + 0.0035) = \$1003.50$. (Remember, we needed to change 0.35% to the decimal 0.0035 in order to perform the arithmetic.) To recap:

$$\text{Quarter 1:} \quad A = \$1000(1.0035) = \$1003.50$$

- At the end of the second quarter, the interest earned is not on the originally \$1000 but the \$1003.50 that was in the account at the beginning of the quarter. Therefore, the amount of money in the account at the end of the second quarter is $A = \$1003.50(1.0035) = \$1007.01$. Here is where things get interesting. The money in the account at the beginning of the second quarter is $1000(1.0035) and the money in the account at the end of the second quarter is $\$1000(1.0035)(1.0035) = \$1000(1.0035)^2$.

$$\text{Quarter 2:} \quad A = \$1000(1.0035)^2$$

- At the end of the third quarter, there will be $A = (\$1000(1.0035)^2)(1.0035) = \$1000(1.0035)^3 = \$1010.54$.
- Continue this pattern:

$$\text{Quarter 4:} \quad A = \$1000(1.0035)^4 = \$1014.07$$

$$\text{Quarter 5:} \quad A = \$1000(1.0035)^5 = \$1017.62$$

$$\text{Quarter 6:} \quad A = \$1000(1.0035)^6 = \$1021.18$$

$$\text{Quarter 7:} \quad A = \$1000(1.0035)^7 = \$1024.76$$

$$\text{Quarter 8:} \quad A = \$1000(1.0035)^8 = \$1028.35$$

In general, if $\$P$ are invested at an annual rate of $r\%$ n times per year, the amount of money, A, in the account after t years is $A = P\left(1 + \frac{r}{100n}\right)^{nt}$. The term $\frac{r}{100}$ converts the percent to a decimal, and dividing this number by n gives the periodic rate of interest (the rate for each period).

EXAMPLE

- Determine the amount of money that will be in an account if $2500 is invested for 10 years in an account that pays 1.2% compounded quarterly.
- $P = 2500$, $r = .003$, and $t = 10$.
- $A = 2500(1 + 0.003)^{4(10)} = 2500(1.003)^{40} = \$2818.24$

The most common interest periods are monthly (12 times each year), semiannually (2 times each year), and annually (1 time a year).

EXAMPLE

▶ Determine the amount of money that will be in an account if \$1250 is invested for 5 years at 2% compounded semiannually:

▶ $A = 1250(1.01)^{10} = \$1380.78$

The formula for population growth is similar to the formula for compound interest. The "credit" of the compound interest problem is increase in the population over one time period. The usual time frame for population growth is 1 year.

EXAMPLE

▶ The population for York County, South Carolina, at the end of 2010 was 226,920 people. If the growth rate for the county is 1.79% per year, what will be the population at the end of 2020?

▶ The time for this problem is 10 years, so the estimated population, P, will be $P = 226{,}920(1.0179)^{10} = 270{,}972$.

Radioactive decay is usually measured in terms of a half-life. A half-life is the amount of time needed for the amount of radioactive material to be cut in half. For example, the half-life of the radioactive isotope of carbon, Carbon-14, is 5730 years. Carbon dating was used to estimate the age of ancient artifacts.

EXAMPLE

- The amount of Carbon-14 (C_{14}) found in an object can be used to estimate the age of ancient objects. The formula for the percent of C_{14} in an object is given by $C = 100\left(\frac{1}{2}\right)^{\frac{t}{5,730}} = 100(2)^{\frac{-t}{5,730}}$. What percent of Carbon-14 should be found in an object that is anticipated to be 10,000 years old?
- The anticipated percentage of C_{14} is $C = 100(2)^{\frac{-10,000}{5,730}} = 29.83$.

Newton's law of heating and cooling is used by forensic scientists to determine the time of death of a corpse and also used by cooks to determine when the Thanksgiving turkey will be ready to be taken out of the oven.

EXAMPLE

- If the temperature of a turkey is 60°F when it is put into a preheated oven set for 350°F, the temperature, T, of the turkey is given by the formula $T = 350 - 290(2.72)^{-0.106t}$, where T is the number of hours the turkey has been in the oven. What will the temperature of the turkey be after 2.5 hours?
- The temperature of the turkey will be $T = 350 - 290(2.72)^{-0.106(2.5)} =$ 127.5°F.
- How much more time must the turkey stay in the oven if the directions state that the turkey should be removed from the oven when its temperature is 155°F?
- There is no way for us to get common bases for this question, so we will have to graph two functions, $T = 350 - 290(2.72)^{-0.106t}$ and $T = 155$, to determine the solution. Clearly the window for the calculator does not need to be too wide, but it does need to reach at least 175 (the extra space gives you the opportunity to read the graph on the calculator).

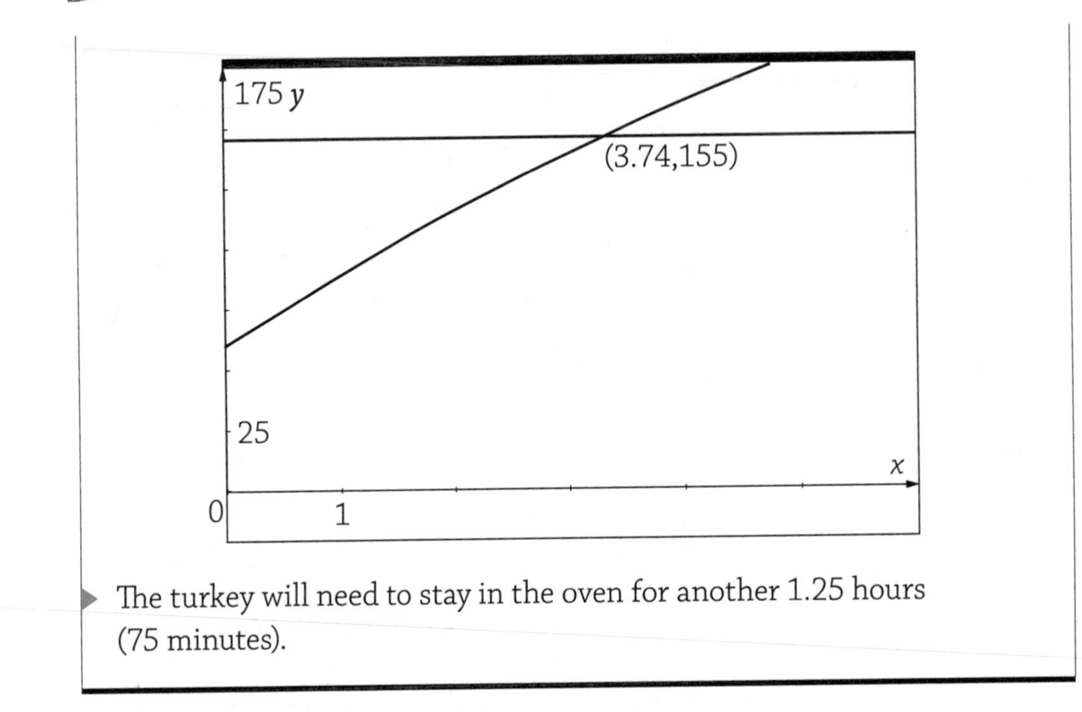

▶ The turkey will need to stay in the oven for another 1.25 hours (75 minutes).

Geometric Sequences as Exponential Functions

Earlier in this book we looked at how arithmetic sequences were a consequence of linear functions. Now we'll take a look at sequences that come from exponential functions.

EXAMPLE

▶ Describe the pattern used to create the sequence 2, 6, 18, 54, . . . and then find the next three terms:

▶ Each term in the sequence is found by multiplying the previous term by 3. The next three terms are 162, 486, and 1458.

Sometimes the common factor is a negative number.

EXAMPLE

▶ Describe the pattern used to create the sequence 3, −6, 12, −24, 48, . . . and then find the next three terms:

▶ Each term in the sequence is found by multiplying the previous term by −2. The next three terms in the sequence are −96, 192, and −384.

Let's take a look at that first sequence again while making use of the description for the pattern. The terms in the sequence are 2, 2(3), $2(3)^2$, $2(3)^3$, $2(3)^4$, $2(3)^5$, and $2(3)^6$. These values are the output for the exponential function $f(n)=2(3)^{n-1}$, where n is a counting number (in the same way that the domain of the function defining arithmetic sequences are counting numbers).

The last sequence needs some explanation to go along with it. Clearly, the function should be $k(n)=3(-2)^{n-1}$. However, we stated earlier that for the exponential function $f(x)=b^n$, the value of b must be greater than 0 and not equal to 1. How is it then that we can have the function $k(n)=3(-2)^{n-1}$ defined this geometric sequence? The answer comes down to domain. The domain of the function $h(n)$ is the set of counting numbers while the domain for $f(x)=b^n$ is the set of real numbers. We do not want to deal with $f\left(\frac{1}{2}\right)$ if $f(x)=(-2)^n$ because that would require we evaluate $\sqrt{-2}$. We know there is no real number that satisfies $\sqrt{-2}$, so we make sure this doesn't occur by putting the restrictions on the values of b. We can allow for the function $k(n)=3(-2)^{n-1}$ to be used to define the **alternating sequence** since we will never be asked to compute the root of a negative number because the domain is the set of counting numbers.

EXAMPLE

- Find the 12th term of the geometric sequence 2, 6, 18, 54, . . .
- We have determined that the function defining this sequence is $f(n) = 2(3)^{n-1}$ so $f(12) = 354{,}294$.

Let's look at another example like that.

EXAMPLE

- Find the 15th term of the geometric sequence 7, 14, 28, 56, . . .
- The constant factor is 2 so the function must be of the form $f(n) = 7(2)^{n-1}$ and $f(15) = 7(2)^{14} = 114{,}688$.

Let's extend the process a little bit.

EXAMPLE

- The fourth term of a geometric sequence is 500, and the seventh term is 62,500. Find the 14th term in this sequence.
- The function that defines this geometric sequence must be of the form $f(n) = a(b)^{n-1}$. The fourth term is 500, so $f(4) = 500 = a(b)^3$ and the seventh term is 62,500, so $f(7) = 62{,}500 = a(b)^6$.
- If we rewrite the second equation to be $62{,}500 = a(b)^3(b)^3$, we can substitute 500 for $a(b)^3$ so the equation becomes $62{,}500 = 500b^3$.
- Divide by 500 to get $b^3 = 125 = 5^3$, so $b = 5$.
- Using $f(4) = 500$, we get $500 = a(125)$ so $a = 4$ and $f(n) = 4(5)^{n-1}$.
- Therefore, $f(14) = 4(5)^{13} = 4{,}882{,}812{,}500$.

Let's do another one like that so we can be sure you have the process.

EXAMPLE

- Find the 20th term of a geometric sequence if the 8th term is 800 and the 13th term is 25.
- Solution: The function that defines the sequence must be of the form $f(n)=a(b)^{n-1}$.
- The 8th term is 800, so $800 = a(b)^7$, and the 13th term is 25, so $25 = a(b)^{12}$.
- Rewrite $a(b)^{12}$ as $a(b)^7 b^5$ and then substitute 800 for $a(b)^7$ to get $25=800b^5$.
- Solve this equation: $b^5=\frac{25}{800}=\frac{1}{32}$ so that $b=\frac{1}{2}$.
- Use $f(8) = 800$ to determine that $a\left(\frac{1}{2}\right)^7=800$ and $a = 102{,}400$.
- Therefore, $f(20)=102{,}400\left(\frac{1}{2}\right)^{19}=\frac{25}{128}$.

Recursive Formulas

Recursion is a process in which one uses the output from one step as the input for the next step. For example, starting with the number 2, each successive term will be three times the previous term. The first term is 2, the second term is 6, the third term is 18, and the fourth term is 54. If you look back to the last section, you will see that this was the first sequence we examined. So, what is different about this than what we have already been doing? For one thing, we can write a nice verbal description of the process. "Start with the number 2" translates to the first term is 2, $a_1=2$. "Each successive term will be three times the previous term" translates to $a_n=3a_{n-1}$ for $n\geq 2$. Our recursion formula for the geometric sequence defined by $f(n)=2(3)^{n-1}$ is $\begin{cases} a_1=2 \\ a_n=3a_{n-1} \end{cases}$.

EXAMPLE

▶ Write the first five terms of the sequence defined recursively by $\begin{cases} a_1 = 7 \\ a_n = 2a_{n-1} \end{cases}$.

▶ The first term is 7. The second term is $a_2 = 2a_1 = 2(7) = 14$. The third term is $a_3 = 2a_2 = 2(14) = 28$. Each term in the sequence is twice the previous term so $a_4 = 56$ and $a_5 = 112$.

Now you have a sense of how to read the notation, but all we did was create another geometric sequence. We can also create arithmetic sequences.

EXAMPLE

▶ Write the first five terms of the sequence defined recursively by $\begin{cases} a_1 = 17 \\ a_n = a_{n-1} + 6 \end{cases}$.

▶ The first term is 17, $a_2 = a_1 + 6 = 17 + 6 = 23$, $a_3 = a_2 + 6 = 23 + 6 = 29$, and if we continue this pattern, we find that the next three terms are 35, 41, and 47.

We can also create some very interesting formulas that are easily (if at all possible) to translate into a formula.

EXAMPLE

▶ Write the first eight terms of the sequence defined recursively by $\begin{cases} a_1 = 1 \\ a_2 = 1 \\ a_n = a_{n-1} + a_{n-2} (n \geq 3) \end{cases}$.

- One obvious difference to this sequence is that the first two terms are defined. The third term in the sequence can be found using the formula $a_3 = a_{3-1} + a_{3-2} = a_2 + a_1 = 1 + 1 = 2$. Continue the process to find the next five terms in the sequence.

$$a_4 = a_{4-1} + a_{4-2} = a_3 + a_2 = 2 + 1 = 3$$

$$a_5 = a_{5-1} + a_{5-2} = a_4 + a_3 = 3 + 2 = 5$$

$$a_6 = a_{6-1} + a_{6-2} = a_5 + a_4 = 5 + 3 = 8$$

$$a_7 = a_{7-1} + a_{7-2} = a_6 + a_5 = 8 + 5 = 13$$

$$a_8 = a_{8-1} + a_{8-2} = a_7 + a_6 = 13 + 8 = 21$$

- For this particular sequence, each term after the first two terms is found by adding the previous two terms together. This sequence is the popular Fibonacci sequence.

IRL

The Fibonacci sequence was first considered by Leonardo Pisano Bigollo during the 12th century. He observed the pattern while counting the number of adult rabbits he was raising over a period of months. It turns out that the sequence has many applications in nature. Two other examples are pretty interesting:

Bees: Each colony of bees has a queen. Female bees are produced when the queen mates with a male so the female bee has two parents. Drones are male produced from an unfertilized egg, so the drone has a mother but no father. Draw a family tree for a male honey bee. There is one parent, two grandparents, three great grandparents, and five great-great grandparents. You can see where this is going.

Plants: Many plants show the Fibonacci numbers in the arrangements of the leaves around their stems. If we look down on a plant, the leaves are often arranged so that leaves above do not hide leaves below. This means that each gets a good share of the sunlight and catches the most rain to channel down to the roots as it runs down the leaf to the stem. The Fibonacci numbers occur when counting both the number of times we go around the stem, going from leaf to leaf, as well as counting the leaves we meet until we encounter a leaf directly above the starting one.*

*www.maths.surrey.ac.uk/hosted-sites/R.Knott/Fibonacci/fibnat.html

Here is an important application called an annuity.

EXAMPLE

▶ In a simple annuity, a client deposits a sum of money into an interest-bearing account. The client agrees to deposit the same amount of money into the account in a time frame that agrees with the interest period. For example, Janine agrees to deposit $100 into an account that pays 2.4% interest compounded monthly at the beginning of each month. (The interest is credited to the account before the new deposit is made.) The rate of interest for each period is $\frac{.024}{12}=0.002$.

The amount of money, A, in the account after n interest periods is defined by

$$A=\begin{cases} a_1=100 \\ a_n=1.002a_{n-1}+100(n\geq 2) \end{cases}$$

▶ How much money will be in the account one year after the account is opened and the last payment is made?

$a_1=100$

$a_2=1.002(100)+100=200.20$

$a_3=1.002(200.20)+100=300.60$

$a_4=1.002(300.60)+100=401.20$

$a_5=1.002(401.20)+100=502.00$

$a_6=1.002(502.00)+100=603.01$

$a_7=1.002(603.01)+100=704.21$

$a_8=1.002(704.21)+100=805.62$

$a_9=1.002(805.62)+100=907.23$

$$a_{10} = 1.002(907.23) + 100 = 1009.05$$

$$a_{11} = 1.002(1009.05) + 100 = 1111.07$$

$$a_{12} = 1.002(1111.07) + 100 = 1213.29$$

$$a_{13} = 1.002(1213.29) + 100 = 1315.71$$

- Janine deposits $1300 and makes more than $15 in interest. Imagine what these numbers might look like if the deposits were $200 per month and they were made over a 25-year period.

EXERCISES

EXERCISE 8-1

Simplify each of the following.

1. $(4x^2)(19x^7)$
2. $(20x^3y^5z)(25x^4y^2z^8)$
3. $(6m^4)^3$
4. $(-4a^3b^2c^4)^3$
5. $(7w^3)^2(4w^5)^3$
6. $\dfrac{(6a^3)^2}{9a^5}$
7. $\dfrac{(15r^6p^4)^2}{(5r^4p^2)^3}$
8. $\left(\dfrac{96k^{-5}g^4}{120k^7g^{-2}}\right)^3$
9. $\left(\dfrac{45c^{-3}d^5}{30c^{-5}d^{-3}}\right)^3$
10. $\left(\dfrac{12w^3v^{-4}}{15w^{-2}v^3}\right)^2\left(\dfrac{5w^{-4}v^4}{6w^5v^{-2}}\right)^3$
11. $27^{\frac{-2}{3}}$
12. $\left(\dfrac{25b^{-4}}{c^{10}}\right)^{\frac{-3}{2}}$

EXERCISE 8-2

Answer the following questions.

1. Write 27,300,000,000,000,000,000 in scientific notation.

2. Write 0.000000000000135 in scientific notation.

3. Compute $(2.56\times10^{7})(3.41\times10^{7})$ and write the answer in scientific notation.

4. Compute $(5.16\times10^{17})(4.1\times10^{9})$ and write the answer in scientific notation.

5. Compute $(6.49\times10^{-7})(1.4\times10^{-10})$ and write the answer in scientific notation.

6. Compute $\dfrac{6.49\times10^{-7}}{1.4\times10^{-10}}$ and write the answer in scientific notation.

7. Compute $\dfrac{4.1\times10^{17}}{5.16\times10^{9}}$ and write the answer in scientific notation.

8. Compute $\dfrac{2.56\times10^{7}}{3.41\times10^{-7}}$ and write the answer in scientific notation.

EXERCISE 8-3

Solve each of the following equations.

1. $2^{x+4}=256$

2. $3^{x+2}=81$

3. $2^{5x-10}=8^{x}$

4. $4^{x-3}=16^{x+1}$

5. $25^{2x+1}=125^{4-x}$

6. $5^{x}=18$

7. $3^{x+2}-7=20$

EXERCISE 8-4

Answer the following questions.

1. Find the 15th term of the sequence 8, 12, 18, 27, . . .
2. Find the 20th term of the sequence 204,800, –102400, 51200, –25600, ...
3. The 4th term of a geometric sequence is 96 and the 9th term is 3072. Find the 17th term in the sequence.
4. The second term is 54,000 of a geometric sequence and the fifth term is 16,000. Find the 10th term.

Find the indicated term of the recursively defined sequence.

5. a_{10} if $\begin{cases} a_1 = 8 \\ a_n = 3a_{n-1} (n \geq 2) \end{cases}$
6. a_8 if $\begin{cases} a_1 = 24 \\ a_n = \frac{3}{2} a_{n-1} (n \geq 2) \end{cases}$
7. a_7 if $\begin{cases} a_1 = 40 \\ a_n = 2a_{n-1} - 20 (n \geq 2) \end{cases}$
8. a_5 if $\begin{cases} a_1 = 2 \\ a_2 = 3 \\ a_n = 5a_{n-2} - 2a_{n-1} (n \geq 3) \end{cases}$

EXERCISE 8-5

Solve each of the following problems.

1. How much money will be in an account that pays 1.6% compounded quarterly if \$3000 is deposited for 4 years?

2. The population of Canada in 2010 was 34.1 million people. If the population grows at a steady rate of 1.2% per year, what will be the 2020 population of Canada?

3. Iodine 123 (I^{123}) is an isotope used to measure thyroid activity. If the half-life of I^{123} is 13.22 hours, what percent of the original dosage injected into a patient's blood stream will remain after 24 hours?

4. Arnold's mother put a container of tomato sauce into the freezer to save for another day. The tomato sauce was at room temperature, 72°F, when she placed the container into the freezer. The temperature, T, of the sauce after t hours is given by the equation $T = 30 + 42(2.782)^{-0.6t}$, where t represents the number of hours since it was put in the freezer. What is the temperature of the sauce after 5 hours?

Quadratic Expressions and Equations

MUST KNOW

A *polynomial* is an expression of more than two algebraic terms containing the sum of several terms that contain different powers of the same variable. A polynomial with the largest exponent 2 is called *quadratic*.

The formulas for the Difference of Squares and Square Trinomials enable us to factor polynomial expressions.

We know if $ab = 0$, then either $a = 0$ or $b = 0$. This simple statement, called the Zero Product property, is the fundamental concept used in solving factorable polynomial equations.

uadratic expressions represent the beginning of "higher" mathematics in your journey through the subject. The manipulations are more complicated than those required with linear expressions. Don't confuse that statement with "the manipulations are complicated" because they really are not. Trust in your ability to do simple arithmetic because statements such as "a negative times a positive is always negative" and "the sum of two signed numbers will take on the same sign as the number with larger magnitude" sound much more confusing than they really are. Let's have some fun with this.

Adding and Subtracting Polynomials

"What do you get when you add three apples with four apples?" I've asked my classes that question many times, and it was almost always the case that the class gave me "that look" that says "What kind of a dolt are you to ask that?" before they responded "Seven apples." Then I asked the next question. "What do you get when you add three apples with four oranges?" The looks on their faces were no longer so smug. How do you answer that question? "Seven pieces of fruit?" often came out and almost always more as a question than a statement. And that, ladies and gentlemen, is the essence of adding and subtracting polynomials!

The point is, combining quantities that clearly have a similar characteristic is really easy and combining quantities that do not have that similarity is not so easy. Instead of apples, we can add $4x$ and $3x$, $4xyz$ and $3xyz$, and $4y^2$ and $3y^2$ easily, but $4x^2 + 3x$, $4x^2 + 3y$, and $4x^2 + 3xy$ are not so easy. Why is that? It comes back to a very simple, but ever-so-important, property of numbers called the **distributive property of multiplication over addition**. Most people have the mindset that you use the distributive property (as it is often called because we try to shorten so many things we say rather than give the concept its full name, and, yes, I am guilty of this as well) to work problems from $2(3x + 5)$ to be $6x + 10$ because it allows us to more easily manipulate the expression. It is a rare day that we ask students to rewrite $6x + 10$ to be $2(3x + 5)$. Yet when we answer the question, "What do you get when you add three apples with four apples?" our mind goes 3 apples + 4 apples = (3 + 4) apples = 7 apples. It also explains why 3 apples + 4 oranges has no response other than 3 apples + 4 oranges.

The initial reaction to the directive "find the sum of $4x^2 + 12x + 9$ and $5x^2 + 7x + 2$" is often "Whatever," because it is fairly instinctual the sum is $9x^2 + 19x + 11$. It takes a little more time to respond to the directive "find the sum of $-4x^2 + 12x - 9$ and $5x^2 - 7x - 2$" because the signed numbers in the problem often cause people to be more cautious, but the response $x^2 + 5x - 11$ is determined. The distributive property is applied mentally when doing the addition and the phrase "combine like terms" takes root.

EXAMPLE

- Find the sum of $8x^3 + 9x^2 - 11x + 10$ and $21 - 7x - 3x^2 - 5x^3$.
- I admit that is cruel to take the second polynomial and write it backwards. It is a common practice that when writing the answer, a polynomial is always written with the highest degree first (that is, the term with the biggest exponent) and the remainder of the terms in the polynomial are written with their exponents in decreasing order.
- Rewrite the problem as $8x^3 + 9x^2 - 11x + 10 + (-5x^3) - 3x^2 - 7x + 21$ to get $3x^3 + 6x^2 - 18x + 31$.
- In reality, you rewrite the problem (in your head) as $(8 - 5)x^3 + (9 - 3)x^2 + (-11 - 7)x + (10 + 21)$ before recording the sum in its final form.

The real challenge turns out to be subtraction. You really need to be careful that the subtraction sign distributes through the entire subtrahend (the number that is being subtracted; the number you are subtracting from is the minuend).

EXAMPLE

- Subtract $3x^2-5x+6$ from $5x^2+7x-3$.
- The minuend is $5x^2+7x-3$, so the problem becomes $5x^2+7x-3-(3x^2-5x+6)$.
- Distribute the minus sign through the parentheses:
$$5x^2+7x-3-3x^2+5x-6$$
- Combine like terms to get $2x^2+12x-9$.
- You might ask: How can one "distribute the minus sign through the parentheses"? If we rewrite the subtraction problem as the addition of a negative number, then $5x^2+7x-3-(3x^2-5x+6)$ becomes $5x^2+7x-3+(-1)(3x^2-5x+6)$ and you are distributing the -1 through the parentheses.

Be careful with the following problem.

EXAMPLE

- Simplify:
$$\begin{array}{r} 5x^2-4x+9 \\ -7x^2+4x-6 \end{array}$$
- Take the time to rewrite the problem horizontally with parentheses around the subtrahend:
$$5x^2-4x+9-(7x^2+4x-6)$$
- Distribute the minus sign:
$$5x^2-4x+9-7x^2-4x+6$$
- Combine like terms: $-2x^2-8x+15$
- The answer is $-2x^2-8x+15$.
- There is a huge temptation to treat this problem as $5-7=-2$, $-4+4=0$, and $9-6=3$, so the answer must be $2x^2+3$.

BTW

I want to be sure that I really drill this point home. Pay careful attention when you subtract polynomials!

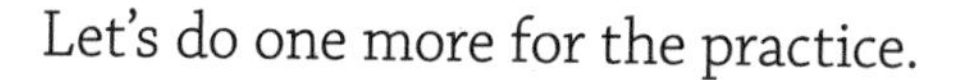

Let's do one more for the practice.

EXAMPLE

- Determine the difference between $-4x^3+5x^2-71$ and $9x^2-6x-12$.
- There are things we need to talk about here. "Determine the difference" is not an ambiguous statement. "Determine the difference between 21 and 9" gives an answer of 12 because $21-9=12$. After all, 21 is larger than 12, so we subtract in that order. There is no way to determine comparative size between the two polynomials, so we do not have that cue to work with. We always treat the first polynomial as the minuend.
- The second issue is that the terms in the polynomials are different. Does it matter? In a word, no.
- We write the problem as $-4x^3+5x^2-71-(9x^2-6x-12)$
- Distribute the minus sign: $-4x^3+5x^2-71-9x^2+6x+12$
- Combine like terms:
  $-4x^3+5x^2-9x^2+6x-71+12 = -4x^3-4x^2+6x-59$

Multiplying a Polynomial by a Monomial

If $2(3x+5) = 6x+10$, then it should not be too difficult to extend the distributive property to include monomials: $2x(3x+5)=6x^2+10x$. Are we limited to only two terms within the parentheses? No, we are not. If your inclination is to say, "Wait! The distributive property says $a\,(b+c)=ab+bc$. There are only two terms inside the parentheses." I will counter with $a((b+c)+d)=a(b+c)+ad=ab+ac+ad$. This extends the distributive property to three terms, and we can continue to expand the number of terms within the parentheses.

EXAMPLE

- Simplify: $3x(2x^2 - 5x + 1)$
- Apply the distributive property:
  $3x(2x^2 - 5x + 1) = 3x(2x^2) - 3x(5x) + 3x(1) = 6x^3 - 15x^2 + 3x$.

Let's increase the level of difficulty a little bit.

EXAMPLE

- Simplify: $-4x^3y^2(5x^2 - 4xy + 3y^2)$
- We have two variables involved here, but the steps are the same.
- Apply the distributive property:

$$-4x^3y^2(5x^2 - 4xy + 3y^2) \\ = -4x^3y^2(5x^2) - (-4x^3y^2)(4xy) + (-4x^3y^2)(3y^2)$$

- Multiply monomials:

$$-20x^5y^2 - (-16x^4y^3) + (-12x^3y^4) = -20x^5y^2 + 16x^4y^3 - 12x^3y^4.$$

Multiplying Polynomials

Multiplication of polynomials is essentially a repeated application of the distributive property. As you will see, the use of the mnemonic FOIL is popular when multiplying binomials but does not extend itself to polynomials with more than two terms.

Multiplying Binomials

We will continue to use the distributive property to multiply algebraic expressions. In this case, the two terms being multiplied will each be a binomial (a polynomial with two terms). Suppose one of the terms is $mx + p$

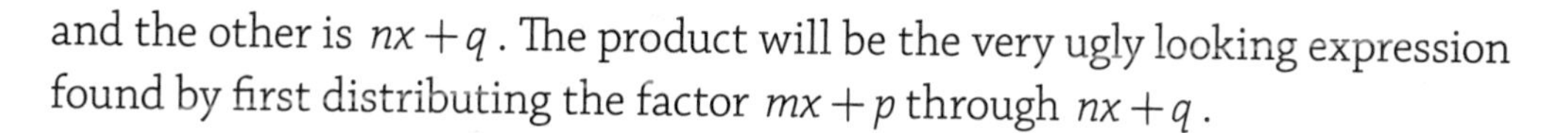

and the other is $nx+q$. The product will be the very ugly looking expression found by first distributing the factor $mx+p$ through $nx+q$.

$$(mx+p)(nx+q)=(mx+p)(nx)+(mx+p)(q)$$

Apply the distributive property again:

$$=(mx)(nx)+(p)(nx)+(mx)(q)+(p)(q)$$

Multiply monomials:

$$=mnx^2+npx+mqx+pq=mnx^2+(np+mq)x+pq$$

Don't try to memorize this, but rather, be comfortable with the double application of the distributive property.

EXAMPLE

- Expand: $(5x+3)(3x+4)$
- First application of the distributive property: $(5x+3)(3x)+(5x+3)(4)$
- Second application of the distributive property:
  $(5x)(3x)+(3)(3x)+(5x)(4)+(3)(4)$
- Multiply the monomials and combine like terms:
  $15x^2+9x+20x+12=15x^2+29x+12$

IRL Most people use the mnemonic FOIL when discussing multiplication of binomials. You begin by multiplying First terms, $(5x)(3x)$, then Outside terms, $(5x)$ (4), then Inside terms, (3) $(3x)$, and finally, Last terms, $(3)(4)$. So, as you can see, FOIL stands for First — Outside — Inside — Last.

Let's practice multiplying binomials.

EXAMPLE

▶ Multiply: $(x+5)(x-4)$

$$(x+5)(x-4) = (x)(x)+(x)(-4)+(5)(x)+(5)(-4)$$
$$= x^2-4x+5x-20 = x^2-x-20$$

Having multiple variables in the expressions does not change the process.

EXAMPLE

▶ Multiply: $(2x+3y)(3x-2y)$

▶ Solution:

$$(2x+3y)(3x-2y) = (2x)(3x)+(2x)(-2y)+(3y)(3x)+(3y)(-2y)$$
$$= 6x^2-4xy+9xy-6y^2 = 6x^2+5xy-6y^2$$

It will be a great benefit to you to write down the pieces as you do the multiplication. You need not write $(2x)(3x)+(2x)(2)+(3)(3x)+(3)(2)$ but really should write $6x^2+4x+9x+6$ as you do each step, especially in the beginning of learning how to perform the multiplication process.

Here are a few more problems (and also a prelude to what is coming later in this chapter).

EXAMPLE

▶ Multiply: $(x+8)(x-8)$

$$(x+8)(x-8) = (x)(x)+(x)(-8)+(8)(x)+(8)(-8)$$
$$= x^2-8x+8x-64 = x^2-64$$

This example is a little different.

EXAMPLE

Multiply: $(5x+8y)(5x-8y)$

$$(5x+8y)(5x-8y)=(5x)(5x)+(5x)(-8y)+(8y)(5x)+(8y)(-8y)$$
$$=25x^2-40xy+40xy-64y^2=25x^2-64y^2$$

Do you see a pattern here? Check your thoughts when you get to the "Differences of Squares" section coming up in this chapter.

EXAMPLE

Multiply: $(x-8)(x-8)$

$$(x-8)(x-8)=(x)(x)+(x)(-8)+(-8)(x)+(-8)(-8)$$
$$=x^2-8x-8x+64=x^2-16x+64$$

This next problem is similar to the last. See if you can detect a pattern. It is not as easy to see as the Difference of Squares.

EXAMPLE

Multiply: $(5x+8)(5x+8)$

$$(5x+8)(5x+8)=(5x)(5x)+(5x)(8)+(8)(5x)+(8)(8)$$
$$=25x^2+40x+40x+64=25x^2+80x+64$$

The expressions $(x-8)(x-8)$ and $(5x+8)(5x+8)$ can be written as $(x-8)^2$ and $(5x+8)^2$, so the results of the multiplication are called square trinomials.

You can also use geometry to help understand the multiplication of binomials (though you really want to be comfortable with either FOIL or the distributive property). The area of a rectangle is the product of the base and height of the rectangle.

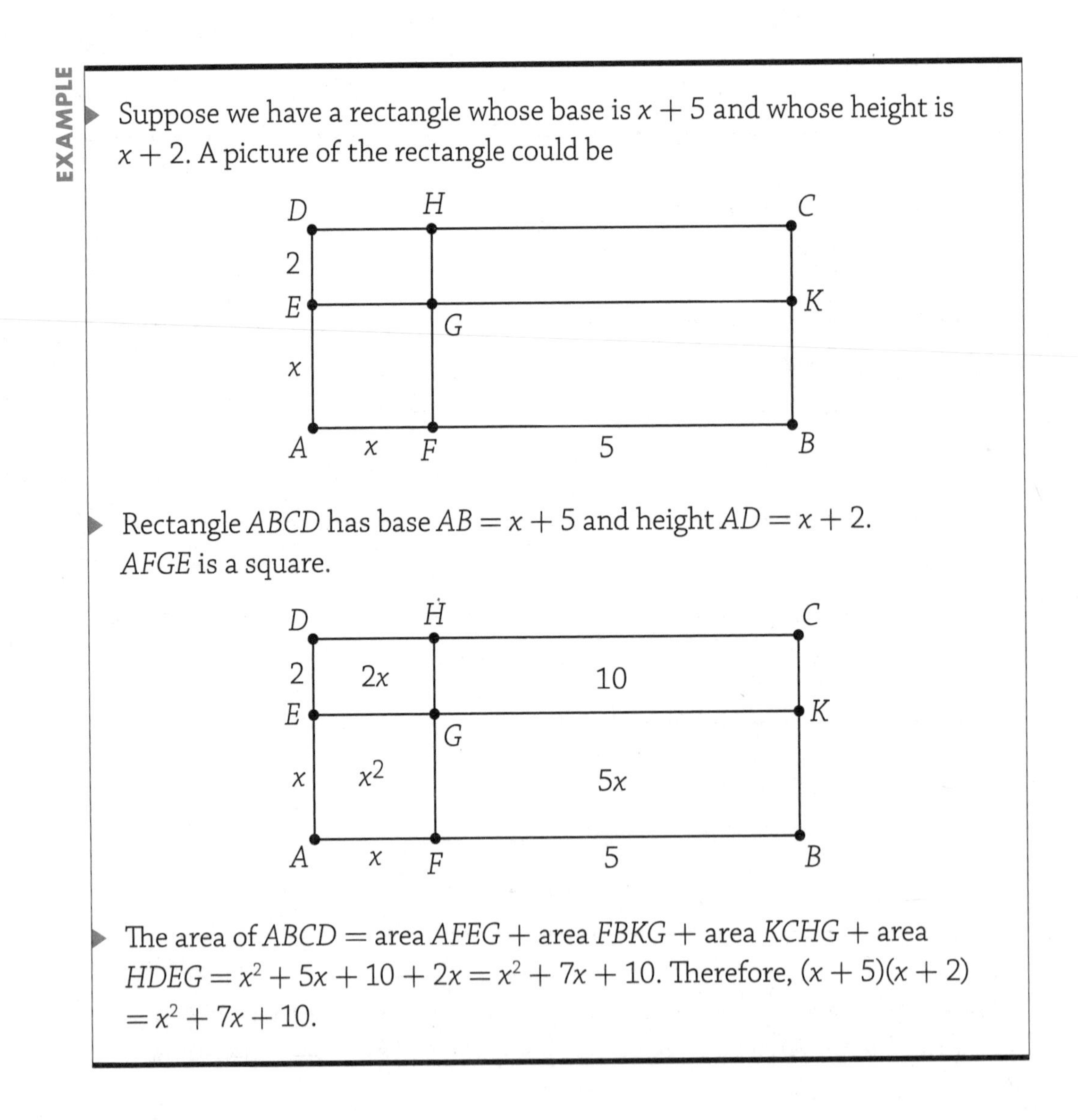

EXAMPLE

▶ Suppose we have a rectangle whose base is $x + 5$ and whose height is $x + 2$. A picture of the rectangle could be

▶ Rectangle *ABCD* has base $AB = x + 5$ and height $AD = x + 2$. *AFGE* is a square.

▶ The area of *ABCD* = area *AFEG* + area *FBKG* + area *KCHG* + area *HDEG* $= x^2 + 5x + 10 + 2x = x^2 + 7x + 10$. Therefore, $(x + 5)(x + 2) = x^2 + 7x + 10$.

Finding the product using a geometric model is a bit trickier if there are negative numbers involved but let's take a look.

EXAMPLE

▶ Suppose we have a rectangle whose base is $x + 5$ and whose height is $x - 2$. Find the area of the rectangle. A picture of the rectangle could be:

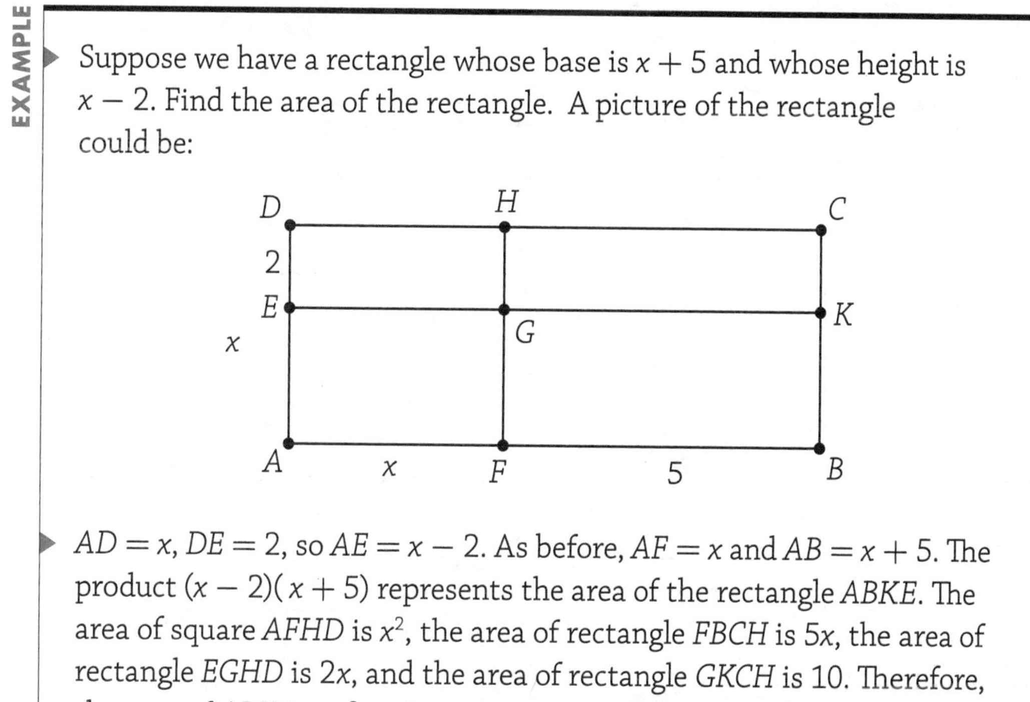

▶ $AD = x$, $DE = 2$, so $AE = x - 2$. As before, $AF = x$ and $AB = x + 5$. The product $(x - 2)(x + 5)$ represents the area of the rectangle $ABKE$. The area of square $AFHD$ is x^2, the area of rectangle $FBCH$ is $5x$, the area of rectangle $EGHD$ is $2x$, and the area of rectangle $GKCH$ is 10. Therefore, the area of $ABKE = x^2 + 5x - 2x - 10 = x^2 + 3x - 10$.

Multiplying Polynomials

The FOIL method works well when multiplying two binomials but only when multiplying two binomials. For any other case, you'll need to use the distributive property twice. I like to refer to this as DD for Double Distributive.

EXAMPLE

- Simplify: $(2x^2+3x-6)(3x^2+9x+2)$
- Distribute the first time:

$$(2x^2+3x-6)(3x^2+9x+2)=(2x^2+3x-6)(3x^2) + (2x^2+3x-6)(9x)+(2x^2+3x-6)(2)$$

- Distribute the second time:

$$(2x^2)(3x^2)+(3x)(3x^2)+(-6)(3x^2)+(2x^2)(9x)+(3x)(9x) + (-6)(9x)+(2x^2)(2)+(3x)(2)+(-6)(2)$$

- Multiply monomials and combine like terms:

$$6x^4+9x^3-18x^2+18x^3+27x^2-54x+4x^2+6x-12 = 6x^4+27x^3+13x^2-48x-12.$$

That is about as hard a problem as you will most likely be asked to do. What makes it so hard? There are quite a few terms that you need to manipulate in order to get the answer. We are going to do this problem again, but we will do it in the style of multi-digit multiplication like we used to do in the lower grade levels.

EXAMPLE

- Simplify: $(2x^2+3x-6)(3x^2+9x+2)$
- Multiply $(2x^2 + 3x - 6)$ by 2: $4x^2+6x-12$
- Multiply by $9x$: $18x^3+27x^2-54x$
- Multiply by $3x^2$: $6x^4+9x^3-18x^2$
- Add the result of the previous three lines:

$$6x^4+27x^3+13x^2-48x-12$$

Which way do you prefer—horizontal or vertical? Then by all means use that method. We'll use the horizontal method here.

EXAMPLE

- Multiply: $(3x^4+5x^2-7x-10)(-2x^3-8x+12)$
- Distribute the first time:

$$(3x^4)(-2x^3-8x+12)+(5x^2)(-2x^3-8x+12)$$
$$+(-7x)(-2x^3-8x+12)+(-10)(-2x^3-8x+12)$$

- Distribute the second time (without writing down all the steps):

$$-6x^7+24x^5+36x^4-10x^5-40x^3+60x^2$$
$$+14x^4+56x^2-84x+20x^3+80x-120$$

- Combine like terms:

$$-6x^7-16x^5+50x^4-20x^3+116x^2-4x-120$$

Special Products

There are two formulas that you should know when multiplying binomials that are special enough to distinguish them from FOIL. They are the *difference of squares* and the *square trinomial*. You should learn these formulas because they will help you with the next topic: factoring polynomials. This is a similar in how you learned your subtraction and division facts way back. You used addition to figure out subtraction and multiplication to figure out division. We'll use the distributive property, FOIL, and these formulas to learn how to factor. But, first things first. Let's learn these specials formulas.

Differences of Squares

It will always be the case that $(a-b)(a+b)=a^2-b^2$. Observe that there are only two terms in the product and that they are both squares. Furthermore, the result is the difference of these squares (and, hence, the name).

EXAMPLE

- Multiply $(2x+3y)(2x-3y)$
- Because the factors fit the form $(a-b)(a+b)=a^2-b^2$, $(2x+3y)(2x-3y)=(2x)^2-(3y)^2=4x^2-9y^2$.

Try this one.

EXAMPLE

- Multiply $(4x^2+7z)(4x^2-7z)$
- Because the factors fit the form $(a-b)(a+b)=a^2-b^2$, $(4x^2+7z)(4x^2-7z) = (4x^2)^2-(7z)^2=16x^4-49z^2$.

Perfect Squares

The second special formula is actually two formulas. Each has to do with squaring a binomial. (You remember that the area of a square whose is length x is x^2. That is why we refer to x^2 as "x squared".)

$$(a+b)^2=a^2+2ab+b^2$$
$$(a-b)^2=a^2-2ab+b^2$$

Observe that the squares of the terms in the parentheses are the first and third terms in the product and also that twice the product of the terms in

the factor is the middle term in the trinomial. You can use a picture to help you remember this:

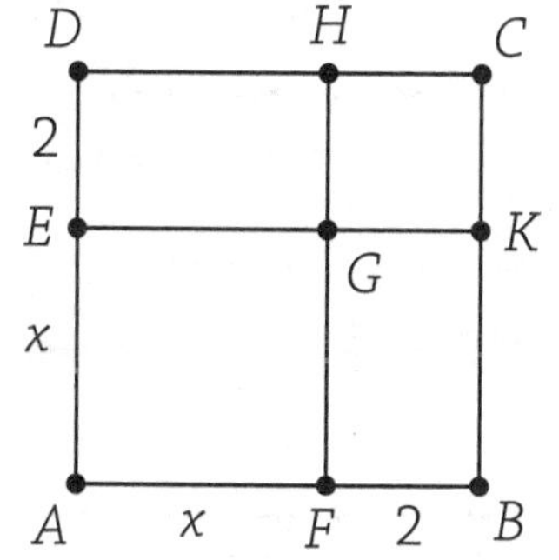

ABCD is a square with each side having length $x + 2$. *AFGE* is also a square and each side has length x. In addition, the length of each side of square *GKCH* is 2. The area of *ABCD* = the area of *AFGE* + the area of *BKGF* + the area of *GKCH* + the area of *EGHD* $= x^2 + 2x + 4 + 2x = x^2 + 4x + 4$. It will always be the case that *AFGE* and *GKCH* will be squares and that the two rectangles *FBKG* and *EGHD* will be exactly the same (congruent is the term used by mathematicians).

In the case when the side has length $a - b$:

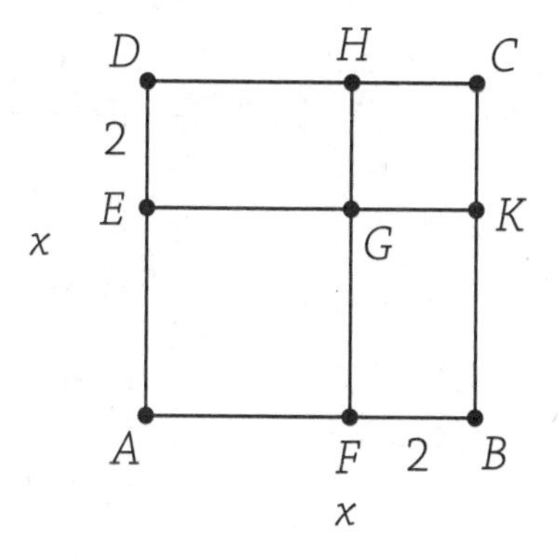

$AB = AD = x$, while $AE = AF = x - 2$. It looks like the area of *AFGE* = the area of *ABCD* – the area of *FBCH* – the area of *EKCD*. However, if we look at that more carefully, we see that we subtracted the area of square *GKCH* twice. Therefore, like the area of *AFGE* = the area of *ABCD* – the area of *FBCH* – the area of *EKCD* + the area of *GKCH*. Consequently, $(x - 2)^2 = x^2 - 2x - 2x + 4 = x^2 - 4x + 4$.

EXAMPLE

Expand $(2x+3)^2$.

$$(2x+3)^2=(2x)^2+2(2x)(3)+3^2=4x^2+12x+9$$

Here is another example for you to try. Notice how the pattern makes finding the product that much easier.

EXAMPLE

Expand $(7w-4v)^2$.

$$(7w-4v)^2=(7w)^2-2(7w)(4v)+(4v)^2=49w^2-56wv+16v^2$$

Factoring Polynomials

Factoring polynomials is a skill that will you need to succeed in mathematics. It is similar to the idea of learning division; once you figured out multiplication, division was doing the problems in reverse. However, there is more to factoring than memorizing a number of facts. Having said that, the process for factoring can be really straightforward.

First Step: Always look for a common factor.

This is simply a way of saying that you should apply the distributive property whenever you can. We know that $3x(4x+1)=12x^2+3x$. In the language of factoring, we say that $3x$ is a common factor for the terms in $12x^2+3x$, so $12x^2+3x=3x(4x+1)$.

EXAMPLE

Factor $18x^2-24x$.

$6x$ is a common factor for each of the terms in $18x^2-24x$, so $18x^2-24x=6x(3x-4)$.

There will be times when the common factor is more than a constant.

EXAMPLE

▸ Factor $3x^2y+11xy^2$.

▸ xy is a common factor for each of the terms in $3x^2y+11xy^2$, so $3x^2y+11xy^2=xy(3x+11y)$.

Second Step: Is the polynomial one of the special polynomials?

That is, if the polynomial only has two terms, is it the difference of squares? If the polynomial has three terms, are the first and third terms squares and is the middle term twice the value of the square roots of the first and third terms? Answer yes to these questions and you have your factors. WARNING: Don't forget to apply step one before you move on to step two!

EXAMPLE

▸ Factor: $36x^2-49$

▸ Because 36, x^2, and 49 are all squares, and the terms are being subtracted from each other (but beware, the sum of squares does not factor with the real numbers), this problem fits the form for the difference of squares formula. Therefore, $36x^2-49=(6x+7)(6x-7)$.

Let's try another problem.

EXAMPLE

▸ Factor: $81x^2y^2-121z^2$

▸ Again, all the terms involved are squares, so we can apply the difference of squares formula to get that $81x^2y^2-121z^2=(9xy+11z)(9xy-11z)$.

Be careful. You may run into a trick in the next problem.

EXAMPLE

▶ Factor: $24x^2 - 54$

▶ We know that 24 and 54 are not squares BUT 24 and 54 do have a common factor of 6. So, $24x^2 - 54 = 6(4x^2 - 9)$ and $4x^2 - 9$ fits the form for the difference of squares. Therefore, $24x^2 - 54 = 6(2x - 3)(2x + 3)$.

We now have a clue as to what to do if there are two terms in the polynomial. Let's take a look at the special trinomials.

EXAMPLE

▶ Factor: $16x^2 + 72x + 81$

▶ The first term can be written as $(4x)^2$ and the last term can be written as 9^2. The question is this: is the middle term equal to the product $2(4x)(9)$? Yes, it is. Therefore, $16x^2 + 72x + 81 = (4x + 9)^2$.

Try this one.

EXAMPLE

▶ Factor: $49x^2 - 112x + 64$

▶ The first term can be written as $(7x)^2$ and the last term can be written as 8^2. The question is this: Is the middle term equal to the product $2(7x)(8)$? Yes, it is. Therefore, $49x^2 - 112x + 64 = (7x - 8)^2$.

Did you pick up on how we know whether to write the answer as $(a + b)^2$ versus $(a - b)^2$? When the terms within the parentheses are being added, the middle term in the expansion is added, but when the terms within the parentheses are being subtracted, the middle term in the expansion is subtracted.

EXAMPLE

▸ Factor: $250x^3 - 400x^2 + 160x$

▸ Clearly, this is not a straightforward application of the square trinomial formula since the first term contains x^3. You can see that x is a common factor, and when you look closer, you will realize that $10x$ is a common factor (but, you already saw that, didn't you). We have that $250x^3 - 400x^2 + 160x = 10x(25x^2 - 40x + 16)$. Both the first and third terms of the trinomials are squares and the middle term is equal to twice the square root of the first and third terms, $40x = 2(5x)(4)$, so we have $250x^3 - 400x^2 + 160x = 10(5x - 4)^2$.

Trial and Error: Part 1

We now tackle the situation that occurs the most often in factoring. The polynomial is not a special polynomial. We saw earlier in this chapter that the product of the binomials $mx + p$ and $nx + q$ is $(mx + p)(nx + q) = mnx^2 + (np + mq)x + pq$. Let's modify this for a moment and assume that m and n both have a value of 1. We are now dealing with the problem $(x + p)(x + q) = x^2 + (p + q)x + pq$. Compare this against the trinomial $x^2 + 8x + 15$. If $(x + p)(x + q) = x^2 + 8x + 15$, then $p + q = 8$ and $pq = 15$. We know that p and q must be 3 and 5 (and we do not care which is which), so we know that $x^2 + 8x + 15$ factors to be $(x + 3)(x + 5)$.

What we know is that if $(x + p)(x + q) = x^2 + bx + c$, then $b = p + q$ and $c = pq$. The trick to factoring trinomials is being able to write the factors of the constant c (either on paper or in your head) while looking for the pair that adds (or subtracts if needed) to equal b.

EXAMPLE

Factor: $x^2+9x+18$

What numbers multiply to be 18 and add to be 9? The numbers 6 and 3 fit, so $x^2+9x+18=(x+6)(x+3)$.

Here's another example.

EXAMPLE

Factor: $x^2+3x-18$

What numbers multiply to be −18 and add to be 3? The numbers 6 and −3 fit, so $x^2+3x-18=(x+6)(x-3)$.

BTW

Arithmetic skills learned in past years are essential to factoring trinomial expressions. Simple questions like "when is the product of two numbers positive?" and "what is the sign of the answer when a negative and positive number are added?" will go a long way in helping you to factor.

Try this example.

EXAMPLE

Factor: $x^2-13x-48$

What numbers multiply to be −48 and add to be −13? The factors of 48 are 1 and 48, 2 and 24, 3 and 16, 4 and 12, and 6 and 8. Of these, the pair that differs by 13 is 3 and 16. Because the middle term of the trinomial is −13, the pair we need must be 3 and −16. Therefore, $x^2-13x-48=(x+3)(x-16)$.

Trial and Error: Part 2

Now that we know how to look for the factors of c that sum to equal b, we can expand the process to the case that the coefficient of x^2 is not 1.

If $(mx+p)(nx+q)=mnx^2+(np+mq)x+pq=ax^2+bx+c$, then $a = mn$, $b = np + mq$, and $c = pq$. This requires that we find the factors of a and the factors of c and then determine the pairs that will give the correct middle coefficient. Sounds hard? It really isn't. It does, however, require some patience and persistence.

Let's start off with some easier problems. We'll limit the value of a to prime numbers. This way we are guaranteed that either m or n must be 1.

EXAMPLE

▶ Factor: $2x^2-3x-5$

▶ We know the factors of 2 are 1 and 2 and the factors of 5 are 1 and 5. We need to arrange these factors so that the difference $np + mq = -3$. Well, $2 - 5 = -3$, so we can have $n = 2$, $p = 1$, $m = 1$, and $q = -5$. Therefore, $2x^2-3x-5=(x+1)(2x-5)$.

This next question is a bit more complicated because the value of c is not a prime number.

EXAMPLE

▶ Factor: $5x^2+11x-12$

▶ The factors of 5 are 1 and 5, while the factors of 12 are 1 and 12, 2 and 6, and 3 and 4. Earlier in this chapter you saw how basic arithmetic facts can be a big help in determining the correct factors. We are looking for the values of m, n, p, and q so that $np + mq = 11$. Let me rephrase that. We are looking for two numbers that add to be an odd number. Does that sound familiar? One of the numbers must be even and the other number must be odd.

We know that m and n must be 1 and 5, so let's set $n = 1$ and $m = 5$. This tells us that p and q cannot both be even numbers because the product of an even number and an odd number is always even. We've already decided that one of the terms, p or $5q$, must be odd. Does it seem reasonable to use the factors 1 and 12 for p and q? $1 + 5(-12) = -59$. Way too big. Use 3 and 4. $3 + 5(-4) = -17$. Nope, still doesn't work. How about $4 + 5(-3) = -11$? We're almost there. We have the right number with the wrong sign. The answer must be $-4 + 5(3) = 11$. Bingo! The factors of $5x^2 + 11x - 12$ must be $(5x - 4)(x + 3)$.

Did that seem long to you? I'll bet it did. It takes much longer to write all that out than it does to think it through. Once you get comfortable with the process, you'll find that it goes much easier. BUT! Write things down—especially in the beginning. You want to make sure that you determined all the possible factors so that as you start to pair them off, you will not have missed a piece.

EXAMPLE

Factor: $7x^2 + 33x - 10$

The factors of 10 are 1 and 10 and 2 and 5. With the middle term being the relatively large number 33, it should be clear that $35 - 2 = 33$, so that $7x^2 + 33x - 10 = (7x - 2)(x + 5)$.

This problem appears to be a bit more challenging.

EXAMPLE

Factor: $13x^2 + 61x - 20$

Having 13 as a factor actually makes the problem easier than harder. We'll need to multiply 13 by a number that will allow us to get in the neighborhood of 61. The factors of 20 are 1 and 20, 2 and 10, and 4 and 5. We know that $13(5) - 4 = 61$, so $13x^2 + 61x - 20 = (13x - 4)(x + 5)$.

That really wasn't too difficult, was it? As I wrote earlier, those arithmetic skills you used in your earlier years *do* come into play here.

Be careful with this next example.

EXAMPLE

Factor: $5x^2 + 5x - 60$

Did you catch the fact that there is a common factor in this problem? When working on a set of exercises, we sometimes get such a case of tunnel vision that we forget to check the first two steps in the factoring process. $5x^2 + 5x - 60 = 5(x^2 + x - 12) = 5(x + 4)(x - 3)$.

Stop and take a breath. That was a lot to think about, and we are about to get into something even more complicated. What happens if the leading coefficient is not a prime number? At the Algebra I level, we need to keep the values reasonable (who decides what is reasonable?), so concentrate on the process rather than concern yourself with the number of steps you might have to perform to get the correct answer.

EXAMPLE

Factor: $6x^2 + 19x + 15$

The factors of 6 (m and n) are 1 and 6 and 2 and 3, while the factors of 15 (p and q) are 1 and 15 and 3 and 5. We need to find the pairs of numbers for which $np + mq = 19$ (an odd result). How quickly can you find the answer?

Pairing 1 and 6 with 1 and 15: $(1)(15) + (6)(1) = 21$. Close but not it. We won't try $(6)(15) + (1)(1)$ because the result is much too large.

- Pairing 1 and 6 with 3 and 5: (1)(5) + (6)(3) = 23 and (1)(3) + (6)(5) = 33.
- We've paired 1 and 6 with the possible values of p and q and none worked.
- Pairing 2 and 3 with 1 and 15: (2)(1) + (3)(15) = 47 and (2)(15)+ (3)(1) = 33. We know 1 and 15 are not part of the answer.
- Pairing 2 and 3 with 3 and 5: (2)(5) + (3)(3) = 19! We've found it.

$$6x^2+19x+15=(2x+3)(3x+5)$$

All the possibilities were written out so that you could see the process. You do not have to start at the beginning as I did if you know, or believe you know, what the pairings of the factors should be. The process gets much faster once you've had some experience.

EXAMPLE

- Factor: $4x^2-4x-15$
- The factors of 4 are 1 and 4 and 2 and 2, while the factors of 15 are 1 and 15 and 3 and 5. We have opposite signs in the factors since 15 is negative and $np + pq = -4$. It will be hard to get −4 as a result with these numbers if we use 15 as a factor. Using 3 and 5 as the factors of 15, we see that 2(5) − 2(3) = 4, so $4x^2-4x-15=(2x-5)(2x+3)$.

We should try a few more for the practice.

EXAMPLE

- Factor: $12x^2+36x+27$.
- Did you catch the common factor of 3 in this problem? $12x^2+36x+27=3(4x^2+12x+9)$. Look closely and you'll see that the trinomial is a square trinomial, so $12x^2+36x+27=3(2x+3)^2$.

Let's do one more for the good of the cause.

EXAMPLE

Factor: $10x^2 - 11x - 6$

The factors of 10 are 1 and 10 and 2 and 5, while the factors of 6 are 1 and 6 and 2 and 3. If 10 was one of the numbers, we would have a relatively large term in the middle of this trinomial, so let's begin with 2 and 5 as the factors of 10. Matching 2 and 5 with 2 and 3, we get (5)(3) − (2)(2) = 11. Therefore, $10x^2 - 11x - 6 = (5x + 2)(2x - 3)$.

Not all polynomials can be factored! Consider $x^2 + 3x + 5$. There are no integers that multiply to equal 5 and add to equal 3. Polynomials such as these are called prime polynomials.

Solving $x^2 + bx + c = 0$

We apply a very simple rule of numbers to help us solve polynomial equations. It is called the **Zero Product property**.

Zero Product Property If the product of two or more numbers is equal to zero, at least one of the factors must be equal to zero.

This tells us that if the product $ab = 0$, then either $a = 0$ or $b = 0$, or possibly both are equal to zero.

EXAMPLE

Solve: $(x + 5)(x - 7) = 0$

Applying the Zero Product property, then either $x + 5 = 0$ and $x = -5$ or $x - 7 = 0$ and $x = 7$. Therefore, $x = -5, 7$.

That seems pretty easy. Try this one.

EXAMPLE

▶ Solve: $x^2 - 2x - 35 = 0$

▶ The Zero Product property tells us what we can do if the product is equal to zero. Consequently, we need to factor $x^2 - 2x - 35$. You can see that $x^2 - 2x - 35 = (x+5)(x-7)$, and since we just solved that problem, we know that $x = -5, 7$.

Let's try another one.

EXAMPLE

▶ Solve $x^2 - 10x + 24 = 0$.

▶ $x^2 - 10x + 24 = (x-4)(x-6)$, so the solution to $x^2 - 10x + 24 = 0$ is $(x-4)(x-6) = 0$ and $x = 4, 6$.

Some problems you can just look at and know what the answer(s) will be.

EXAMPLE

▶ Solve $x^2 - 100 = 0$.

▶ What numbers squared will give the answer 100? The formal solution is to factor $x^2 - 100 = (x+10)(x-10)$ to get the solution $(x-10)(x+10) = 0$ and $x = \pm 10$.

Some problems are not that obvious.

EXAMPLE

▶ Solve $x^2 - 10x - 24 = 0$.

▶ $x^2 - 10x - 24 = (x-12)(x+2)$, so the solution to $x^2 - 10x - 24 = 0$ is $(x-12)(x+2) = 0$ and $x = 12, -2$.

EXERCISES

EXERCISE 9-1

Simplify each of the following.

1. Add $12x^2 + 6x - 9$ and $21x^2 - 17x + 91$.
2. Add $-21x^2 + 17x + 39$ and $11x^2 - 25x - 13$.
3. $(42x^2 + 62x + 52) + (33x^2 - 37x + 41) =$
4. $(42x^2 + 62x + 52) - (33x^2 - 37x + 41) =$
5. Subtract $21x^2 - 17x + 91$ from $12x^2 + 6x - 9$.
6. Subtract $-21x^2 + 17x + 39$ from $11x^2 - 25x - 13$.
7. $-8(3x^2 - 7x - 2)$
8. $3x(2x^3 + 5x - 7)$
9. $9x^2y(7x^2 + 5xy - 8y^2)$
10. $(x + 6)(x + 4)$
11. $(x + 3)(x - 4)$
12. $(x - 8)(x - 5)$
13. $(x + 8)(x - 7)$
14. $(4x + 3)(x - 7)$
15. $(x + 3)(5x - 7)$
16. $(x - 3)(2x - 1)$
17. $(3x + 4)(x - 2)$

18. $(4x + 3)(3x - 2)$

19. $(4x - 3)(5x + 4)$

20. $(4x - 3)(4x + 3)$

21. $(3x + 4y)(3x - 4y)$

22. $(3x + 4)^2$

23. $(4x - 3)^2$

24. $(2x + 5y)^2$

25. $(3w - 7p)^2$

26. $(2x + 3)(x^2 - 3x - 18)$

27. $(2x - 5)(x^2 - 4x - 12)$

28. $(x + 3)(3x^2 - 4x - 2)$

29. $(x + y)(3x^2 - 4xy - 5y^2)$

30. $(x^2 - 3)(3x^2 + 7x + 5)$

EXERCISE 9-2

Factor each of the following.

1. $5x^3 - 10x^2$
2. $12y^3 - 36y^4$
3. $z^2 + 9z + 14$
4. $p^2 + 29p + 100$
5. $r^2 - 20r + 75$

6. $m^2 - 4m - 32$

7. $g^2 - 4g - 45$

8. $k^2 + 8k - 48$

9. $4t^2 - 25$

10. $64v^2 - 169f^2$

11. $36d^2 + 108de + 81e^2$

12. $36q^2 - 180q + 225$

13. $2x^2 + 11x + 15$

14. $7x^2 + 13x - 24$

15. $11x^2 - 20x - 39$

16. $6x^2 - 7x - 20$

EXERCISE 9-3

Solve each equation.

1. $x^2 - 5x - 24 = 0$

2. $y^2 - 12y + 20 = 0$

3. $4d^2 - 49 = 0$

4. $v^2 - 16v + 64 = 0$

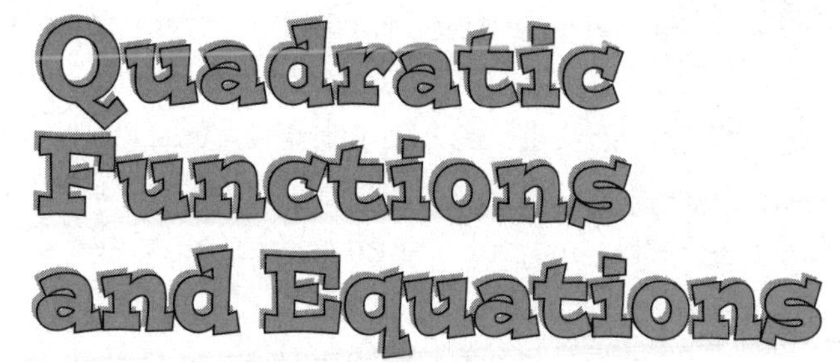

10 Quadratic Functions and Equations

MUST KNOW

- The parabola, in either standard form or vertex form, is the first nonlinear function you will analyze.

- The solutions to all curved functions are tied to where the graph crosses the *x*-axis.

- Once you've determined where the parabola crosses the *x*-axis, you can use the information to solve quadratic equations.

- Knowing the rules for transformations allows you to draw all variations of the parabola $y = x^2$.

- Whereas factoring and graphing approaches are limited as to when they can be used, the quadratic formula enables you to solve all quadratic equations.

aving studied the behavior of linear functions, we now turn our attention to quadratic functions. How high is an object t-seconds after it has been thrown vertically into the air (or dropped from some elevation)? How long does it take a car to stop when the brakes are applied? How much revenue can an operator earn once she has determined the number of units she can sell at a given price? These are some of the applications for quadratic functions.

Completing the Square

In Chapter 9 we saw how rectangles with algebraic dimensions can be used to explain multiplication of binomials. We are again going to use an illustration with algebraic dimensions to change a rectangle into a square. Once we've looked the transformation of rectangle into the square, we'll work algebraically to create a square trinomial.

For example, examine the following illustration:

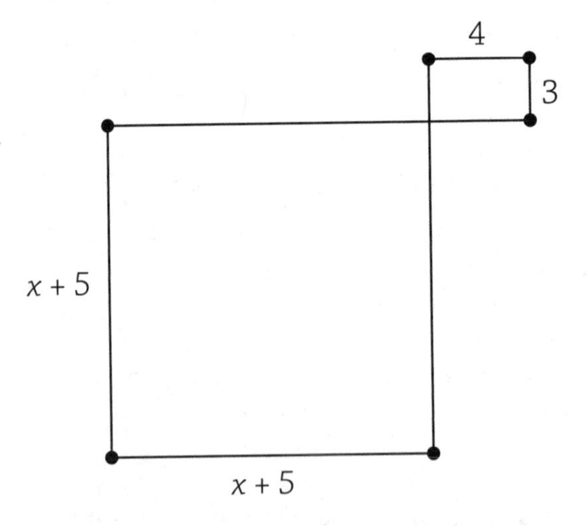

The combined area of the figure is represented by $(x+5)^2+(3)(4) = x^2+10x+25+12 = x^2+10x+37$. So, how do we change a rectangle to a square and another rectangle to keep the area constant?

EXAMPLE

- Rewrite the trinomial $x^2+8x+15$ as a square and a rectangle whose total areas are constant.
- Solution: The key step in performing this task is to take the linear coefficient, in this case 8, cut it in half, and create a square with $x+4$.

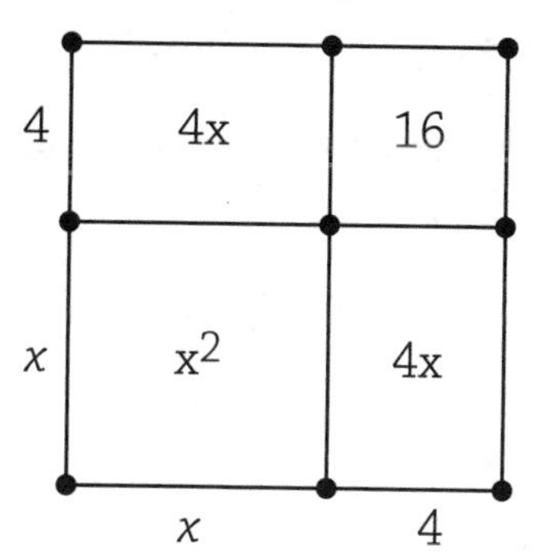

- The problem is that the area of the square is $x^2+8x+16$, which is one unit too big. In order to correct this, we'll need to reduce the area by one. In other words, we need to remove a rectangle with area one from the picture. An easy way to do that is with a square having side of length 1.

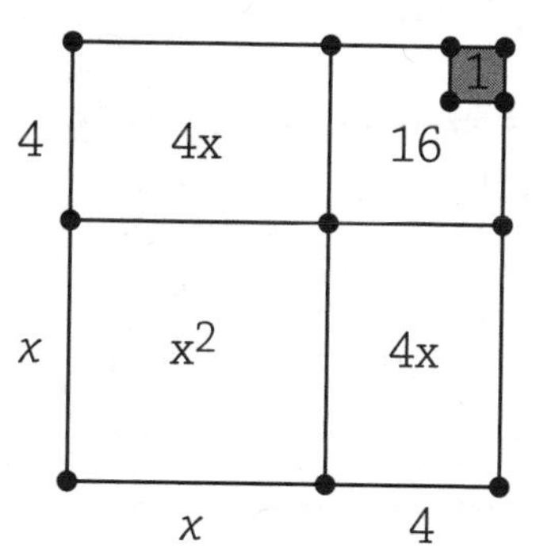

- We have $x^2+8x+15=(x+4)^2-1$.

Let's look at this process using the trinomial $x^2 - 20x + 110$.

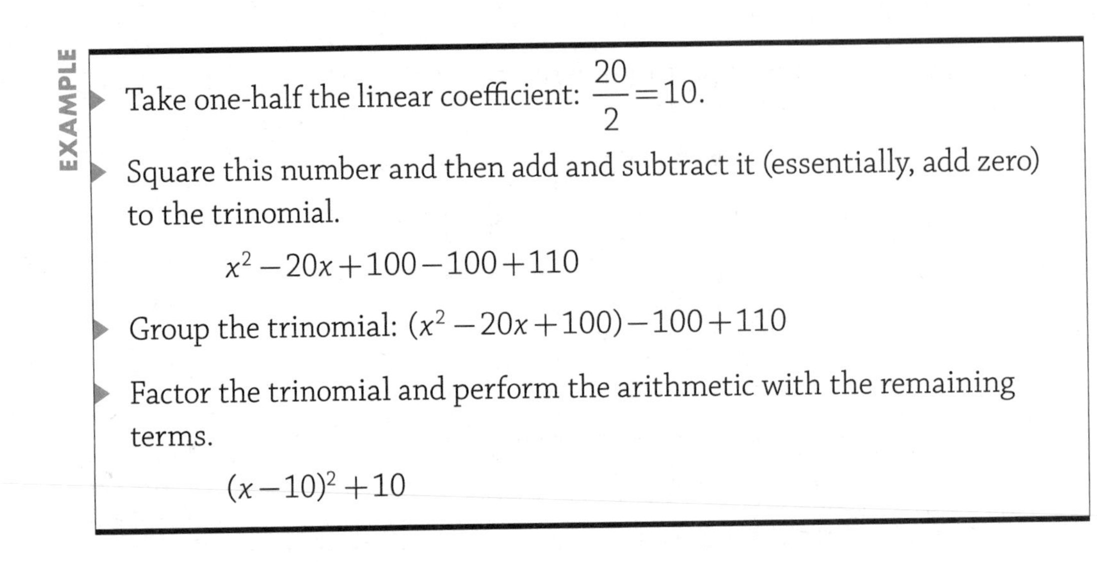
EXAMPLE

- Take one-half the linear coefficient: $\frac{20}{2} = 10$.
- Square this number and then add and subtract it (essentially, add zero) to the trinomial.

$$x^2 - 20x + 100 - 100 + 110$$

- Group the trinomial: $(x^2 - 20x + 100) - 100 + 110$
- Factor the trinomial and perform the arithmetic with the remaining terms.

$$(x - 10)^2 + 10$$

Let's try another problem.

EXAMPLE

- Complete the square with the trinomial $w^2 + 14w - 17$.
- Take one-half the linear coefficient, square it, and then add and subtract it to the trinomial.

$$w^2 + 14w + 49 - 49 - 17$$

- Factor the square trinomial and combine the remaining terms to get: $w^2 + 14w - 17 = (w + 7)^2 - 66$.

There will be times when the middle term is an odd number. Yes, the result will contain fractions but, hey, fractions are numbers too.

EXAMPLE

- Complete the square with the trinomial z^2+9z+7.
- Take one-half the linear coefficient, square it, and then add and subtract it to the trinomial.

$$z^2+9z+\left(\frac{9}{2}\right)^2-\left(\frac{9}{2}\right)^2+7$$

- Factor the square trinomial and combine the remaining terms to get:

$$\left(z+\frac{9}{2}\right)^2-\frac{53}{4}.$$

The process for completing the square can be performed if the quadratic coefficient (the leading coefficient) is not 1. The trick is to factor this number from the first two terms, complete the square with these two terms, and be sure to multiply the value being added and subtracted by the quadratic coefficient.

EXAMPLE

- Complete the square with the trinomial $2x^2+16x-9$.
- Factor the quadratic coefficient from the first two terms: $2(x^2+8x)-9$
- Complete the square inside the parentheses: $2(x^2+8x+16)-32-9$
- Pay attention to the fact that the number added inside the parentheses is really 32, because the 16 is being multiplied by the 2 outside the parentheses. That is why 32 must be subtracted to maintain a statement that is equivalent to the original expression.
- Factor the term inside the parentheses and combine the terms outside: $2x^2+16x-9=2(x+4)^2-41$.

Graphing Quadratic Functions

By now you know that the graph of the function $f(x) = x^2$ is the parabola that opens up, turns at the point (0, 0), and will fold onto itself if the graph paper is folded along the x-axis. From here on, the turning point will be called by its technical name, the vertex.

The parabola has an interesting property. For starters, for each increase to the right (or left) by 1, the y-coordinate will increase by a sequence of odd numbers. Look at the accompanying table of values.

x	y	Difference
-3	9	
		5
-2	4	
		3
-1	1	
		1
0	0	
		1
1	1	
		3
2	4	
		5
3	9	
		7
4	16	

The next value for x will be 5, the corresponding y value will be 25, and 25 is 9 more than 16. Make a table of values for the function $y = 2x^2$. What is the pattern between successive values of y? Extend this to the function $y = ax^2$. The answer appears at the end of the chapter. Don't peek until you have an answer!

Things get a bit more interesting when the equation is expanded to be in standard form, $y = ax^2 + bx + c$, or to the vertex form, $y = a(x-h)^2 + k$.

The Standard Form of the Parabola

The line of symmetry is no longer the y-axis but is the line with the equation $x = \frac{-b}{2a}$. Substitute this value of x into the function to determine the y-coordinate of the vertex. Once the value for the axis of symmetry has been determined, you can create a table of values (five to seven points are usually recommended, with the vertex being the middle value in the table). So long as the value for the axis of symmetry is an integer, or midway between two integers, the values computed for y will prove to be a check for your arithmetic, as they should read the same from top to bottom.

EXAMPLE

Given the function $f(x)=2x^2-8x-3$, determine the equation for the axis of symmetry, the coordinates of the vertex, and the coordinates for six other points, symmetric about the axis of symmetry. Sketch a graph using these results.

The equation for the axis of symmetry is $x=\frac{-(-8)}{2(2)}=2$. The vertex of the parabola has coordinates $(2, f(2)) = (2, -11)$. The table of values is

x	y
−1	7
0	−3
1	−9
2	−11
1	−9
2	−3
3	7

▶ A sketch of the graph is

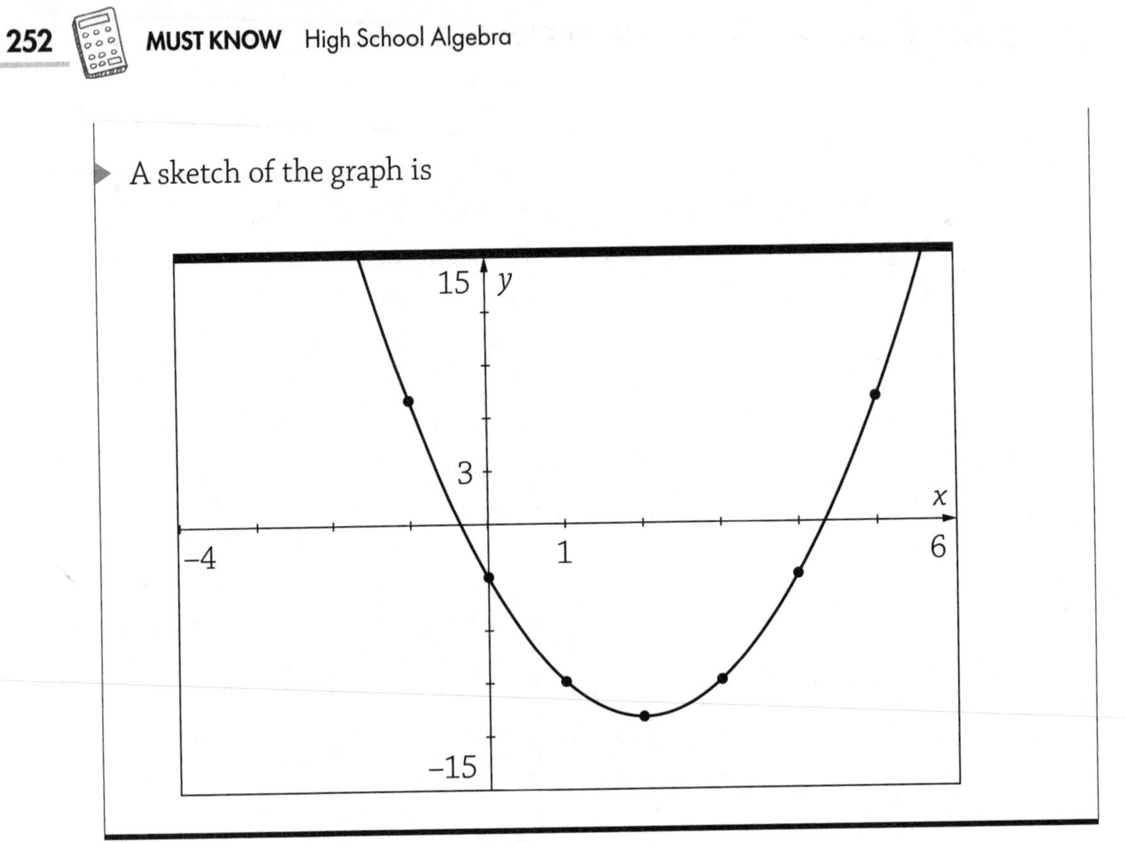

It is critical for you to remember that the graph of the parabola is a smooth curve and not an open polygon whose points are connected by line segments. That is to say, this would not be an acceptable graph for the function $f(x)=2x^2-8x-3$.

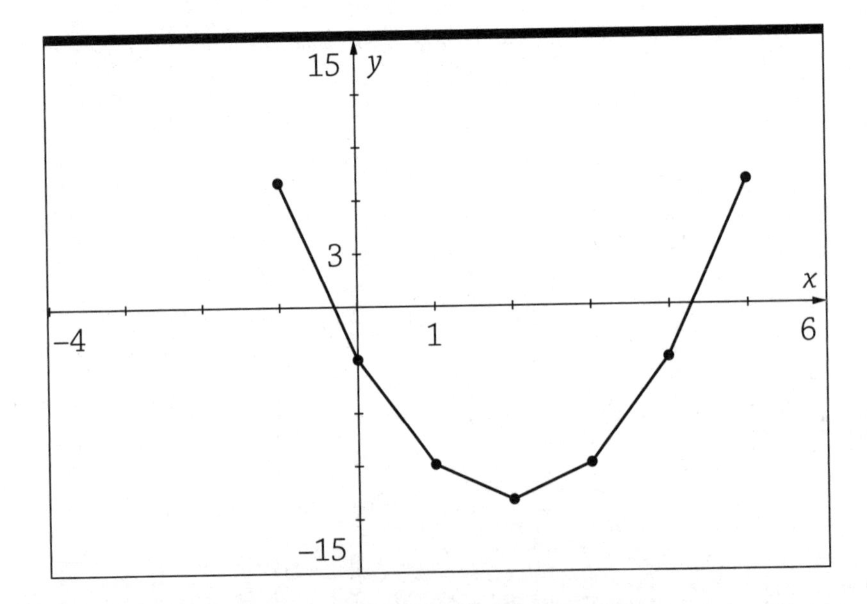

An easy point to find on the parabola is the y-intercept because it will always be the point $(0, c)$. In a pinch, you can find the equation for the axis of symmetry and then the vertex, and if the vertex is not the y-intercept (as it is with $f(x) = x^2$), the reflection of the y-intercept across the axis of symmetry as well.

EXAMPLE

- Given the function $p(x) = -3x^2 - 18x + 4$, determine the coordinates of the y-intercept, the vertex, and the image of the y-intercept over the axis of symmetry. Sketch a graph of the function.
- The y-intercept has coordinates (0, 4). The equation for the axis of symmetry is $x = \dfrac{-(-18)}{2(-3)} = -3$, so the vertex has coordinates
- $(-3, p(-3)) = (-3, 31)$. The image of the (0, 4) after reflection over the line $x = -3$ is the point $(-6, 4)$.

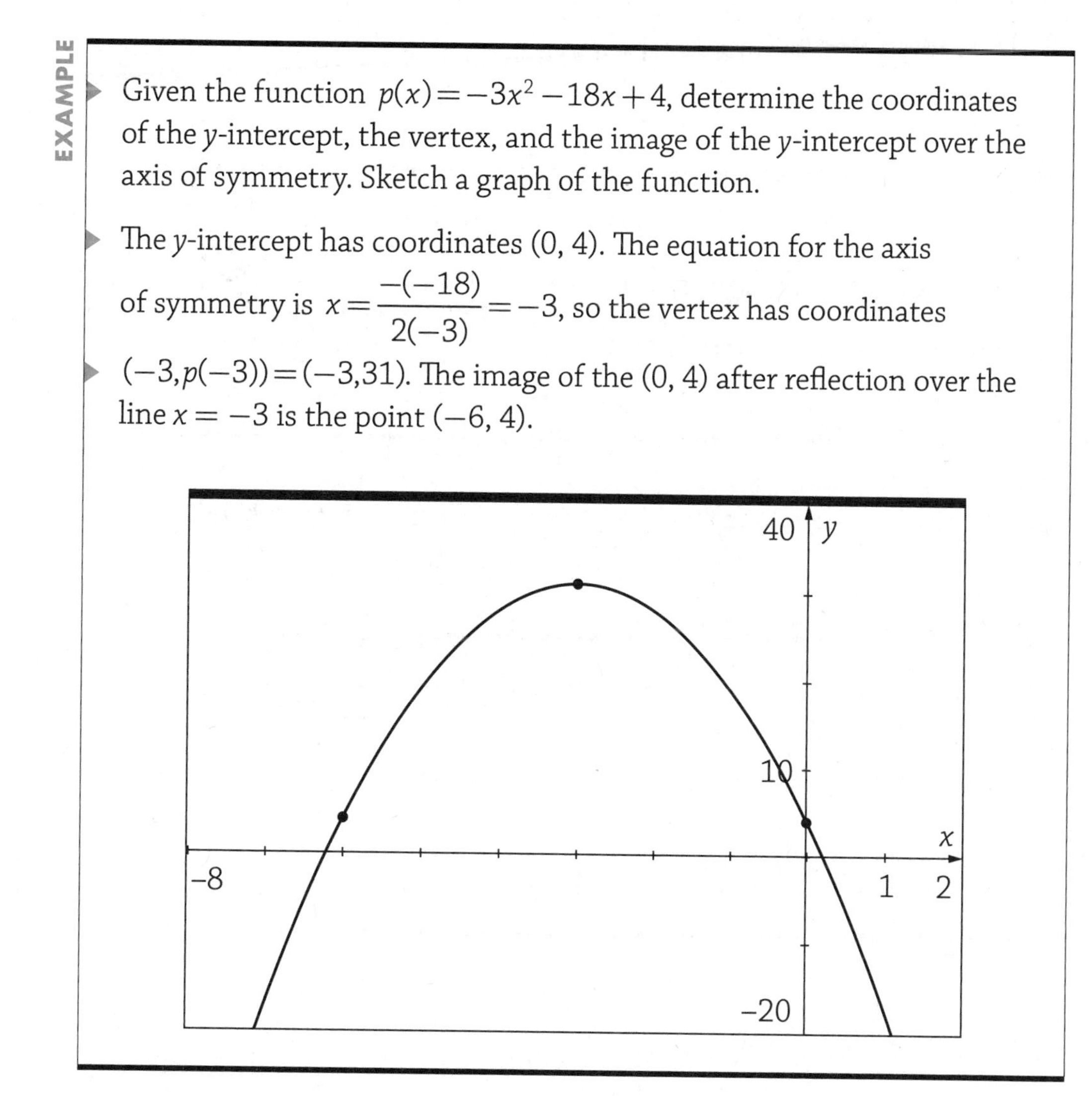

Did you notice that the graph of $f(x)$ opened up while the graph of $p(x)$ opened down? The direction the parabola opens (up or down) is controlled by the quadratic coefficient, a. If $a > 0$, the parabola opens up and if $a < 0$, the parabola opens down. If the parabola opens up, the y-coordinate for the vertex is the minimum value of the function defining the parabola and the range of the parabola will be $y \geq$ this value, while if the parabola opens down, the y-coordinate for the vertex is the maximum value of the function defining the parabola and the range of the parabola will be $y \leq$ this value.

EXAMPLE

▶ Determine the range of the function $q(x) = 5x^2 + 8x + 2$.

▶ The equation for the axis of symmetry for $q(x)$ is $x = \dfrac{-8}{2(5)} = -0.8$.

The coordinates of the vertex are $(-0.8, -1.2)$, so the range for $q(x)$ is $y \geq -1.2$.

In the next problem, don't let the fact that the quadratic coefficient is a fraction bother you. The process is the same.

EXAMPLE

▶ Determine the range of the function $v(x) = -\dfrac{2}{3}x^2 + 4x + 2$.

▶ The equation for the axis of symmetry for $v(x)$ is $x = \dfrac{-4}{2\left(\dfrac{-2}{3}\right)} = 3$. The coordinates of the vertex are $(3, 8)$, so the range for $v(x)$ is $y \leq 8$.

The Vertex Form of the Parabola

The equation for the axis of symmetry for the parabola written in the form $y = a(x - h)^2 + k$ is $x = h$, and the vertex of the parabola has coordinates (h, k).

EXAMPLE

▶ Determine the equation of the axis of symmetry, the coordinates of the vertex, and the range of the parabola with equation $y=3(x-2)^2+4$.

▶ The axis of symmetry is the line with equation $x = 2$, and the vertex has coordinates (2, 4). Since the parabola opens up (the quadratic coefficient is positive), the range is $y \geq 4$.

You need to be careful when working with the equation of the parabola written in vertex form. There is a subtraction and an addition in the formula yet the coordinates of the vertex are (h, k). The coordinates of the vertex of the parabola with equation $y=2(x+3)^2+5$ are (−3, 5) because, when matching this equation to the formula, the equation would need to be written as $y=2(x-(-3))^2+5$. (Why this is will be explained when we talk about transformations of quadratic functions.)

EXAMPLE

▶ Sketch a graph for the parabola with equation $y=-2(x+1)^2+3$.

▶ The coordinates of the vertex are (−1, 3). You can create a table of values by substituting −2, −3, and −4 for values to the left of the vertex and 0, 1, and 2 for values to the right of the vertex. (Or, you can take advantage of the pattern of successive odd numbers as you step away from the vertex.)

x	y
−4	$-5+(-2)(5)=-15$
−3	$1+(-2)(3)=-5$
−2	$3+(-2)(1)=1$
−1	3
0	$3+(-2)(1)=1$
1	$1+(-2)(3)=-5$
2	$-5+(-2)(5)=-15$

▸ The graph is

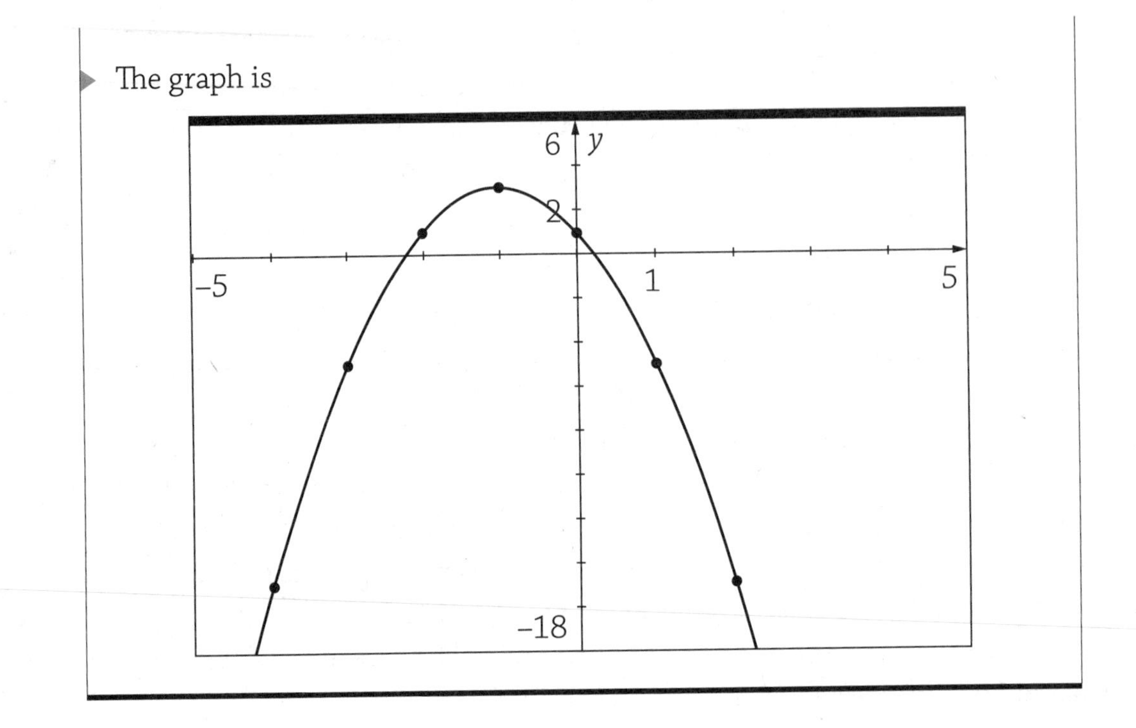

Solving Quadratic Equations by Graphing

The solution to the equation $f(x) = c$, where c is some constant, are those values of x for which $y = c$. That is to say, if we look at the graphs of $y = f(x)$ and $y = c$, the solution are the x-coordinates of the point(s) of intersection.

EXAMPLE

▸ Solve the equation $x^2 - 4x - 5 = 7$ graphically.

▸ Use your graphing utility to graph both $y = x^2 - 4x - 5$ and $y = 7$ on the same set of axes, and use the intersect command to determine the coordinates for the points of intersection.

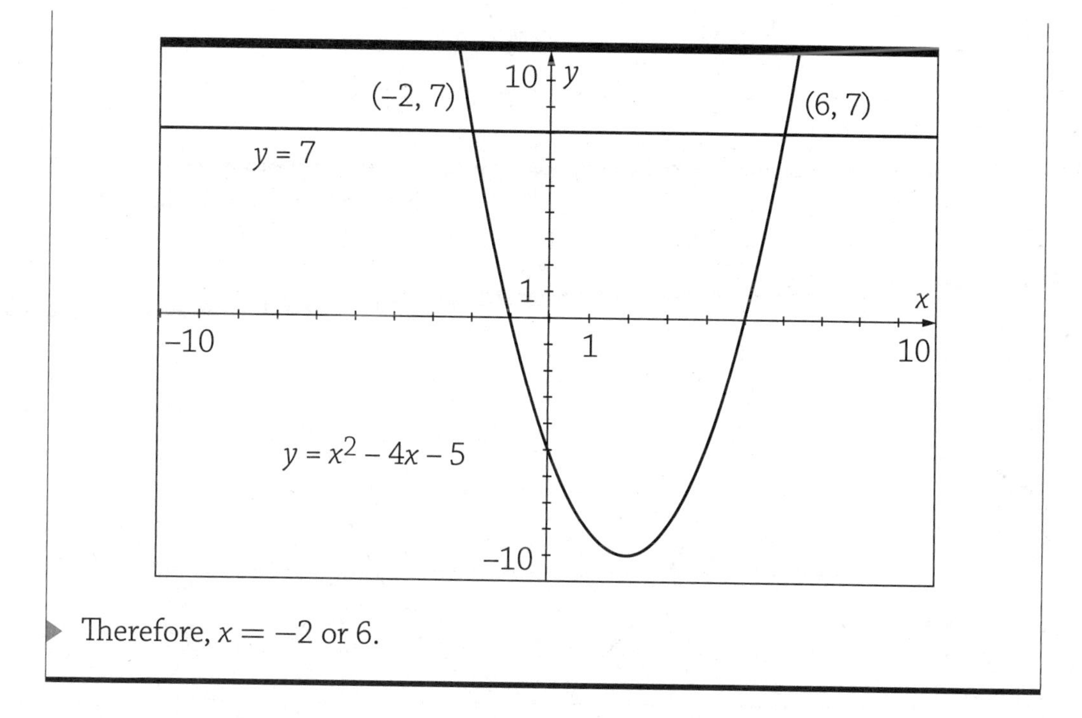

Therefore, $x = -2$ or 6.

Maybe you already did this, but did you notice that if you subtract 7 from the original equation, you get $x^2 - 4x - 12 = 0$ and you can solve this equation by factoring? $x^2 - 4x - 12 = 0$ becomes $(x-6)(x+2) = 0$, so that $x = -2$ or 6.

Using a graphical approach is handy when the numbers in the quadratic make the factoring difficult, if not impossible.

EXAMPLE

▶ Solve the equation $\frac{1}{3}x^2 - x - 18 = 42$ graphically.

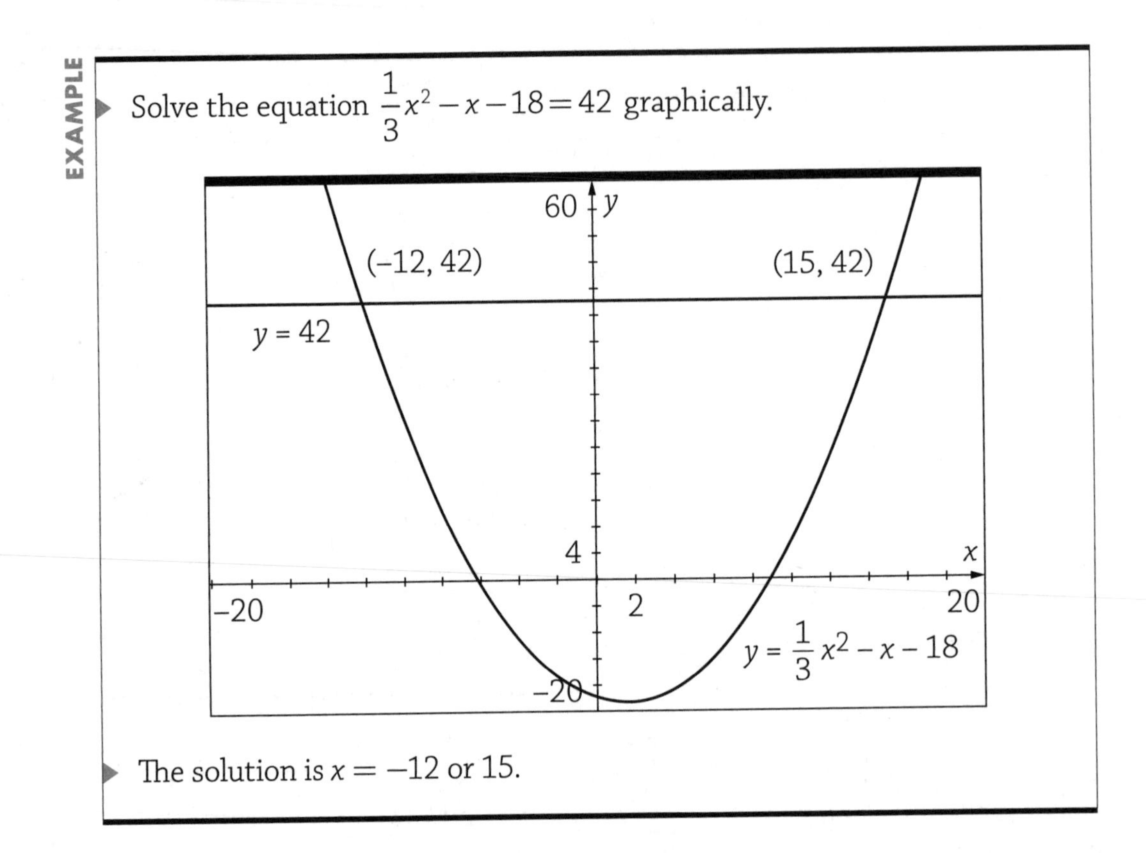

▶ The solution is $x = -12$ or 15.

This process seems a bit easier than trying to factor $x^2 - 3x - 180 = 0$, which is what you would have had to do if asked to solve the problem by factoring. It is not impossible to do, just a bit more tedious.

EXAMPLE

▶ Solve $\frac{4}{5}x^2 - 12x - 8 = 0$.

▶ We can graph the parabola but do not have to graph the line $y = 0$ (although we can). The points where the y values equal 0 are called the x-intercepts of the graph but are also known as the zeros of the function. You can use the Zero feature in the CALC menu to determine these values.

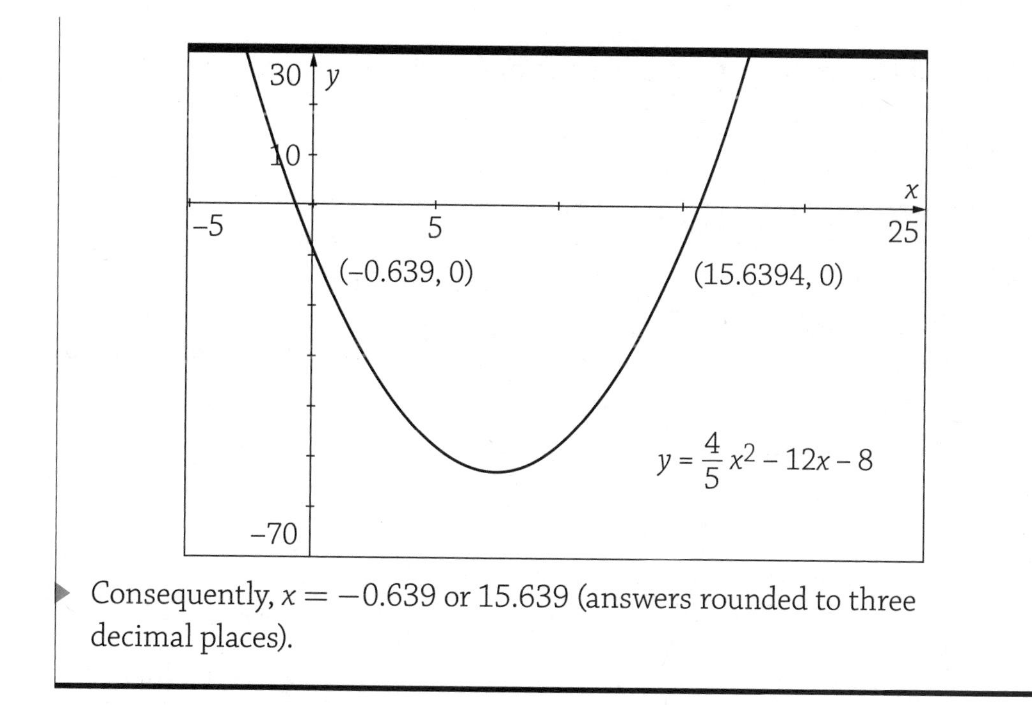

Consequently, $x = -0.639$ or 15.639 (answers rounded to three decimal places).

We should look at two more problems to illustrate what can happen when solving quadratic equations.

EXAMPLE

Solve the equation $x^2 + 3x + 5 = 6$.

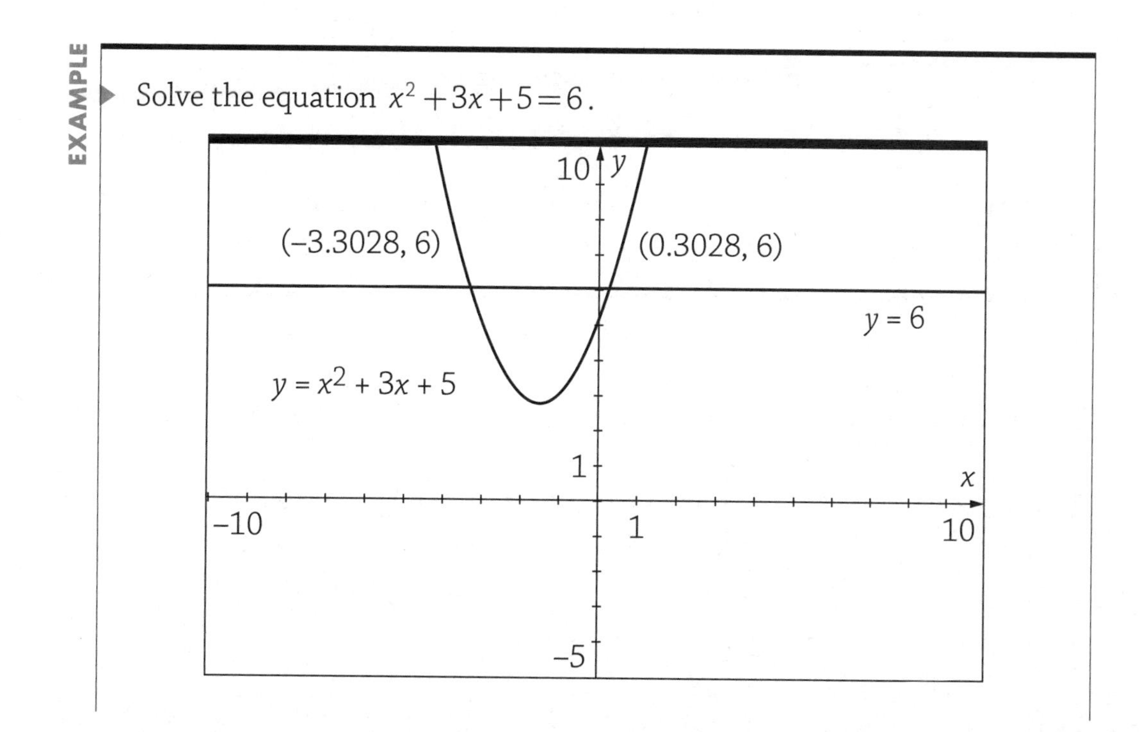

- The solutions to the problem do not appear to be rational numbers, so the calculator/computer gives us the decimal approximations of $x = -3.3028$ or 0.3028.

Have you figured out what the other situation we might encounter is?

EXAMPLE

- Solve the equation $x^2 + 3x + 5 = 1$.

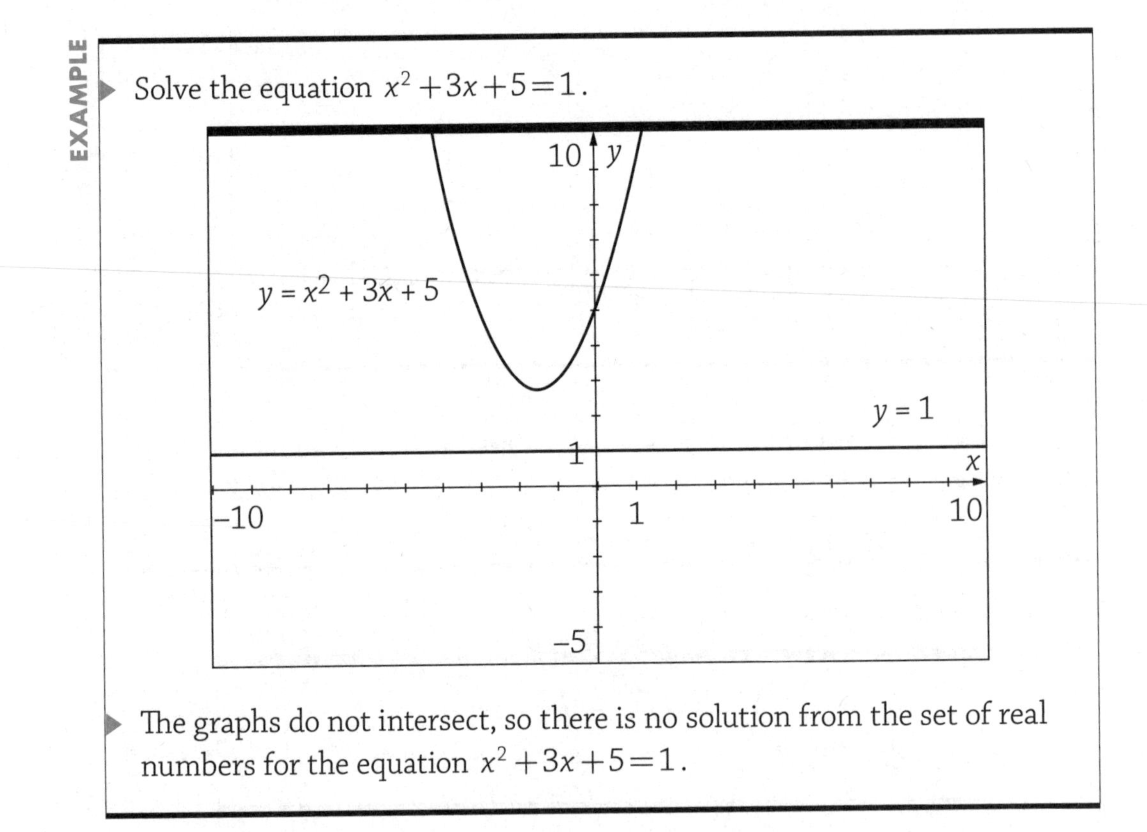

- The graphs do not intersect, so there is no solution from the set of real numbers for the equation $x^2 + 3x + 5 = 1$.

Transformations of Quadratic Functions

Transformations of functions are a big deal. If you know what a basic function, like $y = x^2$, looks like, then you'll know what the graph of the equation $y = a(x - h)^2 + k$ looks like. (Yes, we already talked about the vertex form of the parabola. We'll use this knowledge to set ourselves up for dealing with all transformations—yes, all!)

Let's start off easy. We'll graph the equations of $y = x^2$, $y = (x - 2)^2$, and $y = (x + 5)^2$ and on the same set of axes.

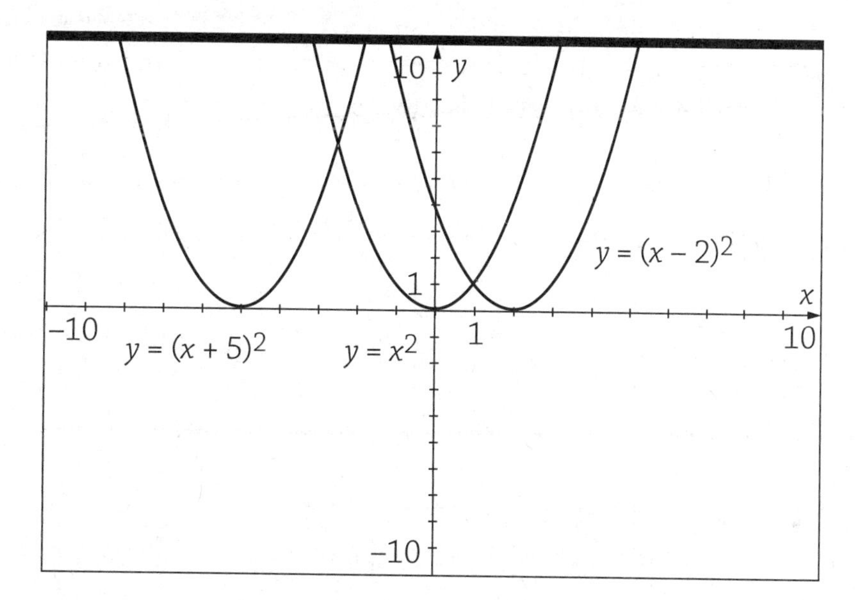

Why is it that when I subtract 2 from x, the graph moves to the right and when I add 5 to x, it moves to the left? That doesn't seem to make sense. The trouble with the logic is that we are thinking about moving the graph of the base function to the transformed function. We need to look at this the other way around. How do we move the transformed function back to the base function?

To move the graph of $y = (x - 2)^2$ back to the graph of $y = x^2$, we move left, and to move the graph of $y = (x + 5)^2$ back to the graph of $y = x^2$, we move right. That makes more sense when we think about positives and negatives.

That seems to be fine, but why, then, must the graphs of $y = x^2 + 3$ and $y = x^2 - 2$ fit our idea of what the graphs should do? After all, the graph of $y = x^2 + 3$ is 3 units above the graph of $y = x^2$ and the graph of $y = x^2 - 2$ is 2 units below. The reasons for this are twofold. First, we're still thinking about moving the graph of the base function to the transformed function rather than the other way around. Second, we've gotten into the habit (and our technology supports this thinking) of writing our equations as $y =$ or $f(x) =$ rather than $y - k =$ or $f(x) - k =$. Look at how you can interpret the equation $y - 3 = x^2$. To send this equation back to the graph of $y = x^2$, you have to move it down 3 units. That is consistent with the interpretation for the horizontal motion. Now that is what mathematical notation is supposed to be—consistent!

EXAMPLE

- How does the graph of $y = f(x - 4) - 3$ compare to the graph of $y = f(x)$?
- The graph of $y = f(x - 4) - 3$ lies 4 units to the right and 3 units below the graph of $y = f(x)$.

The last piece to consider is the scaling factor. What happens when we compare the graphs of $y = x^2$ with the graph of $y = 4x^2$? The points on the graph of $y = 4x^2$ have been stretched from the x-axis by a factor of 4. That is, the x-coordinates do not change, only the y-coordinates do. In general, the graph of the function $y = ax^2$ will expand from the x-axis if $a > 1$, will contract toward the x-axis if $0 < a < 1$, and will flip over the x-axis if $a < 0$.

EXAMPLE

- Describe how the graph of $y = -3(x + 2)^2 + 5$ is determined based on the graph of $y = x^2$.
- The graph of $y = -3(x + 2)^2 + 5$ is found by moving the graph of $y = x^2$ left 2 units, flipping it over the x-axis, stretching it from the x-axis by a factor of 3, and then moving it up 5 units.
- How do we know this? Watch what happens to a value of x. First, 2 is added. We know that is a motion to the left. Then this value is squared. Well, that is just the behavior of the base function we are examining.

Then this value is multiplied by −3. This is a two-step behavior—the flip because the value is negative and the stretch because of the multiplication. Finally, 5 is added to get the y-coordinate.

That's a lot to think about. Let's try one more just to make sure we've got it.

EXAMPLE

▶ Describe how the graph of the function $y = \frac{-1}{4}(x-5)^2 - 2$ is determined based on the graph of $y = x^2$.

▶ The graph of $y = \frac{-1}{4}(x-5)^2 - 2$ is found by moving the graph of $y = x^2$ right 5 units, flipping it over the x-axis, compressing it from the x-axis by a factor of $\frac{1}{4}$, and then moving it down 2 units.

Solving Quadratic Equations by Using the Quadratic Formula

As we saw in the section on solving quadratic equations graphically, there are equations whose solutions are real numbers but not rational numbers. The **quadratic formula** (which is different from a quadratic equation) enables us to find the result. The quadratic formula is derived from the quadratic equation $ax^2 + bx + c = 0$ by completing the square. The algebra itself is not particularly tricky, but having limited experience with literal quadratic equations, I'll suggest you look up the derivation on line if you are interested in seeing it.

The solution to the quadratic equation $ax^2+bx+c=0$ is

$$x=\frac{-b\pm\sqrt{b^2-4ac}}{2a}$$

EXAMPLE

▶ Use the quadratic formula to solve $x^2-5x-8=0$.

▶ We have $a=1$, $b=-5$, and $c=-8$, so the solution is $x=\frac{-(-5)\pm\sqrt{(-5)^2-4(1)(-8)}}{2(1)}=\frac{5\pm\sqrt{25+32}}{2}=\frac{5\pm\sqrt{57}}{2}$. That is, $x=\frac{5+\sqrt{57}}{2}$ or $\frac{5-\sqrt{57}}{2}$. These are the exact values of the solution. (Pay attention to that because there are times when the directions are to give the exact values for the solution rather than a decimal approximation.) If a decimal approximation is required, be sure to evaluate the numerator before dividing. The order of operations strikes again! The roots would be $(5+\sqrt{57})/2$ and $(5-\sqrt{57})/2$. The values of these expressions are 6.275 and −1.275 (answers rounded to the nearest thousandth).

The nice part about the quadratic formula is that it can be used to solve all quadratic equations. The downside to the quadratic formula is that there is a lot of computation to do and there will be many times when factoring the trinomial will get you to the solution more quickly. The reality is that you need to practice factoring so that you become very comfortable with it. Once that is done, you'll find that all you'll really need to do is glance at the equation and you will know whether factoring the trinomial is easy to do.

That old joke, Q: "How do you get to Carnegie Hall?" A: Practice, practice, practice!" has a great deal of truth in it.

EXAMPLE

▶ Solve $24x^2+59x-70=0$.

▶ We have $a=24$, $b=59$, and $c=-70$, so $x=\dfrac{-59\pm\sqrt{59^2-4(24)(-70)}}{2(24)}$. I am so glad I have a calculator to do this. The value of $59^2-4(24)(-70)$ is 10,201. As everyone knows (that is, everyone who used their calculator to figure this out), $\sqrt{10{,}201}=101$. Consequently,

$$x=\frac{-59\pm101}{48}=\frac{-59+101}{48},\frac{-59-101}{48}\text{ so that }x=\frac{7}{8},\frac{10}{3}.$$

This represents a problem where I will use the quadratic formula even though I suspect that there might be trinomial factors because there are a lot of possible factor pairs for 24 (though not that many for 70). I would not consider using the quadratic formula to solve $2x^2-5x+2=0$ because I can "see" the factors but wouldn't think twice about using the quadratic formula to solve $2x^2-5x-2=0$ because there is no way of using the factors of 2 and the negative signs involved to get a middle term of −5.

EXAMPLE

▶ Solve $2x^2-5x-2=0$

▶ We have $a=2$, $b=-5$, and $c=-2$,so

$$x=\frac{-(-5)\pm\sqrt{(-5)^2-4(2)(-2)}}{2(2)}=\frac{5\pm\sqrt{25+16}}{4}=\frac{5\pm\sqrt{41}}{4}.$$

Special Functions

We have examined the linear, exponential, and quadratic functions in this book. There are many, many more functions that you will encounter in your mathematical studies. I want to look at four special functions in this section: the square root function, the absolute value function, and the reciprocal function. Then we will look at a function that makes use of all these other functions. I know I said four functions and listed only three. That is because these three are very specific functions. The last function I want to examine is called a piece-wise or split-domain function and it makes use of all these other functions.

The *square root function*, $f(x)=\sqrt{x}$, is the inverse of the base quadratic function, $y=x^2$. (There is a bit of work behind that statement that will be studied in Algebra II, but for now we are taking advantage of your current mathematical expertise.) The domain of the square root function is $x\geq 0$, and the range is $y\geq 0$.

There is a very important mathematical issue that often gets lost here: the solution to the equation $x^2=4$ is $x=\pm 2$, but $\sqrt{4}=2$ but not -2. Why is that? When solving an equation, you are required to give all values of the variable for which the equation is true. Because 2^2 and $(-2)^2$ both equal 4, there are two solutions to the equation $x^2=4$. When evaluating a function, there is only one output that can come from any input (that is the definition of a function), so the value of $\sqrt{4}$ can only be one number. Keep that in mind as you solve equations and evaluate functions.

EXAMPLE

- The graph of the function $f(x)=\sqrt{x}$ looks like this:

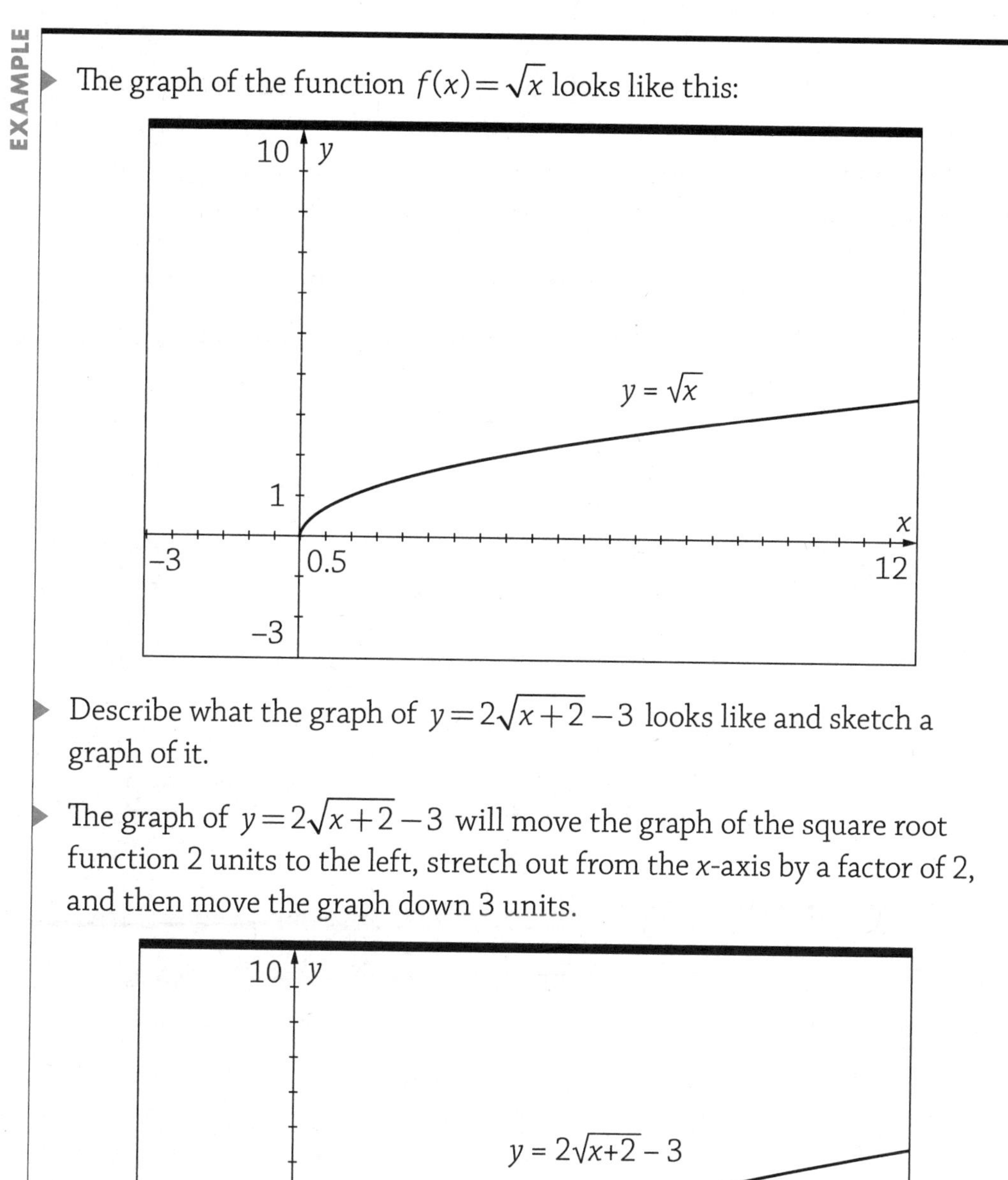

- Describe what the graph of $y=2\sqrt{x+2}-3$ looks like and sketch a graph of it.

- The graph of $y=2\sqrt{x+2}-3$ will move the graph of the square root function 2 units to the left, stretch out from the x-axis by a factor of 2, and then move the graph down 3 units.

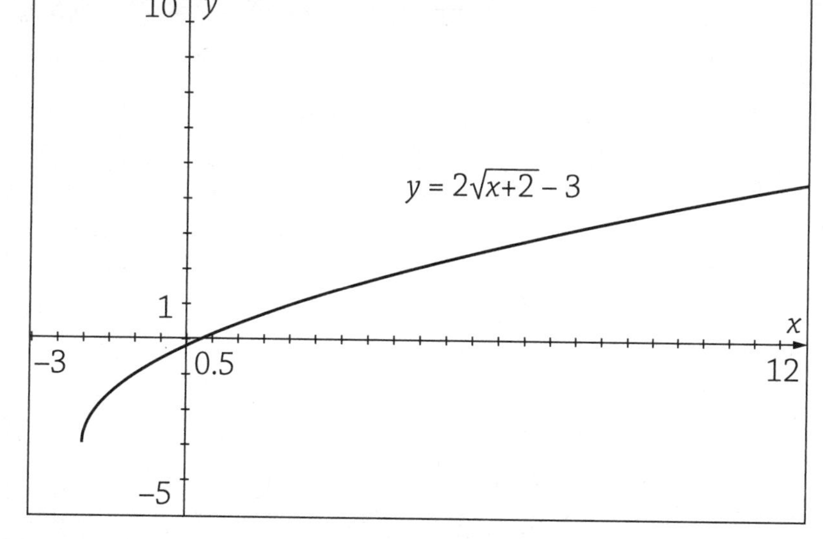

The *absolute value function*, written as $y = |x|$, can be defined a number of different ways. My personal favorite is that the absolute value of x represents the distance that x is from the origin on the number line (as you saw in Chapter 5). What is $|-5|$? The graph of the absolute value function is

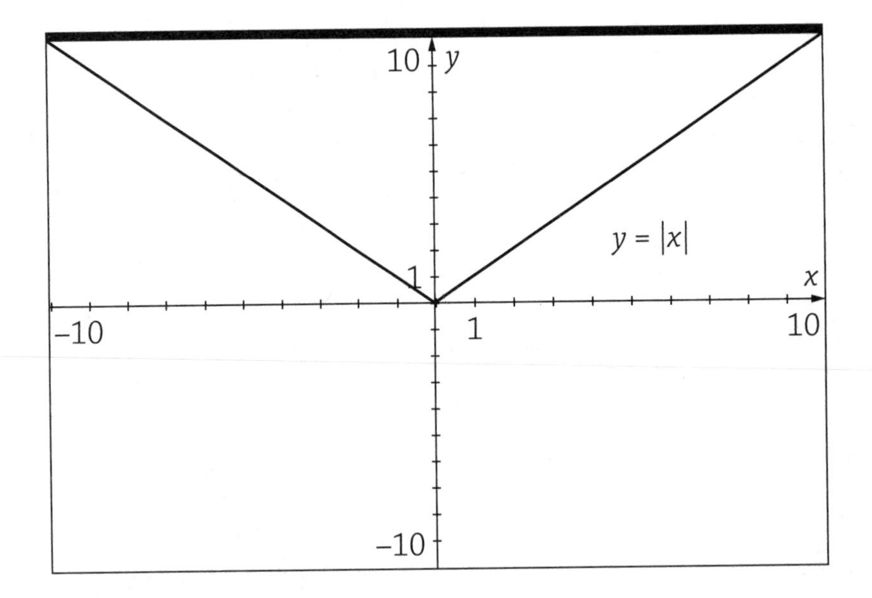

EXAMPLE

▶ Describe how the graph of the function $k(x) = 2|x-3|-4$ is formed from the graph of $y = |x|$.

▶ The graph of $k(x)$ is found by moving the graph of $y = |x|$ three units to the right, stretch from the x-axis by a factor of two, and then slide the graph down four units.

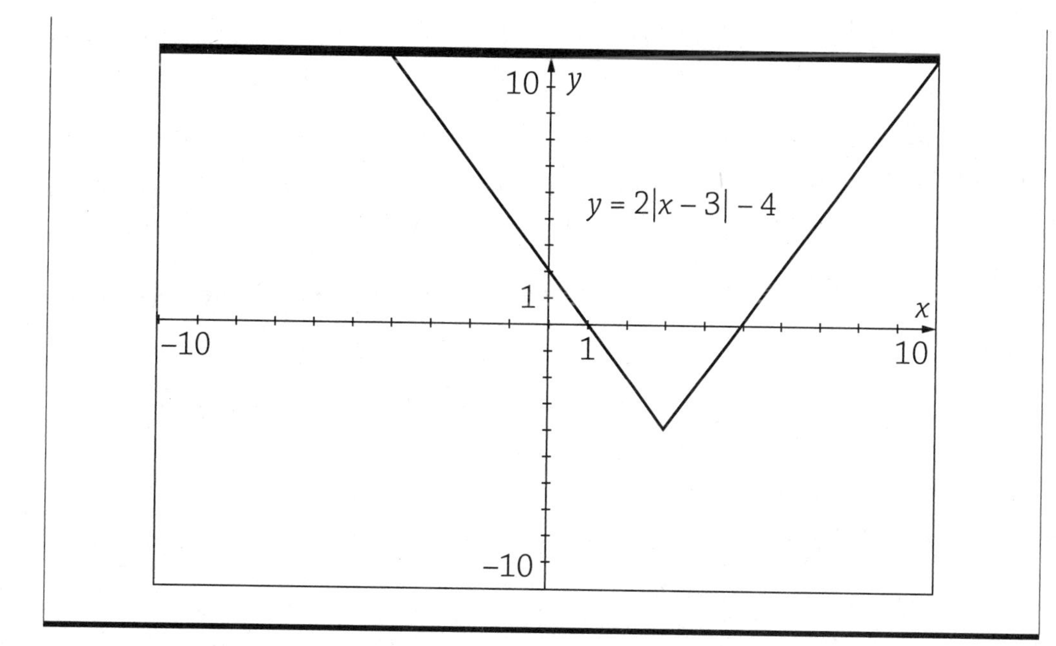

The graph of the *reciprocal function* $g(x) = \frac{1}{x}$ looks like this:

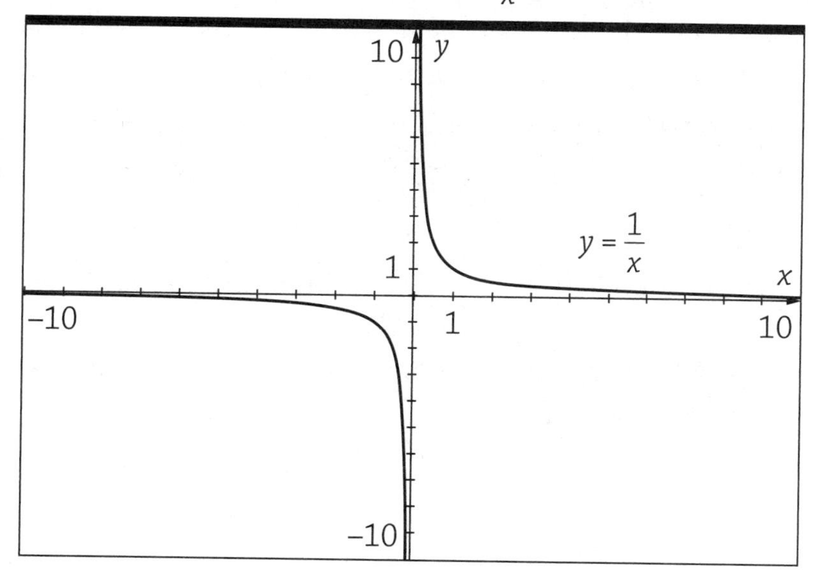

The domain is $x \neq 0$ and the range is $y \neq 0$.

This brings us to the last of the special functions for this section. Examine the following graph.

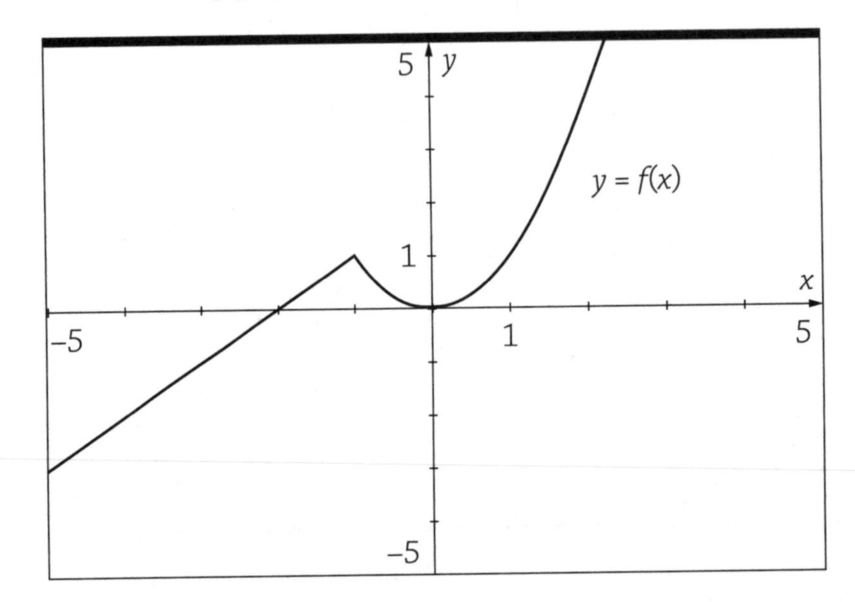

For values of $x < -1$, the graph is the line $y = x + 2$ and for values of $x > -1$, the graph is the parabola $y = x^2$. At $x = -1$, the two sections of the graph meet. This type of graph is referred to as a *piece-wise defined function* (because it consists of pieces of different functions) but is also referred to as a *split domain defined function* because the rule used to create points on the graph is split based on the what part of the domain is being used. This function has the rule $f(x) = \begin{cases} x+2, & x<-1 \\ x^2, & x \geq -1 \end{cases}$. (For this particular function, the sense of equality in the domain could be in either part since at $x = -1$ each rule gives an answer of –2.)

EXAMPLE

▶ Sketch a graph of the function defined by $g(x)=\begin{cases} 2x-1, & x<2 \\ \frac{1}{2}x+2, & x\geq 2 \end{cases}$.

▶ The graph consists of two rays, each emanating from the point (2, 3).

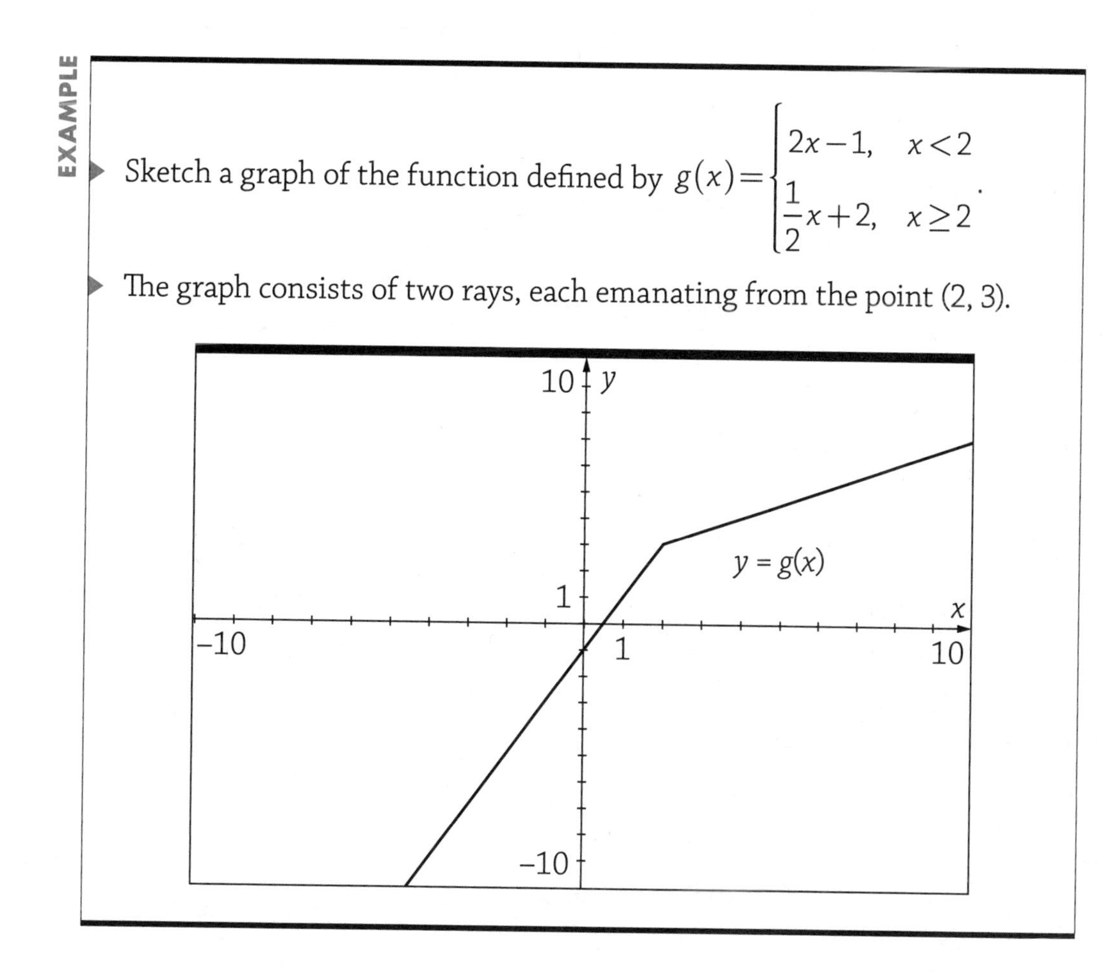

There are many applications for piece-wise functions. The amount you pay to mail a first-class letter in the United States depends on the weight of the letter. At this writing, it costs \$0.49 to mail a letter that weighs at most 1 ounce. For each additional ounce, or part of an ounce, the price increases by \$0.21. For a letter that weighs $1 <$ weight ≤ 2oz, the cost is \$0.70. If the weight is $2 <$ weight ≤ 3 oz, the cost is \$0.91. The graph for this function will look like a set of steps.

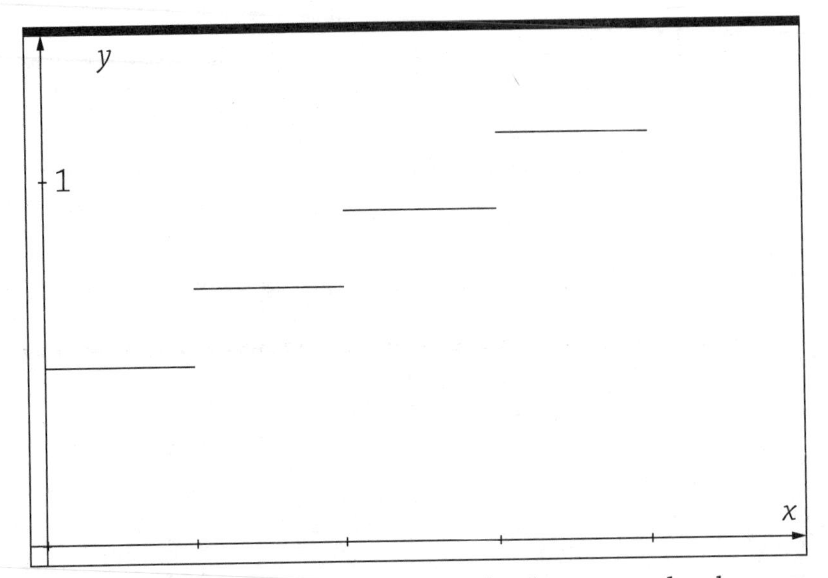

Notice that the pieces that form the graphs do not need to be connected. The graph does not fail the vertical line test because of the way the domain is divided. For example, a letter that weighs exactly 1 ounce requires \$0.49 in postage, not \$0.49 or \$0.70.

EXERCISES

EXERCISE 10-1

Complete the square of each the following.

1. $x^2 - 4x + 2$
2. $p^2 + 10p + 7$
3. $w^2 + 7w - 9$
4. $2m^2 - 12m + 5$
5. $3r^2 + 8r + 4$

EXERCISE 10-2

Determine the equation of the axis of symmetry and the coordinates of the vertex for the parabola defined by the given equation.

1. $y = x^2 - 10x + 3$
2. $y = -x^2 + 8x + 3$
3. $d(x) = 4x^2 - 24x - 5$
4. $v(x) = -3x^2 - 12x + 2$
5. $y = (x - 3)^2 + 1$
6. $y = (x + 5)^2 - 3$
7. $k(x) = -5(x - 2)^2 + 9$
8. $n(x) = \frac{2}{3}(x + 1)^2 + 2$
9. What is the range of the function defined in exercise 8?
10. What is the range of the function defined in exercise 12?

EXERCISE 10-3

Solve each equation. Leave each answer in exact form.

1. $x^2 - 7x - 12 = 0$
2. $3f^2 - 7f - 4 = 0$
3. $8m^2 + 7m - 2 = 0$
4. $-4x^2 + 5x + 3 = 0$
5. $\frac{1}{2}x^2 + 10x + 9 = 0$
6. $\frac{-3}{4}y^2 + 8x + 2 = 0$

EXERCISE 10-4

Solve each equation. Round approximate answers to three decimal places.

1. $12x^2 - 17x - 5 = 0$
2. $-10y^2 + 7y + 8 = 0$
3. $\frac{-3}{2}r^2 - 5r + 2 = 0$

EXERCISE 10-5

Describe the transformation needed to create the graph of the indicated function from its base function.

1. $p(x) = \sqrt{x+2} + 1$
2. $g(x) = 2|x-3| - 1$
3. $d(x) = \frac{1}{x-1} + 2$
4. Sketch the graph of the function defined in exercise 1.

5. Sketch the graph of the function defined in exercise 2.

6. Sketch the sketch of the function defined by $f(x)=\begin{cases} x^2, & x<1 \\ \sqrt{x}, & x\geq 1 \end{cases}$.

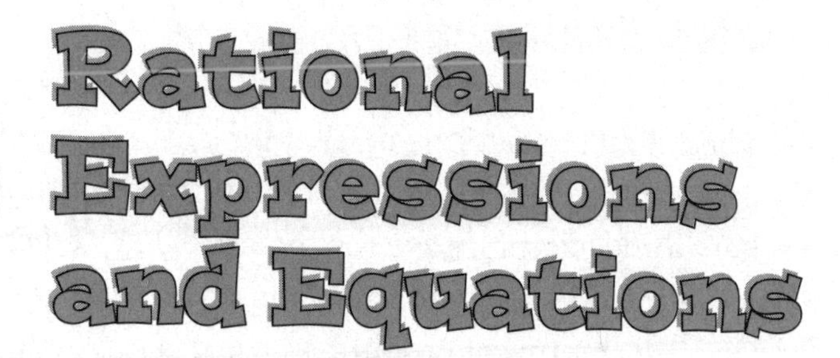

11 Rational Expressions and Equations

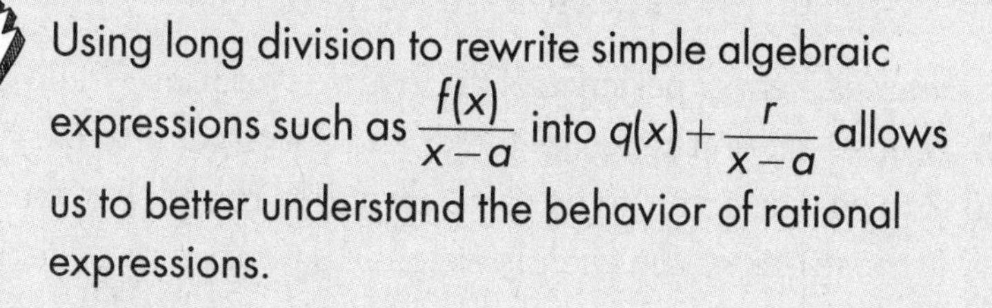

MUST KNOW

- Using long division to rewrite simple algebraic expressions such as $\frac{f(x)}{x-a}$ into $q(x)+\frac{r}{x-a}$ allows us to better understand the behavior of rational expressions.

- The rules for combining algebraic fractions are exactly the same as the rules for adding numerical fractions.

- Inverse variations can be applied to a variety of real-world problems such as determining volume of a gas, intensity of light, and the frequency of a guitar string.

Inverse Variation

We studied direct variations in Chapter 3. The key item to remember about quantities in a direct variation is that as one quantity increases then so does the other. Some examples are the distance traveled compared to the time spent traveling and the amount of money earned as one works more hours. How does the amount of gasoline in the gas tank compare to the distance traveled? The farther you travel, the more gas you use, so there is less gas in the gas tank. That is, as the distance traveled increases, the amount of gasoline in the tank decreases. This is an example of an inverse variation. Whereas in a direct variation, the equation takes the form $y = kx$ or $\frac{y}{x} = k$, where k is the constant of proportionality, in an inverse variation the equation is $xy = k$ or $y = \frac{k}{x}$, where k is a constant.

Perhaps your earliest experience of an inverse variation would be the teeter-totter. Two friends get on the teeter-totter and try to get it to balance. That is, the goal is to get the board to lie parallel to the ground. For the sake of argument, suppose one of the friends weighs 40 pounds and the other 50 pounds. In order to get the board to balance, the child who weighs less must be further from the fulcrum (the point upon which the board rests) than the other. It turns out that the distance one must be from the fulcrum varies inversely as the weight of the rider. That is, $wd = k$, where w is the weight of the person and d is the distance from the fulcrum. So, if the friend with the lighter weight sits 5 feet from the fulcrum, the other friend could sit 4 feet away.

EXAMPLE

▶ What if the child who weighs 40 pounds sits 3 feet from the fulcrum? Where should the child who weighs 50 pounds sit?

▶ Using the equation, $wd = k$, we get that $k = (40)(3) = 120$ foot-pounds. Using the data from the second child, $50d = 120$, so $d = 2.4$ feet.

We can avoid the step of determining the value of k if all we are interested in is learning the distance the second child must sit from the fulcrum. If we consider the data for the first child as w_1 and d_1 and the data for the second child as w_2 and d_2, then the equation we can work with is $w_1 d_1 = w_2 d_2$.

EXAMPLE

▶ When the temperature is held constant, the volume of a gas varies inversely with the pressure of the gas. If the volume of a gas is 15 cubic centimeters when the pressure is 10 Pascals, what is the volume of the gas when the pressure is 25 Pascals?

▶ Using the formula $V_1 P_1 = V_2 P_2$, $(15)(10) = 25V$ gives that the volume is 6 cubic centimeters.

The sound from string instruments can also be described with inverse variations.

EXAMPLE

▶ The frequency of a vibrating guitar string varies inversely as its length. Suppose a guitar string 0.75 meters long vibrates 3.9 times per second. What frequency would a string 0.6 meters long have?

▶ Using the formula $l_1 f_1 = l_2 f_2$, $(0.75)(3.9) = 0.6f$ becomes

$$f = \frac{(0.75)(3.9)}{0.6} = 4.875 \text{ times per second.}$$

Not all examples of inverse variation need to be linear. The relationship between the source of a light and the intensity of that light is not.

EXAMPLE

- The intensity of light produced by a light source varies inversely as the square of the distance from the source. If the intensity of light produced 3 feet from a light source is 750 foot-candles, find the intensity of light produced 7 feet from the same source.

- According to the description, the intensity of the light, I, and the distance from the light source, d, are related by the equation $Id^2 = k$, where k is the constant of variation. We can also write $I_1 d_1^{\,2} = I_2 d_2^{\,2}$. Consequently, we can solve the problem with the equation $(750)(3)^2 = I(7)^2$ so that $I = 137.755$ foot-candles.

Rational Functions

The most basic rational function is $y = \frac{1}{x}$. The domain is $x \neq 0$. If we take a moment to rewrite the equation as $xy = 1$, we see that $y \neq 0$ as well. The graph of $y = \frac{1}{x}$ is interesting as the values of x get very close to zero. For instance, if $x = 0.0000001$ (that is, if $x = \frac{1}{1{,}000{,}000}$), then $y = 1{,}000{,}000$.

Do you see what happens when x is $\frac{-1}{1{,}000{,}000}$? By the same token, if $x = 1{,}000{,}000$, then $y = \frac{1}{1{,}000{,}000}$. The graph will get very close to each of the axes but will never cross it. Lines that the graphs approach but do not cross are called **asymptotes**. The graph of $y = \frac{1}{x}$ is:

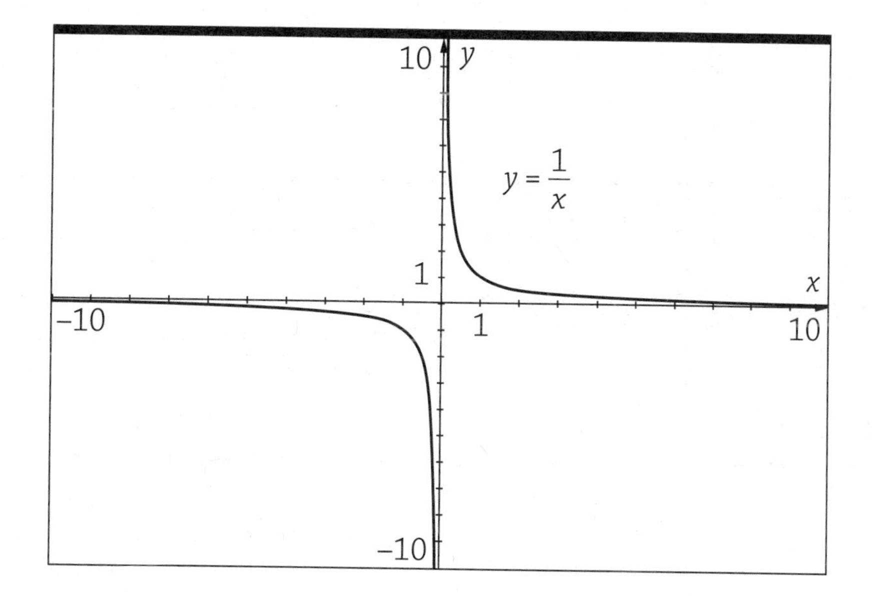

In general, rational functions will take on the form $f(x)=\frac{g(x)}{k(x)}$. The domain for this function are all values of x except for those for which $k(x) = 0$.

EXAMPLE

▶ Determine the domain of the function $f(x)=\frac{3}{4x-6}$.

▶ Set the denominator equal to zero to find that $x = 1.5$. Therefore, the domain for the function $f(x)$ is $x \neq 1.5$.

The vertical asymptote will occur at $x = 1.5$, while the horizontal asymptote remains at $y = 0$. Do you see that the horizontal asymptote for the function $f(x)=\frac{3}{4x-6}+2$ will be at $y = 2$ while the vertical asymptote remains at $x = 1.5$?

Naturally, given that this is a course in algebra, the rule defining the function will get more complicated. But all you have to remember about the domain of the function is that you cannot allow the denominator to be zero. The numerator of the fraction has no influence on the domain.

EXAMPLE

- Determine the domain of the function $r(x)=\dfrac{x^2-5x-6}{4x^2-25}$.
- Set the denominator equal to zero and solve:

$4x^2-25=0 \Rightarrow (2x-5)(2x+5)=0 \Rightarrow x=\dfrac{5}{2},\dfrac{-5}{2}$.

We know that this function has two vertical asymptotes, one at $x=\dfrac{5}{2}$ and one at $x=\dfrac{-5}{2}$. We could look at the graph of $y = r(x)$ to determine what, if any, the equation of the horizontal asymptote is, but we will not need to do an analysis for the horizontal asymptote until Algebra II.

EXAMPLE

- When does the graph of $r(x) = 0$? Said two other ways, what are the x-intercepts for the graph of $r(x)$ or what are the zeros of the function $r(x)$?
- In each case, set $r(x) = 0$ and solve. $\dfrac{x^2-5x-6}{4x^2-25}=0$
- Multiply both sides of the equation by $4x^2-25$: $x^2-5x-6=0$
- Factor: $(x-6)(x+1)=0$
- Solve: $x = 6, -1$

To recap, rational functions are undefined when the denominator is equal to zero and have their zeros when the numerator is equal to zero.

Simplifying Rational Expressions

Just as you do when asked to reduce any fraction, you look for any common factors that the numerator and denominator might have. Once found, you cancel these factors (any number divided by itself is 1). With constants, you will multiply the remaining terms in the numerator and terms in the denominator to denote your answer. With algebraic expressions, we tend to leave the terms in factored form. That is,

$$\frac{42}{385}=\frac{2\times3\times7}{5\times7\times11}=\frac{2\times3\times\cancel{7}}{5\times\cancel{7}\times11}=\frac{6}{55}\text{ but}$$

$$\frac{4(x^2-x-2)}{6(x^2-4)}=\frac{2\times2(x-2)(x+1)}{2\times3(x-2)(x-2)}=\frac{\cancel{2}\times2\cancel{(x-2)}(x+1)}{\cancel{2}\times3\cancel{(x-2)}(x-2)}=\frac{2(x+1)}{3(x-2)}.$$

Be aware that all this work is to prepare you to be able to multiply and divide rational expressions.

EXAMPLE

- Completely simplify: $\frac{4x^4+20x^3+25x^2}{4x^3-25x}$.
- First and foremost, you *cannot* cancel the $4x^4$ in the numerator and the $4x^3$ denominator because these are not factors. They are being added to other terms. The same is true for $25x$. Having said that, the first order of business is to factor the numerator and denominator.
- Common factors: $\frac{4x^4+20x^3+25x^2}{4x^3-25x}=\frac{x^2(4x^2+20x+25x)}{x(4x^2-25)}$
- The numerator contains a square trinomial, while the denominator has the difference of squares.
- $\frac{x^2(4x^2+20x+25x)}{x(4x^2-25)}=\frac{x^2(2x+5)^2}{x(2x+5)(2x-5)}$
- Remove common factors: $\frac{x^2(2x+5)^2}{x(2x+5)(2x-5)}=\frac{x(2x+5)}{2x-5}$

Let's try one more of before we get to the main event.

EXAMPLE

▶ Simplify: $\dfrac{4x^2+5x-6}{5x^2+4x-12}$

▶ Factor and reduce common factors:

$$\frac{4x^2+5x-6}{5x^2+4x-12}=\frac{(4x-3)\cancel{(x+2)}}{(5x-6)\cancel{(x+2)}}=\frac{4x-3}{5x-6}$$

Multiplying and Dividing Rational Expressions

As we said with regard to reducing rational expressions, the process for multiply and dividing rational algebraic expressions is the same as for multiplying and dividing fractions. With multiplication, we factor all numerators and denominators and then reduce when we can find a common factor in the numerator with one in the denominator. With division, we apply "invert and multiply." That is, we change the division problem to a multiplication problem and take the reciprocal of the divisor. We then follow the rules for multiplication.

EXAMPLE

▶ Simplify: $\dfrac{x^2+9x+20}{x^2+x-20}\times\dfrac{x^2-11x+28}{x^2+10x+24}$

▶ Factor all terms:

$$\frac{x^2+9x+20}{x^2+x-20}\times\frac{x^2-11x+28}{x^2+10x+24}=\frac{(x+5)(x+4)}{(x+5)(x-4)}\times\frac{(x-4)(x-7)}{(x+4)(x+6)}$$

▶ Reduce common factors: $\dfrac{\cancel{(x+5)}\cancel{(x+4)}}{\cancel{(x+5)}\cancel{(x-4)}}\times\dfrac{\cancel{(x-4)}(x-7)}{\cancel{(x+4)}(x+6)}=\dfrac{x-7}{x+6}$

The keys are (1) factor first, reduce second, and (2) when factoring, look around to see if the polynomial you are trying to factor might have a factor that you have already seen.

EXAMPLE

▶ Simplify: $\frac{4y^2+17y-15}{16y^2-9} \div \frac{2y^2+5y-12}{4y^2-13y-12}$

▶ Invert and multiply:

$$\frac{4y^2+17y-15}{16y^2-9} \div \frac{2y^2+5y-12}{4y^2-13y-12} = \frac{4y^2+17y-15}{16y^2-9} \times \frac{4y^2-13y-12}{2y^2+5y-12}$$

▶ The first denominator is the difference of squares:

$$16y^2-9=(4y+3)(4y-3).$$

▶ Both numerators start with $4y^2$. Could it be (not that it *has* to be) that these terms are factors in the numerators also? The first numerator will have one addition and one subtraction in the factors, so let's see if $4y+3$ is a factor. The other factor would have to be $y-5$ (so that the first term is $4y^2$ and the constant is 15). Is the middle term $17y$? $(4y+3)(y-5)=4y^2-20y+3y-15=4y^2-17y-15$. No, the middle term is $-17y$. Switch the signs in the factors: $(4y-3)(y+5)=4y^2+17y-15$.

▶ We now have $\frac{4y^2+17y-15}{16y^2-9} \times \frac{4y^2-13y-12}{2y^2+5y-12} =$

$\frac{(4y-3)(y+5)}{(4y-3)(4y+3)} \times \frac{4y^2-13y-12}{2y^2+5y-12}$. Is $4y+3$ a factor of $4y^2-13y-12$?

The other factor would have to be $y+4$ to get the first and third terms of the trinomial. Yes, indeed, $(4y+3)(y-4)=4y^2-13y-12$.

- We know that $y + 5$ is not a factor of $2y^2+5y-12$ because 5 is not a factor of 12. Is $y + 4$ a factor of $2y^2+5y-12$? If so, the other factor needs to be $2y - 3$ (again, in order to get the first and third terms). Yes, $(2y-3)(y+4)=2y^2+5y-12$.
- We're ready to cancel: $\frac{\cancel{(4y-3)}(y+5)}{\cancel{(4y-3)}\cancel{(4y+3)}}\times\frac{\cancel{(4y+3)}\cancel{(y+4)}}{(2y-3)\cancel{(y+4)}}=\frac{y+5}{2y-3}$

There is one result that you need to pay attention to when reducing fractions. So long as $a-b\neq 0$, then $\frac{a-b}{b-a}=-1$. Try some different values for a and b so that you feel comfortable with this.

EXAMPLE

- Simplify: $\frac{x^2-7x+12}{16-x^2}$
- Factor:

$$\frac{x^2-7x+12}{16-x^2}=\frac{(x-4)(x-3)}{(4+x)(4-x)}=\frac{\cancel{(x-4)}^{-1}(x-3)}{(4+x)\cancel{(4-x)}}=\frac{-1(x-3)}{4+x}=\frac{3-x}{4+x}$$

Adding and Subtracting Rational Expressions

It should come as no surprise to you that you need to have a common denominator in order to add and subtract rational algebraic expressions. I believe that it is easier to find the common denominator with algebraic expressions than it is with constants because it is usually the case the factors of the denominators are easier to determine. Yes, the common denominator for $\frac{2}{3}+\frac{3}{4}$ is easy to determine, but you'll no doubt agree that the common denominator for $\frac{7}{165}+\frac{3}{715}$ is not all that obvious.

EXAMPLE

▶ Add $\frac{2}{x+3}+\frac{3}{x+4}$.

▶ The common denominator for the fraction is the product of the denominators. Therefore,

$$\frac{2}{x+3}+\frac{3}{x+4}=\frac{2}{x+3}\left(\frac{x+4}{x+4}\right)+\frac{3}{x+4}\left(\frac{x+3}{x+3}\right)$$

$$=\frac{2(x+4)+3(x+3)}{(x+3)(x+4)}$$

$$=\frac{2x+8+3x+9}{(x+3)(x+4)}=\frac{5x+17}{(x+3)(x+4)}.$$

BTW

There is a HUGE temptation to cancel the $x+3$ and $x+4$ in the expression

$$\frac{2(x+4)+3(x+3)}{(x+3)(x+4)}.$$

Don't do it! *Neither $x+3$ nor $x+4$ is a factor of the numerator, and you can only cancel common factors.*

Let's look at a few more examples. The first of these has a constant as a common factor.

EXAMPLE

▶ Combine: $\frac{3x}{4x+8}+\frac{2x}{3x+6}-\frac{5x}{12x+24}$

▶ Factor each denominator:

$$\frac{3x}{4x+8}+\frac{2x}{3x+6}-\frac{5x}{12x+24}=\frac{3x}{4(x+2)}+\frac{2x}{3(x+2)}-\frac{5x}{12(x+2)}$$

▶ The common denominator for 3, 4, and 12 is 12, so get equivalent fractions for these terms.

$$\frac{3x}{4(x+2)}+\frac{2x}{3(x+2)}-\frac{5x}{12(x+2)}=\frac{3x}{4(x+2)}\left(\frac{3}{3}\right)+\frac{2x}{3(x+2)}\left(\frac{4}{4}\right)-\frac{5x}{12(x+2)}$$

$$=\frac{9x+8x-5x}{12(x+2)}=\frac{12x}{12(x+2)}=\frac{x}{x+2}.$$

The next example has a variable expression as a common factor for the denominator.

EXAMPLE

▶ Combine: $\dfrac{x+1}{x^2+7x+12}-\dfrac{x+2}{x^2-16}$

▶ Factor the denominators:

$$\frac{x+1}{x^2+7x+12}-\frac{x+2}{x^2-16}=\frac{x+1}{(x+3)(x+4)}-\frac{x+2}{(x+4)(x-4)}$$

▶ The product $(x+3)(x+4)(x-4)$ is the common denominator. Get equivalent fractions.

$$\frac{x+1}{(x+3)(x+4)}-\frac{x+2}{(x+4)(x-4)}$$

$$=\frac{x+1}{(x+3)(x+4)}\left(\frac{x-4}{x-4}\right)-\frac{x+2}{(x+4)(x-4)}\left(\frac{x+3}{x+3}\right)$$

$$=\frac{(x+1)(x-4)-(x+2)(x+3)}{(x+3)(x+4)(x-4)}=\frac{(x^2-3x-4)-(x^2+5x+6)}{(x+3)(x+4)(x-4)}$$

$$=\frac{x^2-3x-4-x^2-5x-6}{(x+3)(x+4)(x-4)}=\frac{-8x-10}{(x+3)(x+4)(x-4)}$$

$$=\frac{-2(4x+5)}{(x+3)(x+4)(x-4)}.$$

As you can see, there are plenty of steps that can trip you up if you are not careful. The most critical of these is the transition from $\frac{(x+1)(x-4)-(x+2)(x+3)}{(x+3)(x+4)(x-4)}$ to $\frac{x^2-3x-4-x^2-5x-6}{(x+3)(x+4)(x-4)}$. The intermediate step of doing the multiplication in the numerator and leaving the results in parentheses, $\frac{(x^2-3x-4)-(x^2+5x+6)}{(x+3)(x+4)(x-4)}$, lets us deal with one issue at a time. We can worry about the subtraction separately from the multiplication. Take your time when performing these complex problems by doing one thing at a time. As you get more proficient and experienced, you'll be able to work through these problems more efficiently.

Mixed Expressions and Complex Fractions

Read the number $5\frac{2}{3}$. Whether you said it aloud or in your head, you said "five and two-thirds." What does the "and" mean? Where is this number on the number line? Well, you go to the place marked 5 and then go two-thirds of the way toward the place marked 6. So, what we really mean when we say five and two-thirds is five plus two-thirds. Why am I making a big deal out of this? We have no way of writing a mixed number like this involving algebraic expressions unless we use addition. If I wrote $5\frac{2}{x}$ you should think it means to multiply five and $\frac{2}{x}$. That is consistent with our convention that a constant next to a variable indicates a multiplication problem.

EXAMPLE

- Write $5+\frac{2}{x}$ as an improper fraction.
- Get a common denominator: $5+\frac{2}{x}=5\left(\frac{x}{x}\right)+\frac{2}{x}=\frac{5x+2}{x}$.

This gets to be more interesting when there is another variable expression involved in the problem.

EXAMPLE

- Write $x+4+\dfrac{6}{x-1}$ as an improper fraction.
- Get the common denominator: $(x+4)\left(\dfrac{x-1}{x-1}\right)+\dfrac{6}{x-1}=\dfrac{x^2+3x-4+6}{x-1}$

$$=\frac{x^2+3x+2}{x-1}=\frac{(x+2)(x+1)}{x-1}.$$

Another interesting situation comes from division. If $2\div 3$ is the same as $\dfrac{2}{3}$, does it not make sense that $\dfrac{2}{3}\div\dfrac{5}{7}$ is the same as $\dfrac{\frac{2}{3}}{\frac{5}{7}}$? Yes, it does seem a bit absurd to write $\dfrac{2}{3}\div\dfrac{5}{7}$ as $\dfrac{\frac{2}{3}}{\frac{5}{7}}$. However, it does help us to better understand how to work with the expression $\dfrac{\frac{2}{3}+\frac{5}{12}}{\frac{3}{4}-\frac{1}{6}}$. Such an expression is called a complex fraction. It is complex not because it is difficult but because the numerator and/or the denominator of the fraction is itself a fraction. How do we evaluate $\dfrac{\frac{2}{3}+\frac{5}{12}}{\frac{3}{4}-\frac{1}{6}}$? We could rewrite the problem as a division problem. That is, $\dfrac{\frac{2}{3}+\frac{5}{12}}{\frac{3}{4}-\frac{1}{6}}=\left(\dfrac{2}{3}+\dfrac{5}{12}\right)\div\left(\dfrac{3}{4}-\dfrac{1}{6}\right)$. We need to do the addition and subtraction within the parentheses next.

$$\left(\frac{2}{3}+\frac{5}{12}\right)\div\left(\frac{3}{4}-\frac{1}{6}\right)=\left(\frac{8}{12}+\frac{5}{12}\right)\div\left(\frac{9}{12}-\frac{2}{12}\right)=\frac{13}{12}\div\frac{7}{12}=\frac{13}{12}\times\frac{12}{7}=\frac{13}{7}.$$

That's not too bad. However, you might find the next approach easier. The common denominator for the four component fractions in the term $\dfrac{\frac{2}{3}+\frac{5}{12}}{\frac{3}{4}-\frac{1}{6}}$ is 12. Multiplying the numerator and denominator by 12,

$\dfrac{\left(\frac{2}{3}+\frac{5}{12}\right)}{\left(\frac{3}{4}-\frac{1}{6}\right)}\dfrac{12}{12}=\dfrac{8+5}{9-2}=\dfrac{13}{7}$, requires much less work.

EXAMPLE

▶ Simplify: $\dfrac{\frac{2}{x-2}+\frac{3}{x+2}}{1-\frac{5}{x^2-4}}$

▶ The common denominator for the four component fractions is $(x-2)(x+2)$. Multiply the numerator and denominator by this expression.

$$\frac{\frac{2}{x-2}+\frac{3}{x+2}}{1-\frac{5}{x^2-4}}=\frac{\left(\frac{2}{x-2}+\frac{3}{x+2}\right)}{\left(1-\frac{5}{x^2-4}\right)}\frac{(x-2)(x+2)}{(x-2)(x+2)}$$

$$=\frac{\left(\frac{2}{x-2}\right)(x-2)(x+2)+\left(\frac{3}{x+2}\right)(x-2)(x+2)}{1(x-2)(x+2)-\left(\frac{5}{x^2-4}\right)(x-2)(x+2)}$$

$$=\frac{2(x+2)+3(x-2)}{x^2-4-5}=\frac{2x+4+3x-6}{x^2-9}=\frac{5x-2}{(x+3)(x-3)}$$

While that might appear to be challenging, it really isn't. What did we do?

- We looked at the component fractions and saw that, with the exception of the constant, we had at least one of the factors of the difference of squares expression x^2-4 in the denominator.
- We multiplied the numerator and denominator of the original fraction by the product $(x+2)(x-2)$.
- We distributed this product through the numerator and through the denominator.
- We multiplied the terms, canceling factors as we could.
- We combined the terms in the numerator and in the denominator.

Give this problem a try.

EXAMPLE

▶ Simplify: $\dfrac{\dfrac{2x}{x+3}-\dfrac{x}{x-3}}{\dfrac{x}{x+3}+\dfrac{3x}{x^2-9}}$

▶ The common denominator for the four component fractions is $(x+3)(x-3)$.

▶ Multiply the numerator and denominator by this expression:

$$\frac{\left(\dfrac{2x}{x+3}-\dfrac{x}{x-3}\right)}{\left(\dfrac{x}{x+3}+\dfrac{3x}{x^2-9}\right)}\frac{(x+3)(x-3)}{(x+3)(x-3)}$$

▶ Distribute this product through the numerator and denominator:

$$\frac{\left(\frac{2x}{x+3}\right)(x+3)(x-3)-\left(\frac{x}{x-3}\right)(x+3)(x-3)}{\left(\frac{x}{x+3}\right)(x+3)(x-3)+\left(\frac{3x}{x^2-9}\right)(x+3)(x-3)}$$

- Multiply and cancel terms:

$$\frac{\left(\frac{2x}{x+3}\right)(x+3)(x-3)-\left(\frac{x}{x-3}\right)(x+3)(x-3)}{\left(\frac{x}{x+3}\right)(x+3)(x-3)+\left(\frac{3x}{x^2-9}\right)(x+3)(x-3)}=\frac{2x(x-3)-x(x+3)}{x(x-3)+3x}$$

$$=\frac{2x^2-6x-x^2-3x}{x^2-3x+3x}=\frac{x^2-9x}{x^2}=\frac{x(x-9)}{x^2}=\frac{x-9}{x}$$

Rational Equations

We treat equations containing rational algebraic expressions the same way we treat linear equations with rational coefficients—we'll multiply both sides of the equation by the common denominator to remove the fractions from the problem. However, before we do that, we need to be sure to identify those values of the variable for which the expression is undefined.

EXAMPLE

- Solve: $\frac{12}{2x+1}+\frac{1}{3}=\frac{17}{2x+1}$
- We identify that $x\neq\frac{-1}{2}$ because it would make the denominator of the fraction equal to zero.
- We can subtract $\frac{12}{2x+1}$ from both sides of the equation: $\frac{1}{3}=\frac{5}{2x+1}$

- Cross-multiply (multiply by the common denominator): $2x+1=15$
- Solve to determine that $x=7$.
- Check this result: $\frac{12}{2(7)+1}+\frac{1}{3}=\frac{17}{2(7)+1} \Rightarrow \frac{12}{15}+\frac{5}{15}=\frac{17}{15}$. This is correct.
- Therefore, $x=7$.

The process gets a little more involved when there is more than one algebraic expression in the problem. Here is an example in which the numerator contains constants but the denominators consist of different algebraic expressions.

EXAMPLE

- Solve: $\frac{5}{x+6}+\frac{1}{2}=\frac{1}{x-3}$
- We identify that $x \neq -6,3$ because they would make two of the fraction undefined. The common denominator for these three fractions is $2(x+6)(x-3)$.
- Multiply by the common denominator:
$\left(\frac{5}{x+6}+\frac{1}{2}\right)2(x+6)(x-3)=\left(\frac{1}{x-3}\right)2(x+6)(x-3)$
- Distribute: $\left(\frac{5}{x+6}\right)2(x+6)(x-3)+\left(\frac{1}{2}\right)2(x+6)(x-3)=2(x+6)$
- Cancel common factors: $10(x-3)+(x+6)(x-3)=2(x+6)$
- Expand: $10x-30+x^2+3x-18=2x+12$
- Combine terms: $x^2+13x-48=2x+12$

- Bring all terms to one side of the equation: $x^2+11x-60=0$
- Factor: $(x+15)(x-4)=0$
- Solve: $x=-15,4$
- Neither of these values is restricted from the domain, so they seem like good solutions. Seem? Check to make sure they solve the problem.
- Check $x=4$: $\frac{5}{4+6}+\frac{1}{2}=\frac{1}{4-3}\Rightarrow\frac{1}{2}+\frac{1}{2}=\frac{1}{1}$. That is correct.
- Check: $x=-15$:

$$\frac{5}{-15+6}+\frac{1}{2}=\frac{1}{-15-3}\Rightarrow\frac{-5}{9}+\frac{1}{2}=\frac{1}{-18}\Rightarrow\frac{-10}{18}+\frac{9}{18}=\frac{-1}{18}.$$

That is also correct.

- Therefore, $x=-15,4$.

This next problem also contains an algebraic expression.

EXAMPLE

- Solve: $\frac{x-3}{x-2}-\frac{5}{x+3}=\frac{x-4}{x^2+x-6}$.
- Factor the last denominator before determining the restrictions on the variable:

$$\frac{x-3}{x-2}-\frac{5}{x+3}=\frac{x-4}{(x-2)(x+3)}$$

- We now know that $x\neq-3,2$.
- Multiply by the common denominator:

$$\left(\frac{x-3}{x-2}-\frac{5}{x+3}\right)(x-2)(x+3)=\left(\frac{x-4}{(x-2)(x+3)}\right)(x-2)(x+3)$$

- Distribute: $\left(\frac{x-3}{x-2}\right)(x-2)(x+3)-\left(\frac{5}{x+3}\right)(x-2)(x+3)=x-4$
- Cancel common factors: $(x-3)(x+3)-5(x-2)=x-4$
- Expand: $x^2-9-5x+10=x-4$
- Bring all terms to one side of the equation: $x^2-6x+5=0$
- Factor: $(x-1)(x-5)=0$
- Solve: $x=1,5$
- Neither of these values is restricted from the domain, so they seem like good solutions.
- Check $x=1$: $\frac{1-3}{1-2}-\frac{5}{1+3}=\frac{1-4}{1^2+1-6}\Rightarrow\frac{-2}{-1}-\frac{5}{4}=\frac{-3}{-4}\Rightarrow 2-\frac{5}{4}=\frac{3}{4}$.

  That is correct.
- Check: $x=5$: $\frac{5-3}{5-2}-\frac{5}{5+3}=\frac{5-4}{5^2+5-6}\Rightarrow\frac{2}{3}-\frac{5}{8}=\frac{1}{24}\Rightarrow\frac{16}{24}-\frac{15}{24}=\frac{1}{24}$.

  That is also correct.
- Therefore, $x=1,5$.

This seemingly straightforward equation has a trick that you must be aware of.

EXAMPLE

- Solve: $\frac{2}{x+6}+\frac{2}{x-6}=\frac{24}{x^2-36}$
- We know that x^2-36 factors to $(x+6)(x-6)$, so we now know that the restrictions for this equation is that $x\neq\pm6$.

- Multiply by the common denominator:

$$\left(\frac{2}{x+6}+\frac{2}{x-6}\right)(x+6)(x-6)=\left(\frac{24}{x^2-36}\right)(x+6)(x-6)$$

- Distribute: $\left(\frac{2}{x+6}\right)(x+6)(x-6)+\left(\frac{2}{x-6}\right)(x+6)(x-6)=24$
- Cancel common factors: $2(x-6)+2(x+6)=24$
- Expand: $2x-12+2x+12=24$
- Combine terms: $4x=24$
- Solve: $x=6$
- This cannot be so, because $x=6$ was a restriction to this equation. Therefore, this equation has no solution.

Dividing Polynomials

Dividing polynomials is performed the same way that long division is done with constants. *Be warned:* the most difficult thing about this process is subtraction. "That doesn't seem to hard," you say. I taught math at both the high school and the college level for more than 30 years and the number one mistake (that is, the mistake made most often) was in subtracting algebraic expressions. The distributive property can rear its ugly head at the darnedest times.

EXAMPLE

- Divide $4x^2-11x+6$ by $x-2$.
- We set the problem up as a long division problem: $x-2\overline{\smash{)}4x^2-11x+6}$

- Getting started: We deal with the first term in the dividend (under the division sign) and the first term in the divisor. What is $\frac{4x^2}{x}$? The answer is $4x$.
- Write the quotient above the line as always, aligning the result with the terms in the dividend. That is, write the $4x$ over the $11x$.

$$\begin{array}{r} 4x \\ x-2\overline{)4x^2-11x+6} \end{array}$$

- Multiply $4x$ and $x - 2$.

$$\begin{array}{r} 4x \\ x-2\overline{)4x^2-11x+6} \\ \underline{4x^2-8x} \end{array}$$

- Subtract. Here we go. $4x^2 - 4x^2 = 0$, but what is $-11x - (-8x)$? Careful thought tells us the answer is $-3x$. Bring down the 6.

$$\begin{array}{r} 4x \\ x-2\overline{)4x^2-11x+6} \\ \underline{4x^2-8x} \\ -3x+6 \end{array}$$

- We go back to do division. What is $\frac{-3x}{x}$? The answer is, of course, -3. Write this above the 6, multiply by $x - 2$, and subtract.

$$\begin{array}{r} 4x-3 \\ x-2\overline{)4x^2-11x+6} \\ \underline{4x^2-8x} \\ -3x+6 \\ \underline{-3x+6} \end{array}$$

- Subtraction gives zero as a response. So, we have $\frac{4x^2-11x+6}{x-2} = 4x-3$.

That didn't seem to be too bad. Let's try another.

EXAMPLE

- Divide $6x^3 - 11x^2 - 47x - 20$ by $2x + 1$.
- Set the problem up: $2x+1\overline{)6x^3 - 11x^2 - 47x - 20}$
- First terms: $\dfrac{6x^3}{2x} = 3x^2$
- Record the quotient, multiply by $2x + 1$:

$$\begin{array}{r} 3x^2 \\ 2x+1\overline{)6x^3 - 11x^2 - 47x - 20} \\ \underline{6x^3 + 3x^2} \end{array}$$

- Subtract (you've been warned):

$$\begin{array}{r} 3x^2 \\ 2x+1\overline{)6x^3 - 11x^2 - 47x - 20} \\ \underline{6x^3 + 3x^2} \\ -14x^2 - 47x - 20 \end{array}$$

- That's right, $-11 - 3 = -14$, not -8.
- First terms: $\dfrac{-14x^2}{2x} = -7x$.
- Record the quotient, multiply by $2x + 1$:

$$\begin{array}{r} 3x^2 - 7x \\ 2x+1\overline{)6x^3 - 11x^2 - 47x - 20} \\ \underline{6x^3 + 3x^2} \\ -14x^2 - 47x - 20 \\ \underline{-14x^2 - 7x} \end{array}$$

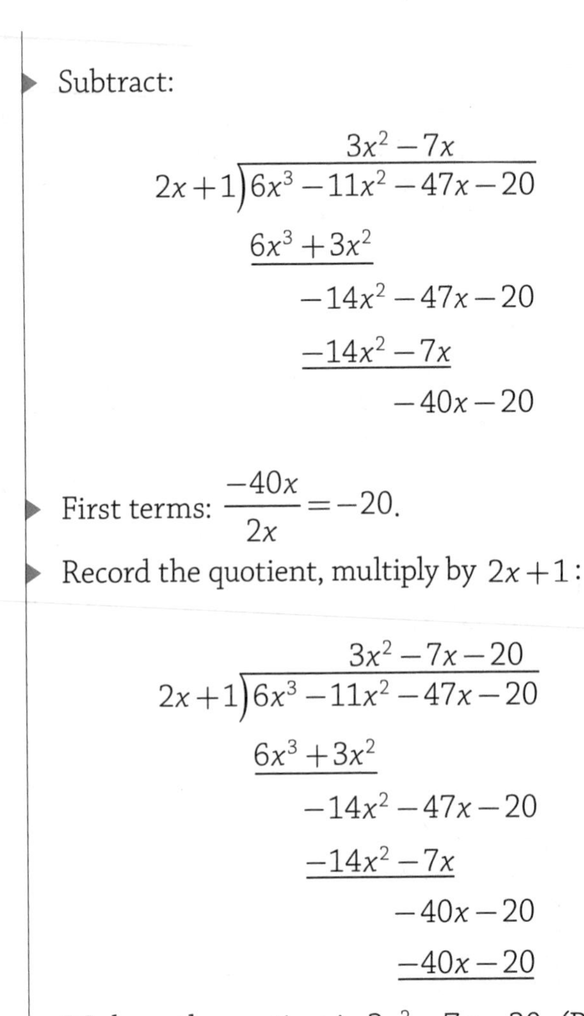

- Subtract:

$$\begin{array}{r} 3x^2-7x \\ 2x+1\overline{)6x^3-11x^2-47x-20} \\ \underline{6x^3+3x^2}\qquad\qquad\quad \\ -14x^2-47x-20 \\ \underline{-14x^2-7x}\qquad \\ -40x-20 \end{array}$$

- First terms: $\dfrac{-40x}{2x}=-20$.
- Record the quotient, multiply by $2x+1$:

$$\begin{array}{r} 3x^2-7x-20 \\ 2x+1\overline{)6x^3-11x^2-47x-20} \\ \underline{6x^3+3x^2}\qquad\qquad\quad \\ -14x^2-47x-20 \\ \underline{-14x^2-7x}\qquad \\ -40x-20 \\ \underline{-40x-20} \end{array}$$

- We have the quotient is $3x^2-7x-20$. (By the way, this quadratic factors to $(x-4)(3x+5)$, so $6x^3 - 11x^2 - 47x - 20 = (2x + 1)(x - 4)(3x + 5)$.)

This final example covers two important features. We'll discuss these when we get to the end of the problem.

EXAMPLE

- Divide $2x^3 - 8x + 7$ by $x + 2$.
- Before we can set up the problem, we need to realize that there is no term with x^2 in the dividend. We will introduce a quadratic term with a coefficient of 0 so that we can align terms in a column.
- Set up:

$$x+2\overline{)2x^3+0x^2-8x+7}$$

- First terms: $\dfrac{2x^3}{x} = 2x^2$
- Record the quotient, multiply by $x + 2$:

$$\begin{array}{r}
2x^2 \\
x+2\overline{)2x^3+0x^2-8x+7} \\
\underline{2x^3+4x^2}
\end{array}$$

- Subtract and bring down the rest of the terms.

$$\begin{array}{r}
2x^2 \\
x+2\overline{)2x^3+0x^2-8x+7} \\
\underline{2x^3+4x^2} \\
-4x^2-8x+7
\end{array}$$

- First terms: $\dfrac{-4x^2}{x} = -4x$
- Record the quotient, multiply by $x + 2$:

$$\begin{array}{r}
2x^2-4x \\
x+2\overline{)2x^3+0x^2-8x+7} \\
\underline{2x^3+4x^2} \\
-4x^2-8x+7 \\
\underline{-4x^2-8x}
\end{array}$$

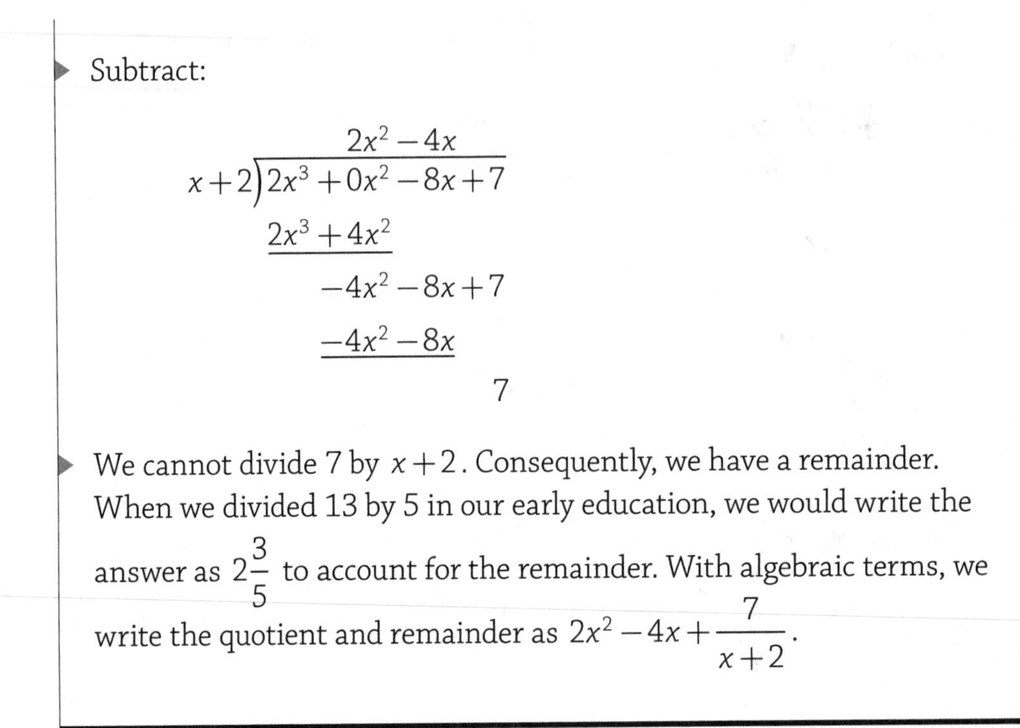

Subtract:

$$
\begin{array}{r}
2x^2 - 4x \\
x+2 \overline{)\, 2x^3 + 0x^2 - 8x + 7} \\
\underline{2x^3 + 4x^2} \\
-4x^2 - 8x + 7 \\
\underline{-4x^2 - 8x} \\
7
\end{array}
$$

We cannot divide 7 by $x+2$. Consequently, we have a remainder. When we divided 13 by 5 in our early education, we would write the answer as $2\frac{3}{5}$ to account for the remainder. With algebraic terms, we write the quotient and remainder as $2x^2 - 4x + \frac{7}{x+2}$.

The two items in this problem were the missing term and the remainder. Now you know how to deal with them.

EXERCISES

EXERCISE 11-1

Find the domain for each function.

1. $f(x) = \dfrac{3x+2}{x^2-4x}$

2. $g(x) = \dfrac{6x-7}{x^2-3x-18}$

3. $p(x) = \dfrac{4x^2-25}{x^2+4}$

EXERCISE 11-2

Answer the following questions.

1. The frequency with which a guitar string vibrates varies inversely with the length of the string. If a string that is 0.3 m long vibrates at 5.2 times per second, how long is the guitar string that vibrates 3.9 times per second?

2. The volume of a gas will vary inversely with the pressure applied to the gas provided that temperature is held constant. Given a constant temperature, if a gas has a volume of 100 ml when the pressure is 5 Pascals, under what pressure will the gas have a volume of 120 ml?

3. The intensity of light produced by a light source varies inversely as the square of the distance from the source. If the intensity of light produced 4 feet from a light source is 1200 foot-candles, find the intensity of light produced 10 feet from the same source.

EXERCISE 11-3

Use $f(x) = \frac{5x-6}{2x+3}$ to answer these questions.

1. What is the equation of the vertical asymptote for the graph of $f(x)$?
2. Determine the value of the x-intercept for the graph of $f(x)$.

EXERCISE 11-4

Use $g(x) = \frac{x^2-5x-6}{2x^2+3x-5}$ to answer these questions.

1. What are the equations of the vertical asymptotes for the graph of $g(x)$?
2. Determine the zeros of $g(x)$.

EXERCISE 11-5

Completely simplify each of the following expressions.

1. $\frac{4x-x^2}{x^2-16}$
2. $\frac{2x^2-x-1}{x^2-6x+5}$
3. $\frac{4x-12}{6x+18} \times \frac{9x^2-81}{x^2-4x+3}$
4. $\frac{x^2+10x+25}{x^2+7x+10} \times \frac{x^2-4x-12}{2x^2+11x+5}$
5. $\frac{4x^2-1}{2x^2+x-1} \div \frac{2x^2+3x+1}{x^2+5x+4}$

6. $\dfrac{x^2-3x-28}{2x^2-2x-40} \div \dfrac{6x^2-66x+168}{18x-72}$

7. $\dfrac{\frac{2}{3}+\frac{5}{6}}{\frac{11}{12}-\frac{3}{4}}$

8. $\dfrac{\frac{x}{y}-\frac{y}{x}}{\frac{1}{y}-\frac{1}{x}}$

9. $\dfrac{\frac{2}{x+1}-\frac{3}{x-1}}{1-\frac{24}{x^2-1}}$

10. $\dfrac{\frac{x+1}{x-2}-\frac{x-1}{x+2}}{1+\frac{4}{x^2-4}}$

EXERCISE 11-6

Simplify the sums and differences in the following questions.

1. $\dfrac{5}{3x-15}+\dfrac{7}{2x-10}$

2. $\dfrac{8}{x-4}-\dfrac{4}{x+3}$

3. $\dfrac{x+1}{x-1}-\dfrac{x-4}{x+4}$

4. $\dfrac{x+5}{2x-2}-\dfrac{x-3}{2x+8}$

5. $\dfrac{2x+5}{x-3}-\dfrac{3x-4}{x-4}$

6. $\dfrac{x+1}{x^2-2x}+\dfrac{x-4}{x^2+5x}$

7. $\dfrac{7}{x^2-x-2}+\dfrac{5}{x^2-5x+6}$

8. $\dfrac{x+7}{x^2-x-2}-\dfrac{x-5}{x^2-5x+6}$

9. $\dfrac{2x+1}{x^2-10x+25}-\dfrac{x+4}{x^2-25}$

10. $\dfrac{2x+5}{x^2-10x}-\dfrac{x+1}{x^2-20x+100}$

EXERCISE 11-7

Solve these equations.

1. $\dfrac{4}{x+10}+\dfrac{10}{x+6}=\dfrac{2}{3}$

2. $\dfrac{3}{p-4}-\dfrac{1}{2}=\dfrac{1}{p+1}$

3. $\dfrac{2}{g-6}-\dfrac{5}{g+4}=\dfrac{1}{g+4}$

4. $\dfrac{m+8}{m+10}+\dfrac{m+6}{m+12}=\dfrac{2}{15}$

5. $\dfrac{n}{n+5}-\dfrac{n-10}{n-5}=\dfrac{1}{4}$

6. $\dfrac{r-2}{r}+\dfrac{r-4}{r+2}=\dfrac{13}{r}$

EXERCISE 11-8

Determine the quotient and remainder, if there is one, in these questions.

1. $\dfrac{3x^2+13x-30}{x+6}$
2. $\dfrac{4x^3+8x^2-9x-18}{2x-3}$
3. $\dfrac{x^4+4x^3-8x-32}{x-2}$
4. $\dfrac{x^3-5x^2+4x-19}{x-5}$

Radical Functions and Geometry

MUST KNOW

- The Pythagorean theorem illustrates the numerical relationship among the lengths of the sides of right triangles.
- The Pythagorean theorem is used by carpenters, plumbers, and mathematicians. It is still as relevant today as it was 2,500 years ago.
- Right triangle trigonometry allows us to see the relationship between angles and the lengths of sides of polygons.

he set of **rational numbers** are defined as the numbers that can be written in the form $\frac{a}{b}$, where both a and b are integers and b does not equal 0. Those numbers that cannot be written in this form are called **irrational**. Perhaps the most famous of the irrational numbers is π, the ratio of the circumference of a circle to the length of the circle's diameter. We'll study how to simplify irrational expressions, do arithmetic with them, and solve equations involving them. Finally, we'll look at two major applications of the irrational numbers, the Pythagorean theorem and right triangle trigonometry.

BTW

The ancient Greek philosopher and mathematician Pythagoras is known for the theorem regarding the relationship among the lengths of the sides of a right triangle, a relationship that was known to both the Chinese and Egyptians long before Pythagoras was born (around 569 BCE). He believed that there was a relationship between numbers and nature and that things like earth, water, wind, and fire had close ties with certain natural numbers. There is story that says when a disciple of his group showed him that the hypotenuse of a right triangle did not have a length that was a natural number, Pythagoras had the man killed. Now that's irrational!

Simplifying Radical Expressions

The most important numerical relationship among the irrational numbers is $\sqrt{ab} = \sqrt{a}\sqrt{b}$. A consequence of this rule is that $\sqrt{\frac{a}{b}} = \frac{\sqrt{a}}{\sqrt{b}}$. The terminology associated with irrational expressions is that is called the *radical* and the term within the radical is called the *radicand*.

We can apply these relationships to implement the two basic rules for irrational expressions:

- Do not leave perfect powers within the radical.
- Do not leave fractions within the radical nor have a radical in the denominator of a fraction.

EXAMPLE

▶ Simplify: $\sqrt{12}$

▶ To perform the action of simplifying this radical, we are asked to remove from the radical any factors of 12 that are perfect squares. Given that $12 = 4 \times 3$, we rewrite $\sqrt{12}$ as $\sqrt{4\times3} = \sqrt{4}\sqrt{3} = 2\sqrt{3}$.

The initial reason for the development of this topic was to get reasonable decimal approximations for irrational numbers in the days prior to electronic calculating devices. Painstaking work was done to find decimal approximations for many irrational numbers. However, it was understood that the approximations did not need to be done for all numbers. If we know that $\sqrt{3}$ is approximately equal to 1.732, then $\sqrt{12}$ is approximately $2(1.732) = 3.464$. "Well then," you argue, "why are we still studying this, since we have calculators that can give decimal approximations for any irrational number we need?" That would seem like a fair question. The reason is that we are studying algebra and not doing arithmetic. Yes, we begin the conversation doing arithmetic, but we will soon move the discussion to algebra. So, I ask for your indulgence while we work through the preliminary material.

EXAMPLE

▶ Simplify: $\sqrt{75}$

▶ The number 75 factors to be 25×3, so $\sqrt{75} = \sqrt{25}\sqrt{3} = 5\sqrt{3}$.

We now can look at the other two operations from arithmetic. Observe:

$$\sqrt{75} + \sqrt{12} = 5\sqrt{3} + 2\sqrt{3} = 7\sqrt{3} \text{ and } \sqrt{75} - \sqrt{12} = 5\sqrt{3} - 2\sqrt{3} = 3\sqrt{3}.$$

But $7\sqrt{3} = \sqrt{49}\sqrt{3} = \sqrt{147}$ and $3\sqrt{3} = \sqrt{9}\sqrt{3} = \sqrt{27}$, so $\sqrt{75} + \sqrt{12} \neq \sqrt{87}$ and $\sqrt{75} - \sqrt{12} \neq \sqrt{63}$.

Consequently, we now know

$$\sqrt{a}+\sqrt{b} \neq \sqrt{a+b} \text{ and } \sqrt{a}-\sqrt{b} \neq \sqrt{a-b}.$$

EXAMPLE

▶ Simplify: $\dfrac{3\sqrt{72}}{2\sqrt{32}}$

▶ We simplify each of the irrational numbers to be

$$\frac{3\sqrt{72}}{2\sqrt{32}} = \frac{3\sqrt{36}\sqrt{2}}{2\sqrt{16}\sqrt{2}} = \frac{3(6\sqrt{2})}{2(4\sqrt{2})} = \frac{9}{4}.$$

Let's look at part of that again. We reduced $\sqrt{32}$ to be $\sqrt{16}\sqrt{2}$. What if you did not see it that way? What if you saw $\sqrt{32}=\sqrt{4}\sqrt{8}$? That's fine so long as you realize that 8 has 4 as a square factor and you continue to reduce the radical. That is, $\sqrt{32}=\sqrt{4}\sqrt{8}=2\sqrt{4}\sqrt{2}=2(2\sqrt{2})=4\sqrt{2}$. You don't have to get to this result in the shortest number of steps. You just need to get there eventually.

EXAMPLE

▶ Simplify: $\dfrac{8\sqrt{5}}{3\sqrt{2}}$

▶ We need to take advantage of the fact that $(\sqrt{2})^2=2$ to "rationalize the denominator" of this fraction. In English, we need to make sure that the denominator of the fraction has no irrational numbers in it. To do so, we will multiply both the numerator and denominator of the fraction by $\sqrt{2}$.

▶ $\dfrac{8\sqrt{5}}{3\sqrt{2}}\dfrac{\sqrt{2}}{\sqrt{2}} = \dfrac{8\sqrt{10}}{3(2)} = \dfrac{4\sqrt{10}}{3}.$

We will often need to reduce radicals when we solve quadratic equations using the quadratic formula.

EXAMPLE

▶ Solve $x^2 - 16x + 48 = 0$

▶ Use the quadratic formula with $a = 1$, $b = -16$, and $c = 46$.

$$x = \frac{16 \pm \sqrt{(-16)^2 - 4(1)(46)}}{2(1)} = \frac{16 \pm \sqrt{256 - 184}}{2} = \frac{16 \pm \sqrt{72}}{2}$$

$$= \frac{16 \pm \sqrt{36}\sqrt{2}}{2} = \frac{16 \pm 6\sqrt{2}}{2} = \frac{16}{2} \pm \frac{6\sqrt{2}}{2} = 8 \pm 3\sqrt{2}$$

We've spent the entire section examining square roots. As we saw when we were working with exponents, $\sqrt{x} = x^{1/2}$. We also saw terms like $x^{1/3}$ and $x^{1/4}$. These also represent radicals. In general, $x^{1/n} = \sqrt[n]{x}$, the nth root of x. For example, $\sqrt[3]{27} = 3$ because $3^3 = 27$. The rules for simplify all radicals is the same. Look for a factor that is a perfect n^{th} power so that that factor can be simplified.

EXAMPLE

▶ Simplify: $\sqrt[3]{24}$

▶ $\sqrt[3]{24} = \sqrt[3]{8}\sqrt[3]{3} = 2\sqrt[3]{3}$.

The vast majority of problems you will do in Algebra I will involve square roots. You have the basics you need should need to work with a cube root.

Operations with Radical Expressions

Let's make things more interesting. We've examined some basic irrational numbers. Now let's look at some more involved numbers.

EXAMPLE

▶ Simplify: $(12+3\sqrt{50})+(7-5\sqrt{8})$

▶ Begin by simplifying the radical statements:

$$(12+3\sqrt{50})+(7-5\sqrt{8})=(12+3\sqrt{25}\sqrt{2})+(7-5\sqrt{4}\sqrt{2})$$
$$=(12+15\sqrt{2})+(7-10\sqrt{2})$$

▶ Combine like terms:

$$(12+15\sqrt{2})+(7-10\sqrt{2})=19+5\sqrt{2}$$

We can even include variable expressions in the problems if we choose. We need to be careful, though, that if we are working with square roots, the variable in the radicand must represent a non-negative (that is greater than or equal to zero) number.

EXAMPLE

▶ Simplify: $(5a+5b\sqrt{20})-(3a-2b\sqrt{80})$

$$(5a+5b\sqrt{20})-(3a-2b\sqrt{80})=(5a+5b\sqrt{4}\sqrt{5})-(3a-2b\sqrt{16}\sqrt{5})$$
$$=(5a+10b\sqrt{5})-(3a-8b\sqrt{5})=2a+18b\sqrt{5}$$

Do you see how the arithmetic with irrational numbers is just like the arithmetic of monomials and binomials (with the exception of having to first deal with the radical expression)? Multiplication with radicals will also be like the multiplication with binomials. We'll need to use the distributive property, and we'll need to combine like terms after.

EXAMPLE

Simplify:

$(5+3\sqrt{2})(4-2\sqrt{2})$

$$(5+3\sqrt{2})(4-2\sqrt{2}) = (5)(4)+(5)(-2\sqrt{2})+(3\sqrt{2})(4)+(3\sqrt{2})(-2\sqrt{2})$$
$$= 20-10\sqrt{2}+12\sqrt{2}-6(\sqrt{2})^2$$
$$= 20-2\sqrt{2}-6(2) = 20-2\sqrt{2}-12 = -2-2\sqrt{2}$$

Not all problems will have the radical expressions already reduced. In that case, it is best to first simplify the radicals before performing the multiplication.

EXAMPLE

Simplify: $(5+6\sqrt{12})(4-2\sqrt{27})$

The first thing for us to do here is to simplify the radicals $\sqrt{12}$ and $\sqrt{27}$.

$$(5+6\sqrt{12})(4-2\sqrt{27}) = (5+6(2\sqrt{3}))(4-2(3\sqrt{3})) = (5+12\sqrt{3})(4-6\sqrt{3})$$
$$= (5)(4)+(5)(-6\sqrt{3})+(12\sqrt{3})(4)+(12\sqrt{3})(-6\sqrt{3})$$
$$= 20-30\sqrt{3}+48\sqrt{3}-72(\sqrt{3})^2 = 20+18\sqrt{3}-72(3)$$
$$= 20+18\sqrt{3}-216 = -196+18\sqrt{3}$$

The special formulas for square trinomials and the difference of squares will come into play. Use the formula to simplify the expression rather than applying the distributive property.

EXAMPLE

▶ Simplify: $(5+3\sqrt{2})^2$

▶ Use the formula $(a+b)^2 = a^2 + 2ab + b^2$ to simplify this expression.

$$(5+3\sqrt{2})^2 = 25+2(5)(3\sqrt{2})+(3\sqrt{2})^2 = 25+30\sqrt{2}+9(\sqrt{2})^2$$
$$= 25+30\sqrt{2}+9(2) = 43+30\sqrt{2}$$

The expression $(a\sqrt{b})^2$ will always equal a^2b. You can use this when simplifying expressions with numbers that allow for mental arithmetic.

EXAMPLE

▶ Simplify: $(4+2\sqrt{2})(4-2\sqrt{2})$

▶ Use the formula $(a+b)(a-b) = a^2 - b^2$ to answer this problem.

▶ $(4+2\sqrt{2})(4-2\sqrt{2}) = 4^2 - (2\sqrt{2})^2 = 16 - 4(2) = 8$

The difference of squares formula is especially important with fractional expressions whose denominators are of the form $a+\sqrt{b}$. You are guaranteed that the denominator will be a rational number. Terms such as $a + b$ and $a - b$ are called **conjugates**. They always add to equal $2a$ and multiply to equal $a^2 - b^2$.

EXAMPLE

▶ Simplify: $\dfrac{5-\sqrt{3}}{2+\sqrt{3}}$

▶ Multiply the numerator and denominator of this fraction by the conjugate of the denominator.

$$\left(\frac{5-\sqrt{3}}{2+\sqrt{3}}\right)\left(\frac{2-\sqrt{3}}{2-\sqrt{3}}\right) = \frac{5(2)+5\left(\sqrt{3}\right)-\sqrt{3}(2)-\sqrt{3}\left(-\sqrt{3}\right)}{4-3}$$
$$= 10+2\sqrt{3}+3 = 13+2\sqrt{3}$$

It will not always be the case that the denominator will equal 1, but it is true that the denominator will always be a rational number.

EXAMPLE

▶ Simplify: $\frac{5-2\sqrt{7}}{6-4\sqrt{7}}$

▶ Multiply numerator and denominator of this fraction by the conjugate of the denominator.

$$\left(\frac{5-2\sqrt{7}}{6-4\sqrt{7}}\right)\left(\frac{6+4\sqrt{7}}{6+4\sqrt{7}}\right)=\frac{5(6)+5\left(4\sqrt{7}\right)-\left(2\sqrt{7}\right)(6)-\left(2\sqrt{7}\right)\left(4\sqrt{7}\right)}{36-16(7)}$$

$$=\frac{30+20\sqrt{7}-12\sqrt{7}-56}{-76}=\frac{-26+8\sqrt{7}}{-76}=\frac{26-8\sqrt{7}}{76}$$

$$=\frac{2\left(13-4\sqrt{7}\right)}{76}=\frac{13-4\sqrt{7}}{38}$$

Radical Equations

Solving radical equations is a straightforward process. Isolate the radical on one side of the equation and then raise both sides of the equation by the appropriate power to remove the radical. Remember, $(x^m)^n = x^{mn}$, so $(x^{1/n})^n = x^{n/n} = x^1 = x$.

EXAMPLE

▶ Solve: $\sqrt{2x+1}=5$

▶ Square both sides of the equation: $\left(\sqrt{2x+1}\right)^2 = 5^2$ to get $2x+1=25$ so that $x=12$.

That was a fairly easy example to consider. This next problem is a bit more challenging.

EXAMPLE

- Solve: $\sqrt{5-x}+7=12$
- Why is the problem more challenging? We need to remember to isolate the radical expression before we raise both sides of the equation to a power.
- $\sqrt{5-x}+7=12$ becomes $\sqrt{5-x}=5$. Square both sides of the equation: $5-x=25$ so that $x=-20$.

You have to admit, that problem was not particularly challenging. Things get very interesting when there is a variable both inside and outside the radical.

EXAMPLE

- Solve: $\sqrt{2x+1}=x-1$
- The radical expression is isolated, so we square both sides of the equation.

  $\left(\sqrt{2x+1}\right)^2=(x-1)^2$ becomes $2x+1=x^2-2x+1$.
- This is a quadratic equation, so we follow the protocol by moving all terms to one side of the equation: $0=x^2-4x$. The quadratic expression has a common factor. Use this to solve the equation: $x(x-4)=0 \Rightarrow x=0,4$.
- That didn't seem to be too bad. Except—it's not correct. No, we didn't make any arithmetic or algebraic mistakes during the solution process. The error is that 0 does not solve the problem. Is $\sqrt{1}=-1$? No, it isn't. (This solution to the equation is called an **extraneous root** because it is extra and erroneous.) Be sure to check your answers when solving these equations. The solution to this equation is $x=4$.

There will be times when the problem will look much more difficult than it really is. What you need to remember is the process.

EXAMPLE

- Solve: $\sqrt{2x^2+11x-2}-2x=1$
- The first step is to isolate the radical expression. Add $2x$ to both sides of the equation.

$$\sqrt{2x^2+11x-2}=2x+1$$

- Square both sides of the equation to remove the radical.

$$\left(\sqrt{2x^2+11x-2}\right)^2=(2x+1)^2$$

- Simplify this statement:

$$2x^2+11x-2=4x^2+4x+1$$

- Set one side of the equation equal to zero.

$$0=2x^2-7x+3$$

- Factor the quadratic (or use the quadratic formula) to solve the equation.

$$(2x-1)(x-3)=0$$

$$x=\frac{1}{2},3$$

- Check the validity of these answers in the original equation.
- Check $x=\frac{1}{2}$.

$$\sqrt{2\left(\frac{1}{2}\right)^2+11\left(\frac{1}{2}\right)-2}-2\left(\frac{1}{2}\right)=1 \Rightarrow \sqrt{2\left(\frac{1}{4}\right)+\frac{11}{2}-2}-1=1$$

$$\Rightarrow \sqrt{\frac{1}{2}+\frac{11}{2}-2}-1=1 \Rightarrow \sqrt{\frac{12}{2}-2}-1=1 \Rightarrow \sqrt{6-2}-1$$

$\Rightarrow \sqrt{4}-1=1 \Rightarrow 2-1=1$ so that $1=1$. $x=\frac{1}{2}$ is a solution.

- Check $x=3$.

$$\sqrt{2(3)^2+11(3)-2}-2(3)=1 \Rightarrow \sqrt{2(9)+33-2}-6=1$$

$$\Rightarrow 18\sqrt{2(9)+33-2}-6=1 \Rightarrow \sqrt{49}-6=1 \Rightarrow 7-6=1$$

- $x=3$ is also a solution.
- Therefore, the roots to the equation are $x=\frac{1}{2}, 3$.

It is not always the case that there is only one solution, so you need to take the time to check your solutions. (And yes, it is possible that the equation has no solutions.)

The Pythagorean Theorem

It is very likely that you have already studied the Pythagorean theorem. If asked to state the theorem, most people will say, "The sum of the square of the legs is equal to the square of the hypotenuse." That would be a correct statement, but "The sum of the squares of the two smaller sides is equal to the length of the square of the longest side" would not be correct. Do you understand the difference in the two statements? There is only one kind of triangle that has an hypotenuse and that is a right triangle. The Pythagorean theorem only applies to right triangles! (Spoiler alert: When you get to the advanced trigonometry course, you'll learn an extension to this relationship that applies to all triangles. It is called the Law of Cosines.)

There are certain right triangles whose sides have lengths that are readily known among the students. The most well-known of these is the 3–4–5 triangle. There are a few other right triangles whose lengths you would benefit from knowing.

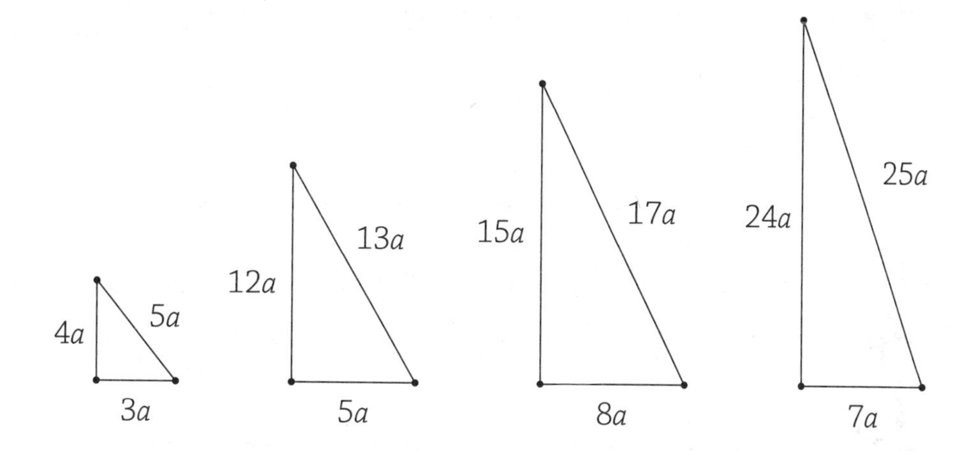

Why did I include an *a* as a factor for each of the sides? If we take the basic 3-4-5 triangle and enlarge it by a factor of 2, we get a 6-8-10 triangle, and this is also a right triangle. (Do you remember similar triangles—triangles with the same shape but proportional lengths?) There are many more right triangles than those listed here, but these four are the most commonly used right triangles that have sides whose lengths are integers. These sets of numbers are referred to as **Pythagorean triples**.

EXAMPLE

- The lengths of the sides of a right triangle measure $x + 2$, $x + 9$, and $x + 10$. What are the lengths of the sides of the triangle?
- Use the Pythagorean theorem, with the longest side having length $x + 10$ being the hypotenuse.

$$(x+2)^2 + (x+9)^2 = (x+10)^2$$

- Expand the terms: $x^2 + 4x + 4 + x^2 + 18x + 81 = x^2 + 20x + 100$
- Bring all terms to one side of the equation: $x^2 + 2x - 15 = 0$
- Solve the equation: $(x+5)(x-3) = 0$

$$x = -5, 3$$

- Realize that if $x = -5$, then the smallest side of the triangle would have length −2. So, $x = -5$ is rejected.
- If $x = 3$, then the lengths of the sides of the triangle will be 5, 12, and 13.

Allow me to repeat something I wrote earlier. The most common lengths of sides for a right triangle whose lengths are *integers* are 3-4-5, 5-12-13, 8-15-17, and 7-24-25.

EXAMPLE

- Determine the length of the hypotenuse of a right triangle if the legs have lengths 3 and 5.

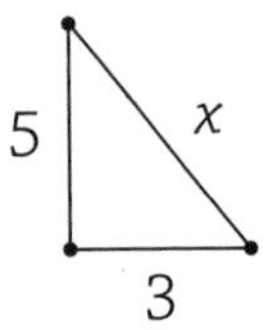

- Use the Pythagorean Theorem to get $3^2 + 5^2 = x^2$. Solve this to get $x^2 = 24$, and thus $x = \sqrt{24} = 2\sqrt{6}$.

A moral to this problem is to not assume that once you see right triangle, 3, and 5, the third side must have length 4.

Let's try another.

EXAMPLE

▶ One leg of a right triangle has length 20, and the hypotenuse has length 40. Determine the length of the third side of the triangle.

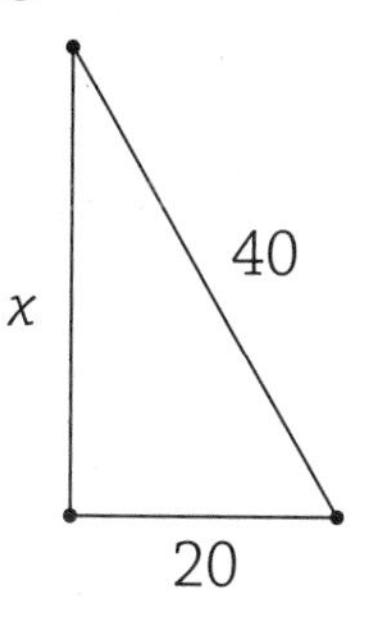

▶ Solve $x^2+20^2=40^2$. This becomes $x^2+400=1{,}600$, so that $x^2=1{,}200$ and $x=\sqrt{1{,}200}=\sqrt{100}\sqrt{12}=10\sqrt{4}\sqrt{3}=20\sqrt{3}$.

IRL

The right triangle whose sides have lengths a, $a\sqrt{3}$, and $2a$ has angles with measurements 30, 60, 90. It is a very important triangle in the study of trigonometry.

I had been teaching for about 12 years when I had this experience: The school day was over and I was working at my desk when I heard a man who was standing in the doorway ask, "Do you remember me?" I looked at the man and couldn't identify him. It turns out Andy was in my Algebra I class my first year teaching and, as I was the only math teacher in the high school at that time, he was also in my Geometry and Algebra II classes. He told me he came by to tell me a story about one his classmates who was a plumber at the time.

A pipe needed to be placed in the basement of the house they were working in, and there was an obstruction in the way, so a direct measurement could not be made. "George measured the diagonal to be 40 inches and the horizontal measure was 20 inches. He looked at me and told me to cut a pipe with length 34.6 inches." Andy asked, "How did you know that's the length?" "Didn't you learn anything in Monahan's class—it's a 30-60-90 triangle." Andy cut the pipe to the prescribed length and it fit perfectly. I have to tell you, it made my day!

The converse to the Pythagorean theorem is a useful statement:

> If the sum of the squares of the lengths of the smaller sides of triangle is equal to the square of the length of the longest side, then the triangle is a right triangle.

Let's look at an example:

EXAMPLE

- The lengths of the sides of a triangle are 9, 40, and 41. Is this triangle a right triangle?
- If $9^2 + 40^2 = 41^2$, then the triangle is a right triangle. $81 + 1600 = 1681 = 41^2$. Yes, the triangle is a right triangle (and you have a new Pythagorean triple).

Trigonometric Ratios

One of the great finds in Spain after the Spanish expelled the Moors in early 1492 was in a library in Seville. This piece of work illustrated the relation between the measure of the angles of a triangle with the length of its sides.

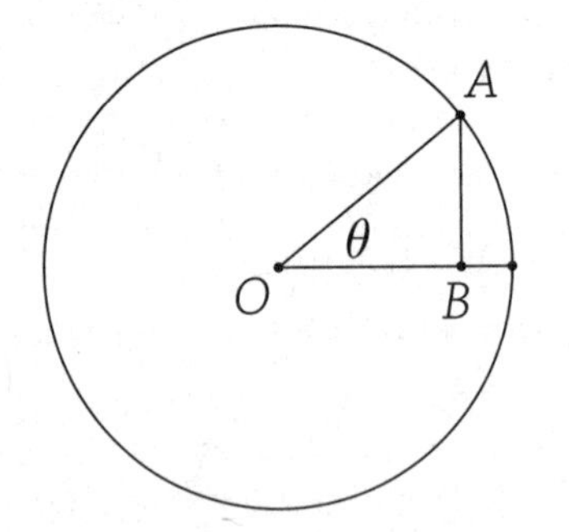

Right triangle *OBA* is drawn in a circle whose radius is 1. The segment $\overline{AB}$ was designated by a word in Arabic that no one knew how to translate. The closest interpretation that could be made was the word that meant "pocket." In Latin, the language in which all scholarly work was done in those days in Europe was *sinus*, which is anglicized to *sine*. We now know that the Arabic word actually meant "half-chord." That makes sense since $\overline{AB}$ is half the chord drawn from A perpendicular to the radius containing O and B.

The modern world tends to ignore the circle, so we work with the definition that the sine value is the ratio of the length of the side opposite the angle to the length of the hypotenuse. The angle in the diagram is represented by the Greek letter theta. We write $\sin(\theta)=\frac{\text{AB}}{\text{OA}}$. The ratio of the adjacent side of the triangle, $\overline{OB}$, to the hypotenuse is called the *cosine of theta*, abbreviated $\cos(\theta)$, and the ratio of the opposite side to the adjacent side is called the *tangent of theta*, abbreviated $\tan(\theta)$. There is a mnemonic for remembering the trigonometric ratios—SOHCAHTOA.

$$\text{sine}=\frac{\text{opposite}}{\text{hypotenuse}} \quad \text{cosine}=\frac{\text{adjacent}}{\text{hypotenuse}} \quad \text{tangent}=\frac{\text{opposite}}{\text{adjacent}}$$

The cosine represents the sine of the complementary angle. If the angle has a measure of 27°, then the complementary angle is 63° and $\sin(27°) = \cos(67°)$. Let's look at a few problems.

EXAMPLE

Determine the value of the three trigonometric ratios for the angle marked by θ.

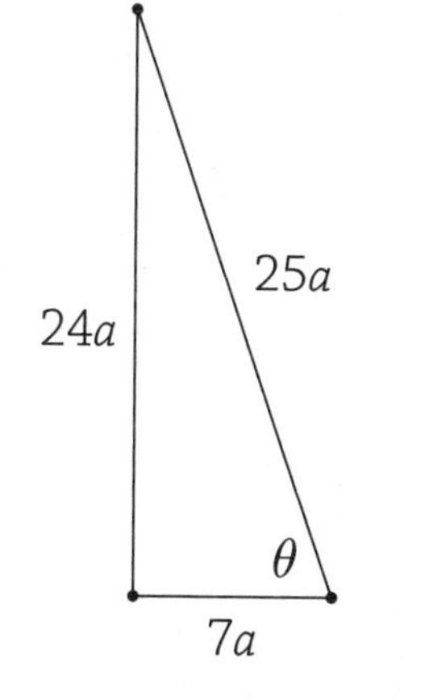

$$\sin(\theta)=\frac{24a}{25a}=\frac{24}{25} \quad \cos(\theta)=\frac{7a}{25a}=\frac{7}{25} \quad \tan(\theta)=\frac{24a}{7a}=\frac{24}{7}$$

You'll need a calculator for the rest of the material in this section. You want to be sure that you have the calculator set for Degree mode. (There is another mode for measuring angles and that is Radian mode. This is a topic you will study in the future. For now, suffice it to say that, like temperature, which can measured in degrees Fahrenheit or degrees Celsius, angles can be measured in different units as well.)

NORMAL FLOAT AUTO REAL DEGREE MP
MATHPRINT CLASSIC
NORMAL SCI ENG
FLOAT 0 1 2 3 4 5 6 7 8 9
RADIAN DEGREE
FUNCTION PARAMETRIC POLAR SEQ
THICK DOT-THICK THIN DOT-THIN
SEQUENTIAL SIMUL
REAL a+bi re^(θi)
FULL HORIZONTAL GRAPH-TABLE
FRACTION TYPE: n/d Un/d
ANSWERS: AUTO DEC
STAT DIAGNOSTICS: OFF ON
STAT WIZARDS: ON OFF
SET CLOCK 01/01/15 12:00 AM
LANGUAGE: ENGLISH

EXAMPLE

▶ The diagonal of a rectangle has length 25 cm and forms an angle with the base of the rectangle that measures 42°. Determine the length of the base and the height of the rectangle. Round these answers to the nearest tenth.

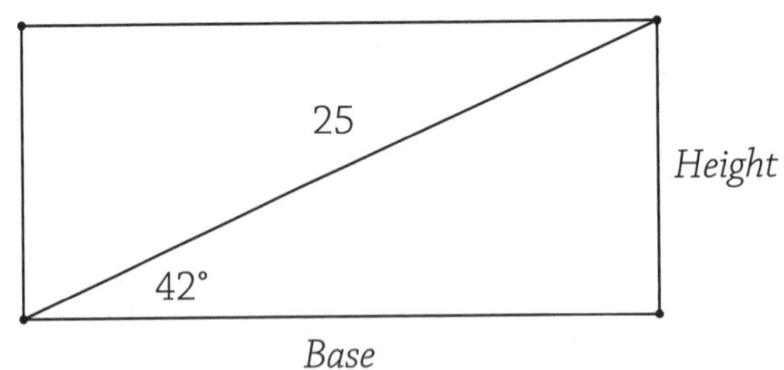

▶ The height of the rectangle is the side opposite the marked angle in the triangle. The ratio of the opposite side to the hypotenuse is the *sine ratio*. Therefore, $\sin(42) = \frac{h}{25}$, where h represents the height of the rectangle. Multiply by 25 to get $h = 25\sin(42)$. Enter 25*sin(42) into your calculator to get the answer 16.7 cm.

- The base of the rectangle is adjacent to the marked angle in the triangle. Therefore, $\cos(42) = \frac{b}{25}$, where b represents the length of the base. Multiply by 25 to get $b = 25\cos(42) = 18.6$ cm.

Let's look at an example that has a real-world application.

EXAMPLE

- The sun casts a shadow of a building. Alice measures the distance from the base of the building to the tip of the shadow. She finds that the shadow is 45 feet long. She then uses a clinometer to measure the angle of elevation to the top of the building. She records the angle to be 59.3°. Determine the height of the building. (Round your answer to the nearest tenth of a foot.)

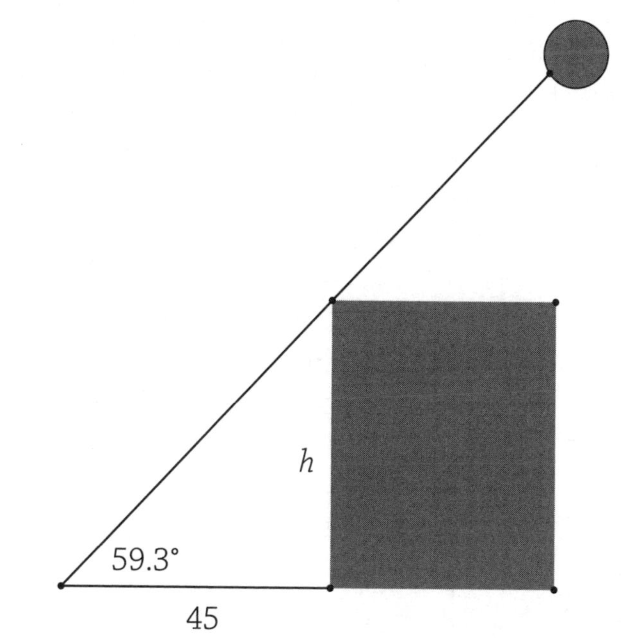

- The two pertinent sides for this triangle are the opposite and adjacent sides, so we use the tangent ratio.
- $\tan(59.3) = \dfrac{h}{45}$, so $h = 45\tan(59.3) = 75.8$ ft

The clinometer is a device used to measure the angle of elevation. What is the angle of elevation? Assuming you are sitting in a chair while reading this, hold your head still and look ahead of you so that your line of vision is parallel to the floor. Without moving your head, look up at the ceiling. The angle that your line of sight changed is the **angle of elevation**. Return so that your line of sight is parallel to the floor. Again, do not move your head. Now look at the floor. The angle through which you changed your line of sight is the **angle of depression**. Imagine that you are standing at the edge of a cliff looking at the view around you and you look down and see that someone is looking at you. The angle of depression needed for you to see that person is equal to the angle of elevation through which the second person needed to raise her line of sight to see you. Why is that? The original angles for both of you were parallel to the ground. The line of sight for each of you is a transversal to those parallel lines, so the alternate interior angles must be equal.

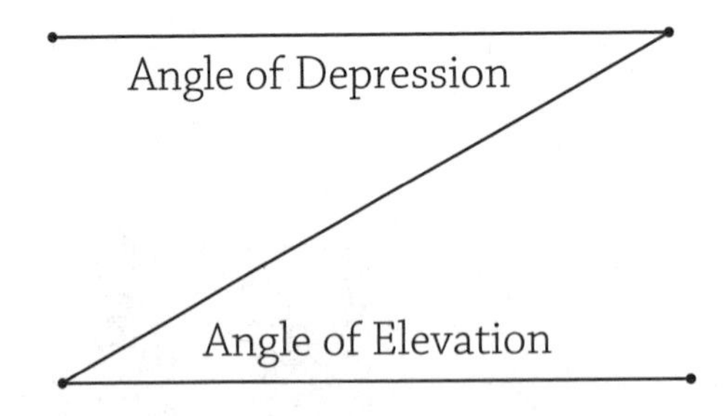

EXAMPLE

▶ Max is standing at the edge of the roof of a building that is 80 feet tall. He hears his name being called, and he looks down to see his friend George waving at him. Max estimates that the angle of depression for him to see George is approximately 30°. How far is George from the base of the building that Max is standing upon?

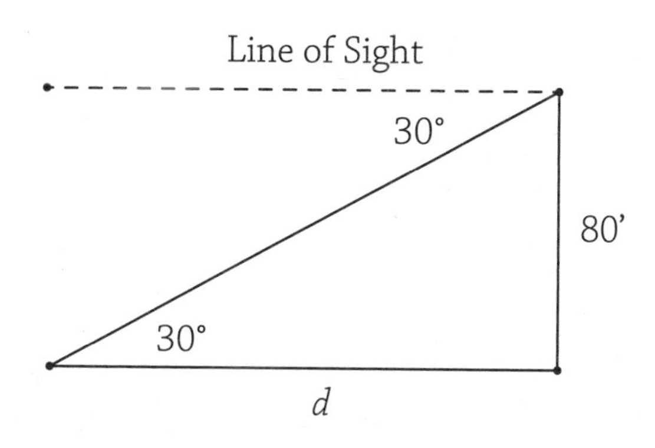

▶ As you can see from the diagram, the angle of depression is outside the triangle. Because the angle of elevation is inside the triangle, we are able to write the equation $\tan(30)=\frac{80}{d}$. Solving this equation for d, we get $d\tan(30) = 80$ and then $d=\frac{80}{\tan(30)}=138.6$.

Let's use Max and George to look at a similar problem. All the examples we have looked at so far involve finding the length of a missing side of the triangle. We can also use trigonometry to find the measure of a missing angle.

EXAMPLE

▶ Max is standing at the edge of the roof of a building that is 80 feet tall. He hears his name being called, and he looks down to see his friend George waving at him. Max estimates that George is about 150 feet from the base of the building. Through what angle of depression did Max's line of vision change so that he could see George?

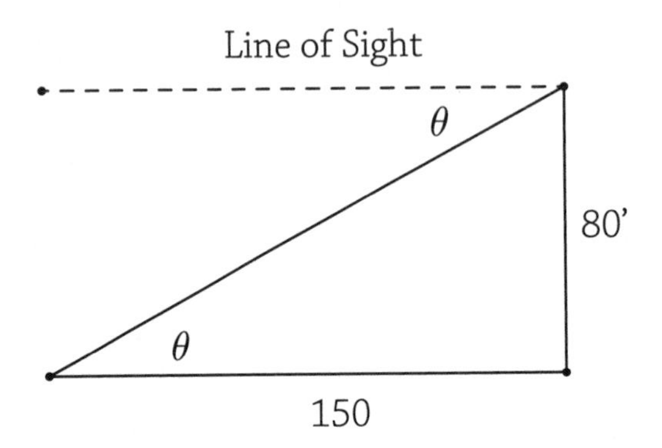

We can use the equation $\tan(\theta) = \frac{80}{150}$ to solve this problem. The question is, how do we find the value of θ? Look at your calculator. You see the buttons sin, cos, and tan. On the plate above each of these buttons, notice that $\sin^{-1}$, $\cos^{-1}$, and $\tan^{-1}$ are written. Recall from Chapter 2 that if A represents a relation, then A^{-1} represents its inverse. These three items, each accessed by first pressing the 2ND button, are the inverse sine, inverse cosine, and inverse tangent functions. Given a ratio, using these buttons determines the angle needed. The solution to our problem is $\theta = \tan^{-1}\left(\frac{80}{150}\right) = 28.1°$.

Here is another problem that asks you to find the measure of an angle.

EXAMPLE

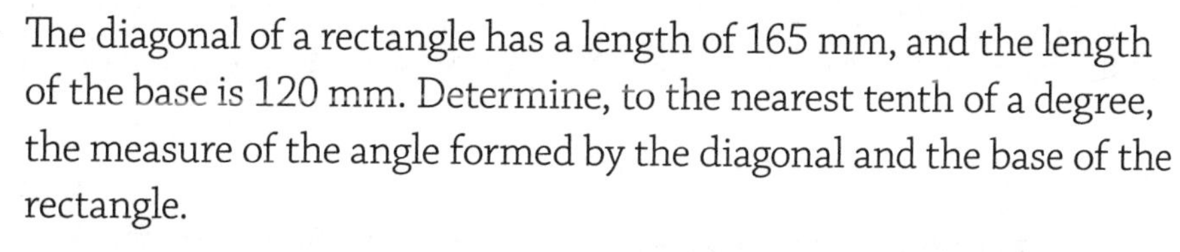

The diagonal of a rectangle has a length of 165 mm, and the length of the base is 120 mm. Determine, to the nearest tenth of a degree, the measure of the angle formed by the diagonal and the base of the rectangle.

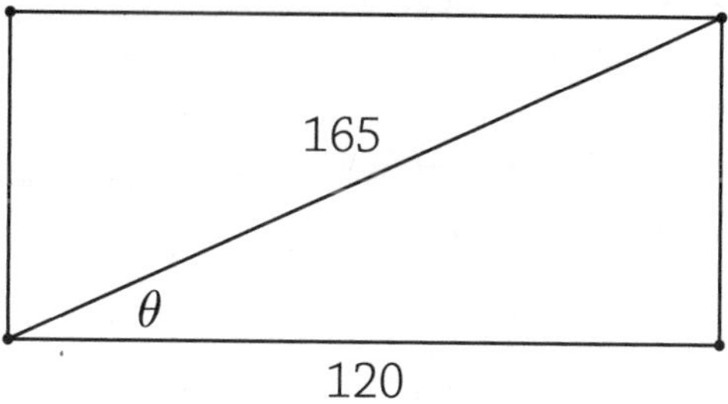

The base is adjacent to the designated angle, and the diagonal is the hypotenuse. Therefore, the equation needed to solve this problem is $\cos(\theta)=\frac{120}{165}$ so that $\theta=\cos^{-1}\left(\frac{120}{165}\right)=43.3°$.

EXERCISES

EXERCISE 12-1

Simplify each of the radical expressions. Assume variables found in the radicand represent non-negative values.

1. $\sqrt{24}$
2. $\sqrt{54}$
3. $\sqrt{162}$
4. $\sqrt{450}$
5. $7\sqrt{24}+4\sqrt{54}$
6. $4\sqrt{98}-2\sqrt{162}$
7. $4\sqrt{600a}-7\sqrt{150a}$
8. $(4\sqrt{6})(5\sqrt{3})$
9. $(4\sqrt{15})(5\sqrt{30})$
10. $(8\sqrt{2ab})(3\sqrt{32ac})$
11. $(5\sqrt{7})^2$
12. $(-6\sqrt{5})^2$
13. $(4\sqrt{5r})^2$
14. $\dfrac{5\sqrt{128}}{4\sqrt{50}}$
15. $(18+3\sqrt{27})+(-7+2\sqrt{108})$
16. $(18+3\sqrt{27})-(-7+2\sqrt{108})$

17. $5(8+5\sqrt{216})+7(4-8\sqrt{54})$

18. $5(8+5\sqrt{216})-7(4-8\sqrt{54})$

19. $(8+2\sqrt{3})(7-4\sqrt{3})$

20. $(9+3\sqrt{12})(2-4\sqrt{48})$

21. $(8+2\sqrt{3})(8-2\sqrt{3})$

22. $(7-4\sqrt{3})^2$

23. $\dfrac{2+\sqrt{5}}{3-\sqrt{5}}$

24. $\dfrac{4+\sqrt{5}}{4-\sqrt{5}}$

25. $\dfrac{7-2\sqrt{8}}{3-\sqrt{18}}$

EXERCISE 12-2

Use the quadratic formula to find the exact solution to each equation. Answers in reduced radical form.

1. $x^2-6x+7=0$

2. $x^2-4x-41=0$

3. $x^2+4x-3=0$

4. $16x^2+40x+19=0$

5. $49x^2-14x-1=0$

EXERCISE 12-3

Solve each equation.

1. $\sqrt{2x+11}=5$
2. $2\sqrt{x+25}-3=9$
3. $\sqrt{x^2+9}+x=1$
4. $\sqrt{3x+1}=x-3$
5. $\sqrt{12x+13}-x=4$

EXERCISE 12-4

Find the length of the missing side of each triangle in questions 1 to 5—and don't forget question 6!

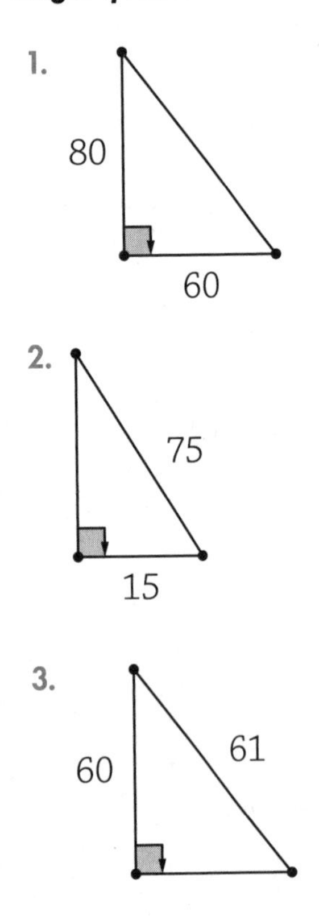

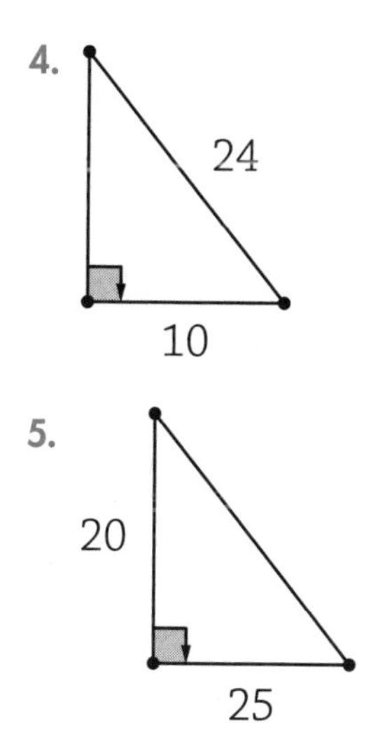

6. Are there integer values of x for which the expressions $x - 2$, $x + 5$, and $x + 6$ represent the lengths of the sides of a right triangle?

EXERCISE 12-5

Find the length of the identified side of each triangle. Round answers to the nearest hundredth.

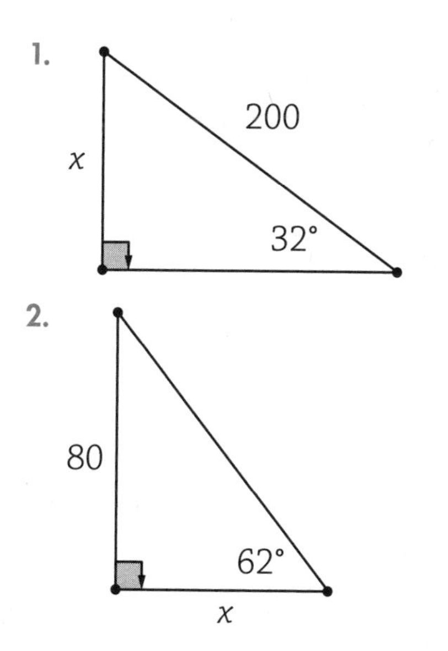

3.

4.

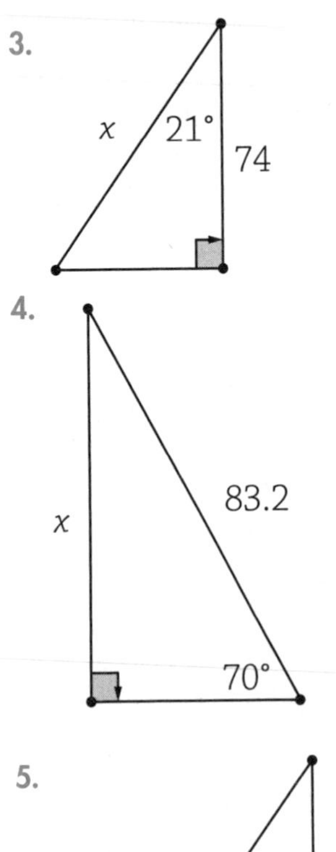

5.

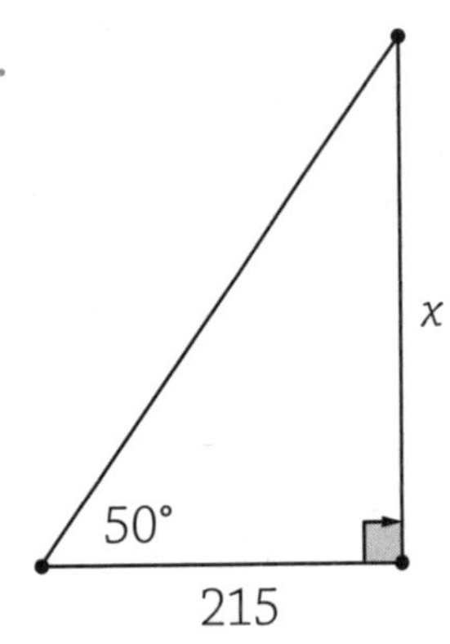

EXERCISE 12-6

For questions 1 to 5, find the length of the identified angle of the triangle. Round answers to the nearest tenth. Again, don't overlook question 6!

1\. 2.

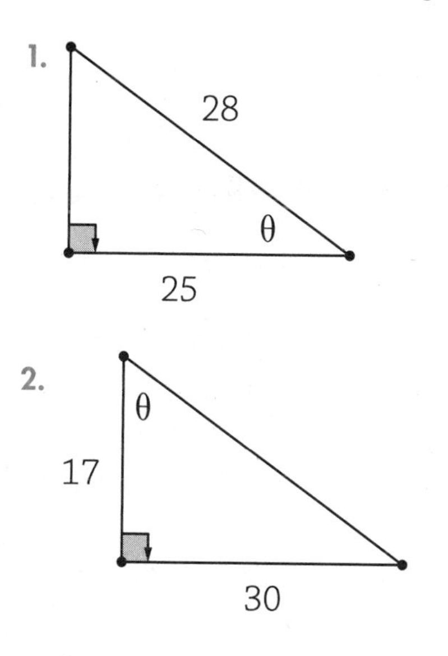

3\.

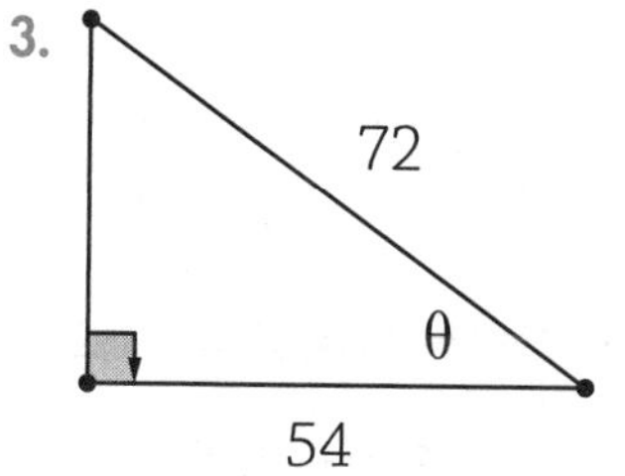

4\.

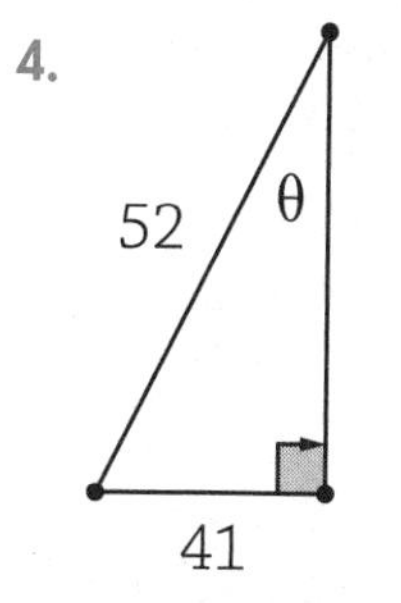

5. 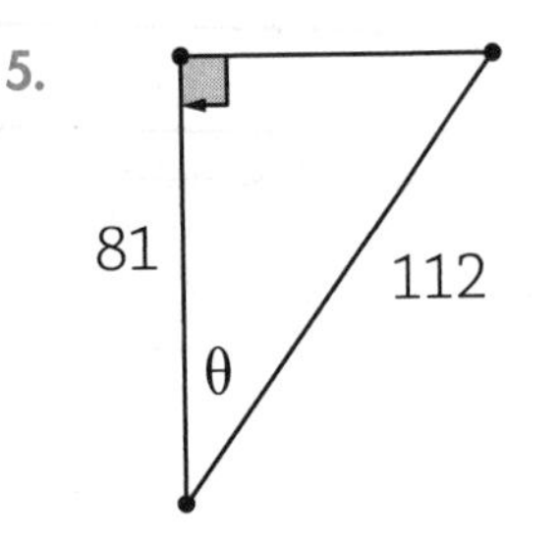

6. The American Disabilities Act (ADA) requires that ramps have a maximum slope of $\frac{1}{12}$. That is, the ramp must travel a horizontal distance of 12 feet for each 1-foot rise. To the nearest tenth of a degree, what is the maximum angle at which a ramp can be constructed and still meet the ADA guidelines?

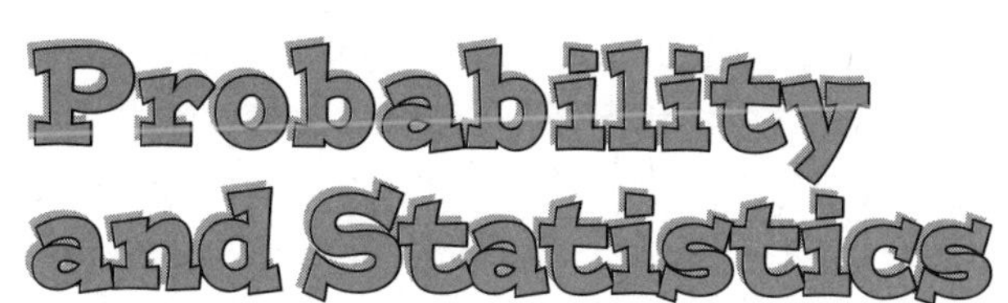

13 Probability and Statistics

MUST KNOW

- Probability, a measure of relative frequency, enables us to predict the likelihood of future outcomes.
- The concept of average indicates where the center of a set of data is located.
- Determining a statistically reliable relationship between two quantities enables us to predict the outcome for input values that have yet to be encountered experimentally.

It's a fact; you cannot avoid reading or hearing about statistics. They are present in the economic world, the political world, and the entertainment industry (this includes music and sports). Which movie has the best sales? Which athlete makes the most money? How many points per game does Steph Curry score? What are the chances that the Saints will get a first down when they have third down and less than 5 yards to go?

We will look at two basic factors in the study of descriptive statistics—center and spread. We will then take a look at predictive statistic—equations for regression.

Statistics and Parameters

To begin studying statistics, we need to know some of the terminology. The first two terms at are population and sample. The *population* is the set of all possible items in a study. For example, "What type of music do the students at your school like to listen to?" is a question of interest. The population is the set of all students in your school. Rather than ask every student in the school for his or her opinion, a researcher designs a plan to randomly select 200 students what type of music each listens to and will then use this information to project what is true of all students. The 200 students are the sample. They are part of the population but not necessarily the entire population.

You might ask the question, "Why not ask all the students? There aren't that many people." That is a fair question. Let's change the topic. The question is now, "Is the global climate change a danger to our country?" The population could well be all the people who live in your country. Is reasonable to think that it would be possible to contact every person in the country to get a response. First, that is going to be an expensive proposition. Second, it could take a lot of time. Most importantly, as you will learn as your study of statistics is extended, your findings will likely be no different than if you worked with just a portion of that population.

The beginning of the study of statistics doesn't seem to be able to trace back to an individual. However, states (get it? *statistics*) and governments have long been collecting data for the purpose of taxation. A more formal study of the subject started in the 16th century when mathematicians such as Blaise Pascal and Gerolamo Cardano began studying probability in games of chance.

Following are the three most often used quantities used by statisticians:

- **Mean** A measure of where the center of the data is located
- **Proportion** A relative measure of the population
- **Standard deviation** A measure of how the data is distributed

These three terms are used in the discussion of the population or a sample of the population. Numbers that describe a population are called **parameters**, while numbers used to describe a sample are called **statistics**. It's easy to remember: *p* for parameter and population; *s* for statistics and sample. Lowercase Greek letters are used to represent parameters, and lowercase English (actually from the Roman, or Latin, alphabet) letters are used to represent statistics:

	Statistic	Parameter
Mean	$\bar{x}$	μ
Proportion	p	ρ
Standard Deviation	s	σ

The three Greek letters are mu (μ), rho (ρ), and sigma (σ). The symbol for the mean of a sample is read as x-bar.

Measures of Central Tendency

You've had plenty of examples in your life of "average." The average grade on a test, your average grade for your report card, baseball averages, average salary for the middle class, average number of miles per gallon for a particular vehicle are just some of these examples. How do you find the average? Isn't that a silly question to be asking you at this point of your education? "Everyone" knows that you add up your data values and then divide by the number of data values. For example, if your grades on a report card are 93, 89, 87, 94, 85, 80, and 91, your average is

$$\frac{93+89+87+94+85+80+91}{7} = 88.43 \text{ (nice job – keep up the good work).}$$

That's what "everyone" would do, isn't it? And "everyone" would be partially correct. This measure for the average (center) is called the **mean** in statistics. There are two other measures for the center of a set of data: the median and the mode. (Full disclosure—the mode isn't used all that often, but it still is a measure of center.)

What is the mode? The **mode** is the piece (or pieces) of data that occur with the highest frequency (most often). For example, given the data set

19, 21, 21, 21, 23, 24, 25, 26, 29, 30, 32, 32, 38

the mode is 21 because this value occurs three times and no other data values occur more often.

However, the mode for the data set

19, 21, 21, 21, 23, 24, 25, 26, 29, 30, 32, 32, 32, 38

is 21 and 32 because they both occur three times. Note that it is possible that a data set not have a mode if all values occur the same number of times.

The **median** for a set of data is the number (which might not be one of the data values) that lies in the center of the data AFTER the data has been sorted from smallest to largest (or largest to smallest). The median of the data set

19, 21, 21, 21, 23, 24, 25, 26, 29, 30, 32, 32, 38

is 25 because there are six values smaller than 25 and six values larger than 25. This process of picking the middle number works well when there are an odd number of data points. When there is an even number of data points as in the data set,

19, 21, 21, 21, 23, 24, 25, 26, 29, 30, 32, 32, 32, 38

we take the two values in the middle of the data set (in this case 25 and 26) and we find the number midway between them, $\frac{25+26}{2}=25.5$.

The mean for the data set 19, 21, 21, 21, 23, 24, 25, 26, 29, 30, 32, 32, 38 is 26.23, while the mean for the data set 19, 21, 21, 21, 23, 24, 25, 26, 29, 30, 32, 32, 32, 38 is 26.64.

EXAMPLE

- Find the three measures of central tendency for the data set:

  30, 32, 35, 39, 36, 34, 36, 22, 22, 22, 30, 25

- Let's begin by arranging the data from lowest to highest numbers:

  22, 22, 22, 25, 30, 30, 32, 34, 35, 36, 36, 39

- The mode for the data is 22.
- There are 12 pieces of data, so the median will be midway between the sixth and seventh piece of data. The median is $\frac{30+32}{2}=31$.
- The mean is

$$\frac{22+22+22+25+30+30+32+34+35+36+36+39}{12}=30.25.$$

Frequency tables are used when there is a large number of pieces of data. We'll consider just those cases in which data points are repeated as opposed to creating intervals of scores.

EXAMPLE

Find the three measures of central tendency for the data set:

Score	Frequency
100	3
97	6
92	10
88	12
83	20
79	15
73	11
68	4

The data is already arranged in sequential order. The mode for this data is 83, since it occurs with the highest frequency.

Before we can determine the median, we'll need to sum the Frequency column to determine the number of data values. There are 81 data points for this set, so the median will be the 41st data value. Add the numbers in the frequency column until 41 is reached, $3 + 6 + 10 + 12 = 31$, so the 41st data value will be in the next group, and that is 83.

To compute the mean manually, we'll add a third column to this table and label it Product. The entries for this column will be the product of the score and frequency on each row.

Score	Frequency	Product
100	3	300
97	6	582
92	10	920
88	12	1056
83	20	1660
79	15	1185
73	11	803
68	4	272

The mean is found by dividing the sum of the column labeled Product with the 81 data values, $\overline{x} = \frac{6778}{81} = 83.68$.

We need to consider a rather drastic scenario before we continue the discussion of central tendency. In the next example, we include a piece of data that is very much different from all the other data.

EXAMPLE

What happens to the measures of central tendency when "unusual" numbers are introduced into the data set? You may have inferred that the data set used in the last example were test scores. Well, I led you to believe that by labeling the first column as scores. So not to make you feel like you were being tricked, I'll change the label to "Values" and will introduce one more piece of data.

Values	Frequency
1000	1
100	3
97	6
92	10
88	12
83	20
79	15
73	11
68	4

- There are now 42 pieces of data. The mode and mean are still 83 (do you agree?). Is the mean still 83.68?

Values	Frequency	Product
1,000	1	1000
100	3	300
97	6	582
92	10	920
88	12	1056
83	20	1660
79	15	1185
73	11	803
68	4	272

- The mean for this data is $\overline{x} = \frac{7778}{82} = 94.85$.

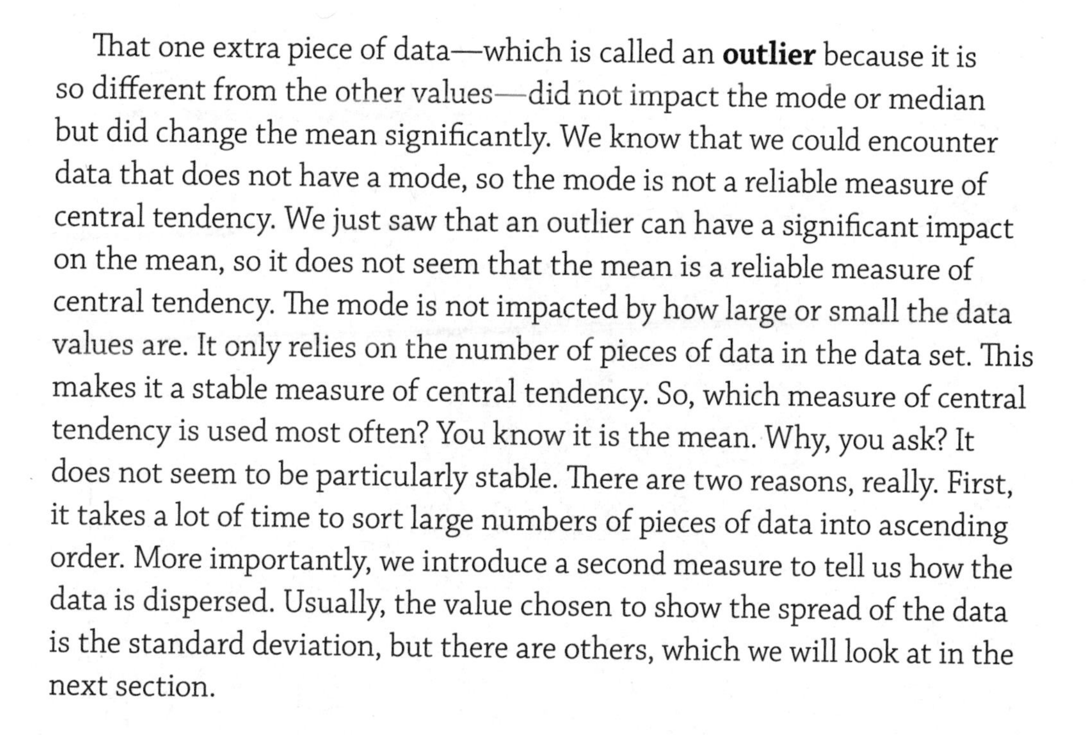

That one extra piece of data—which is called an **outlier** because it is so different from the other values—did not impact the mode or median but did change the mean significantly. We know that we could encounter data that does not have a mode, so the mode is not a reliable measure of central tendency. We just saw that an outlier can have a significant impact on the mean, so it does not seem that the mean is a reliable measure of central tendency. The mode is not impacted by how large or small the data values are. It only relies on the number of pieces of data in the data set. This makes it a stable measure of central tendency. So, which measure of central tendency is used most often? You know it is the mean. Why, you ask? It does not seem to be particularly stable. There are two reasons, really. First, it takes a lot of time to sort large numbers of pieces of data into ascending order. More importantly, we introduce a second measure to tell us how the data is dispersed. Usually, the value chosen to show the spread of the data is the standard deviation, but there are others, which we will look at in the next section.

Distributions of Data

"A picture is worth a thousand words." It's an old saying, but still one that applies, particularly with regard to statistics. There are a number of different ways to represent day graphically. We will look at **dot plots**, **histograms**, and **box and whisker plots** in this section. Later, when we look at bivariate data (a fancy way of saying ordered pairs), we'll examine scatter plots.

We'll use the set of data from earlier:

19, 21, 21, 21, 23, 24, 25, 26, 29, 30, 32, 32, 38

The data is displayed in a box and whisker plot.

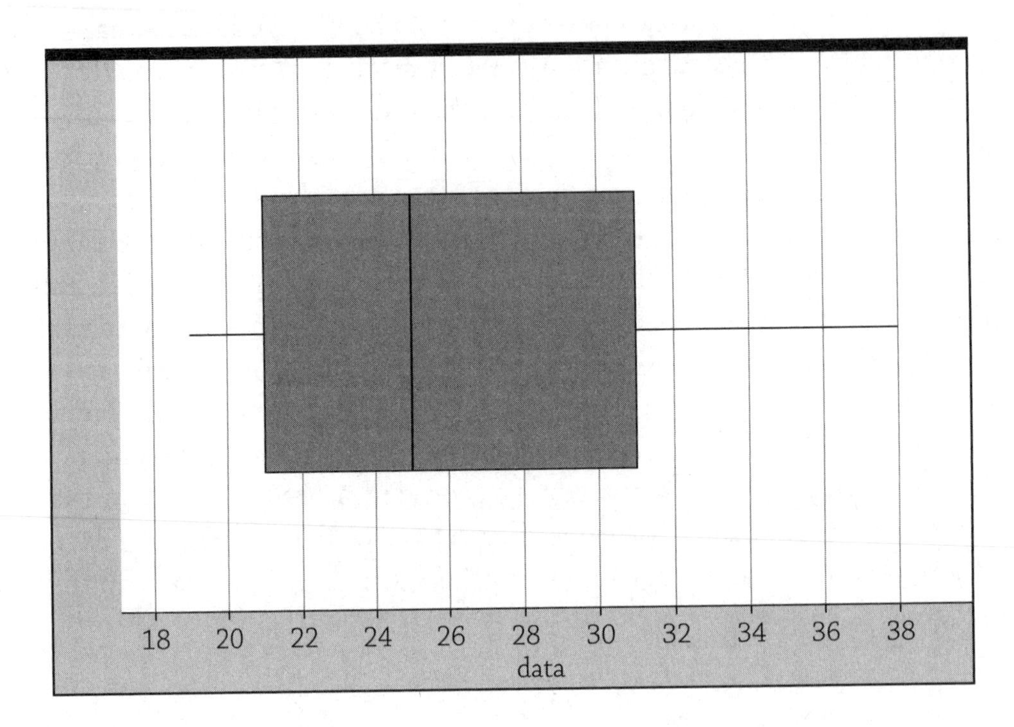

What do you see? The "whisker" to the left is shorter than the whisker to the right. The median of the data is 25. Can you see where this value is located on the graph? You can see that there is a 6-unit difference between the smallest value in the data and the median, while there is a 12-unit difference between the median and the largest value in the data set. Can you see where these numbers are located on the graph? (That accounts for the different lengths of the whiskers.) What about the box? How is it created? The median is the point in the data in which 50% of the data lies below and 50% above. This is also referred to as the 50th percentile. Midway between the smallest value and the median is the 25th percentile. One fourth of the data lies below this value. For that reason, it is also known as the first quartile (Q_1). Midway between the median and the maximum value is the 75th percentile, or third quartile (Q_3). For this data $Q_1 = 21$ and $Q_3 = 31$. Examine the graph again and notice where these values fit on the graph.

To summarize, the box and whisker graph is determined by five points: the minimum, the first quartile, the median, the third quartile, and the maximum. These values compose what is called the **five-number summary**. It gives a measure of center, the median, as well as two measures of spread. The range of a set of data is the difference between its largest and smallest values. For this data set, the range is 19. The other measure of spread is the difference between the third quartile and the first quartile. This is called the **inter-quartile range** (IQR). For this data set, the IQR is 10.

Our calculators have the ability to take data sets and give us these summary values.

$\bar{x}$	26.2308
Σx	341.
Σx^2	9323.
sx := sn-...	5.61477
σx := σn-...	5.3945
n	13.
MinX	19.
Q_1X	21.
MedianX...	25.
Q_3X	31.
MaxX	38.

What does this table tell us? We know the value of the mean, $\bar{x} = 26.2308$. There are 13 pieces of data ($n = 13$). The minimum value, MinX, is 19. The first quartile, Q_1X, is 21. The median is 25. The third quartile, Q_1X, is 31. The maximum value MaxX is 38. What about these other values? How does one calculate a mean? You add all the data values and then divide by the number of pieces of data. The sum of the data points is 341. The oddly shaped *E* is actually the uppercase Greek letter sigma and $\sum x$ indicates that we should add all the data points, whereas $\sum x^2$ indicates that we should add the squares of all the data points. We'll discuss $\sum x^2$ more in a moment. The last two items on the list are the measures of the standard deviation. The sample standard deviation, s_x, is 5.61477, while the population standard deviation, σ_x, is 5.3945.

What is a standard deviation? The best interpretation for you (though this is not completely accurate) is that the **standard deviation** gives the average difference between the mean and each piece of data.

Why are there two measures of standard deviation? The sample standard deviation is used to provide an estimate for the population value. It was determined that the best predictive value for the approximation has a slightly different formula than the procedure used for computing the population value. Sound difficult? It is. However, all you need to concern yourself with at this level is to be able to determine whether the data represents a sample or a population so that you use the appropriate value for the standard deviation.

With regard to $\sum x^2$, this is a number that can be used to compute the standard deviation. It is displayed when you execute the One-Variable Statistics calculation on your calculator.

Now that we have this under control, let's add one more piece of data to the data set, 50. This value is significantly different from all the other data values.

$\bar{x}$	27.9286
Σx	391.
Σx^2	11823.
sx := sn-...	8.33403
σx := σn-...	8.03087
n	14.
MinX	19.
Q_1X	21.
MedianX...	25.5
Q_3X	32.
MaxX	50.

The range for this new data set is 31. The first quartile is still 21, and third quartile is now 32, so the IQR = 11. The box and whisker plot has a new twist to it.

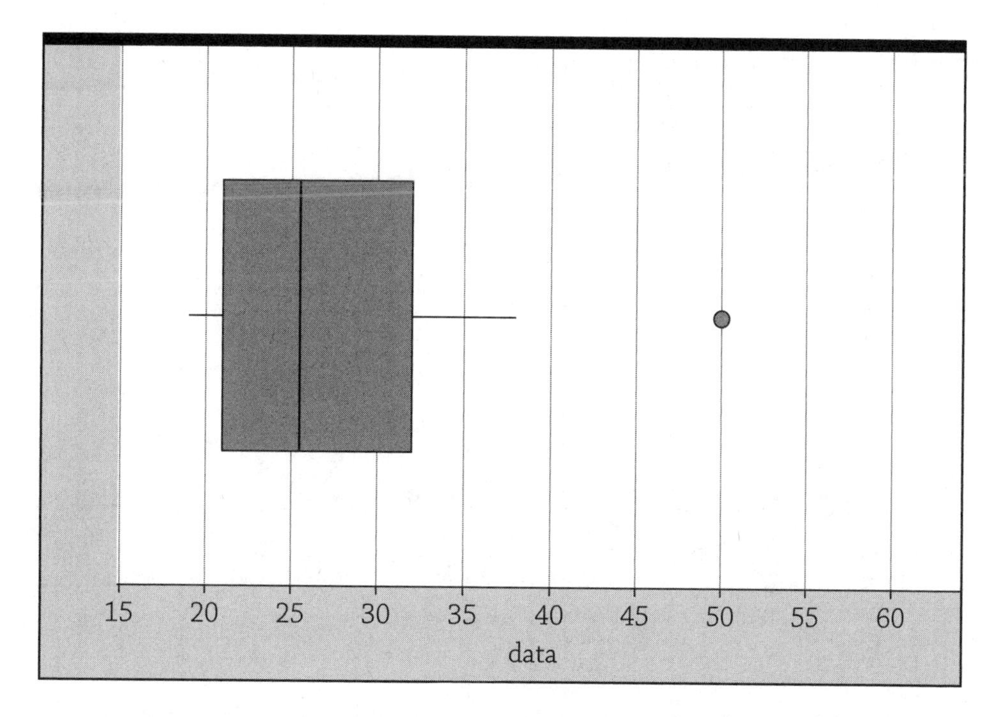

The data value 50 is isolated by itself. We call this data point an outlier because it is that much different from the rest of the data. The rule of thumb for an outlier is any data value is more than 1.5 times the IQR from Q_3 (or less than 1.5 times the IQR from Q_1). The end of the whisker is the largest data value before the outlier.

You can see that the range is greatly impacted by an outlier but the IQR is not. Notice that if the extra data value had been 500 rather than 50, the range would be 481 but the IQR would still be 11.

The dot plot for the original set of data is

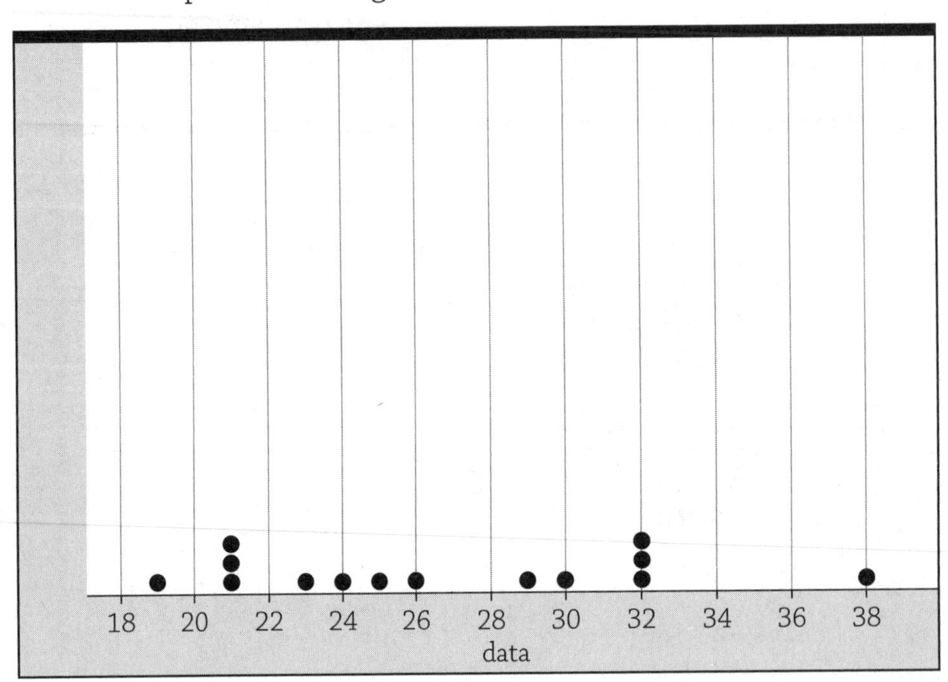

while the histogram for the same data is

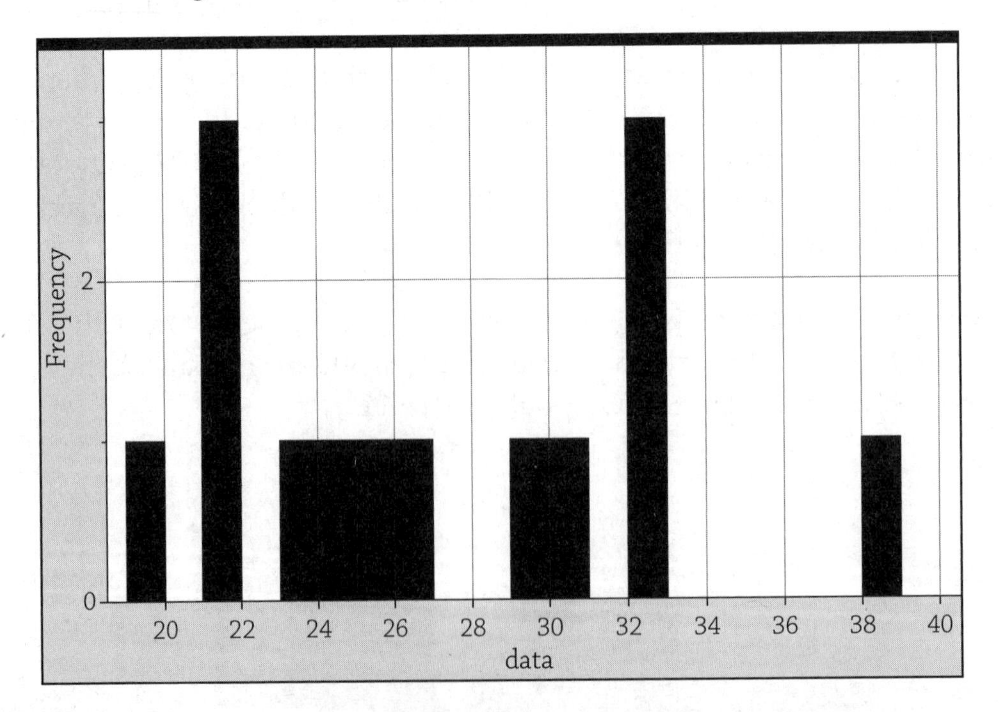

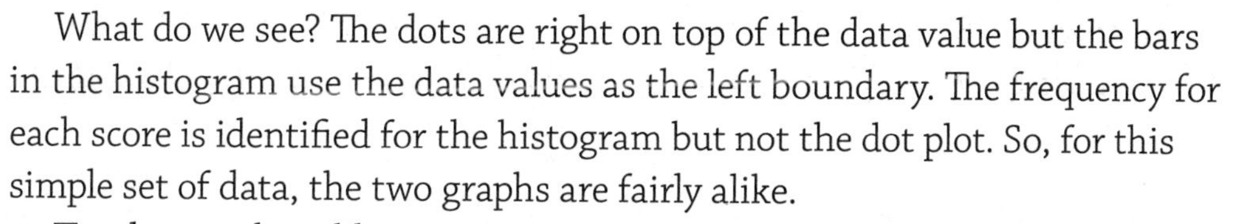

What do we see? The dots are right on top of the data value but the bars in the histogram use the data values as the left boundary. The frequency for each score is identified for the histogram but not the dot plot. So, for this simple set of data, the two graphs are fairly alike.

The data in the table is a summary of test scores for a final exam.

Score	Frequency
100	3
97	6
92	10
88	12
83	20
79	15
73	11
68	4

The results of the One-Variable Statistics command is

$\bar{x}$	83.679
$\sum x$	6778.
$\sum x^2$	572532.
sx := sn-...	8.18203
σx := σn-...	8.13137
n	81.
MinX	68.
Q_1X	79.
MedianX...	83.
Q_3X	88.
MaxX	100.

The mean of the 81 pieces of data is 83.679, and the median is 83. The range of the data is 32, the IQR is 9, and, assuming this is a sample, the standard deviation is 8.18.

The box and whisker plot is

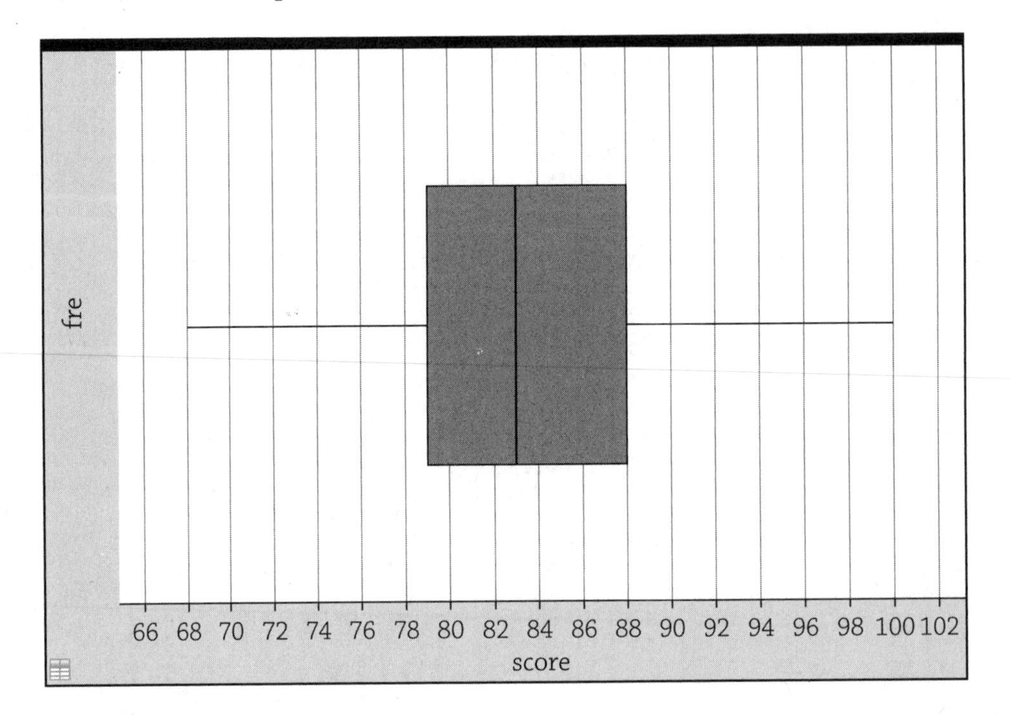

The dot plot for this data gives a sense of the shape of the distribution of the data, as does this histogram:

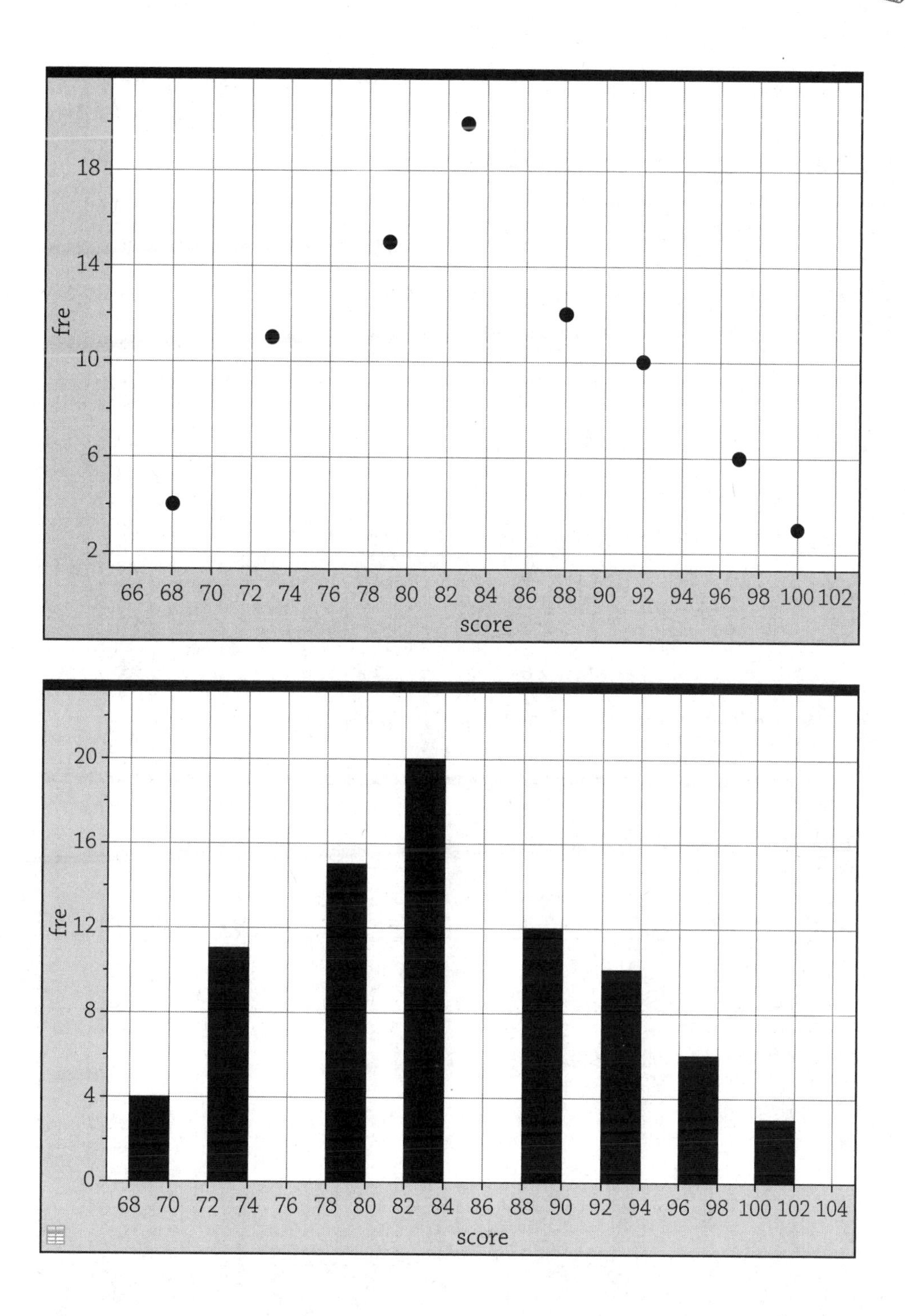
fre
18
14
10
6
2
66 68 70 72 74 76 78 80 82 84 86 88 90 92 94 96 98 100 102
score
fre
20
16
12
8
4
0
68 70 72 74 76 78 80 82 84 86 88 90 92 94 96 98 100 102 104
score

As before, we'll add one more piece of data to the data set. The new number is 200. Examine the results of the One-Variable Statistics calculation and the box and whisker plot.

Before

$\bar{x}$	83.679
Σx	6778.
Σx^2	572532.
sx := sn-...	8.18203
σx := σn-...	8.13137
n	81.
MinX	68.
Q_1X	79.
MedianX...	83.
Q_3X	88.
MaxX	100.

After

$\bar{x}$	85.0976
Σx	6978.
Σx^2	612532.
sx := sn-...	15.2028
σx := σn-...	15.1098
n	82.
MinX	68.
Q_1X	79.
MedianX...	83.
Q_3X	88.
MaxX	200.

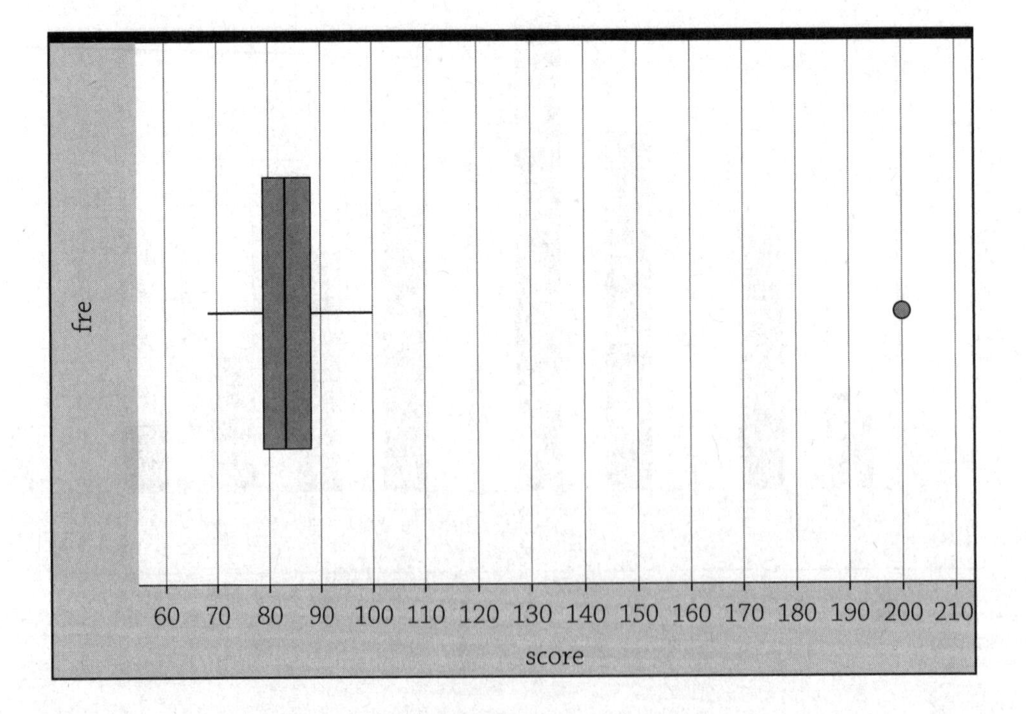

The mean increased by 2 points, the standard deviation is almost doubled, yet the median and IQR remain unchanged. Such is the impact of an outlier. The large standard deviation (large in comparison to the data values) is an indication that the data is widely spread and a strong hint that there might be an outlier in the data. If there is an outlier, it would be prudent to look at the data and the data collection process to determine the cause of the outlier. Was it a data entry error? Is there an exceptional source to the data? Is the process faulty? These questions, and others, are reasonable concerns for the statistician.

All the examples used to this point show a distribution that is fairly evenly distributed about the center. That is, there are a comparable number of data points below and above the measure of central tendency (even if there is an outlier). This occurs when the mean and median are approximately equal to each other. Not all data sets behave as such. Consider the two graphs shown below:

Left Skewed

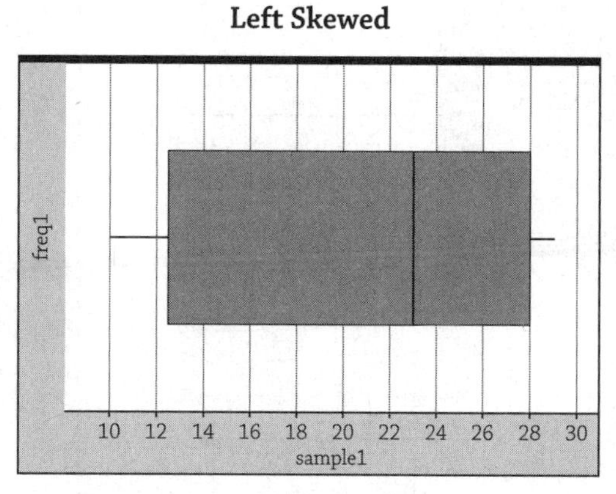

Tail is pulled to the left.

Mean < Median

Right Skewed

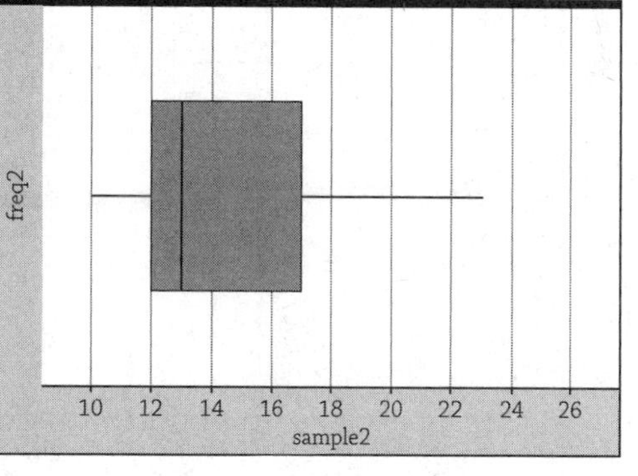

Tail is pulled to the right.

Mean > Median

EXAMPLE

The heights, in inches, of 14 students are listed below:

68, 63, 58, 66, 74, 60, 70, 79, 72, 61, 67, 72, 75, 67

Which of the following statements are true?

(1) The set of data is evenly spread.

(2) The IQR of the data is 21.

(3) 59 is an outlier, which would affect the standard deviation of these data.

Statement 1 is true.

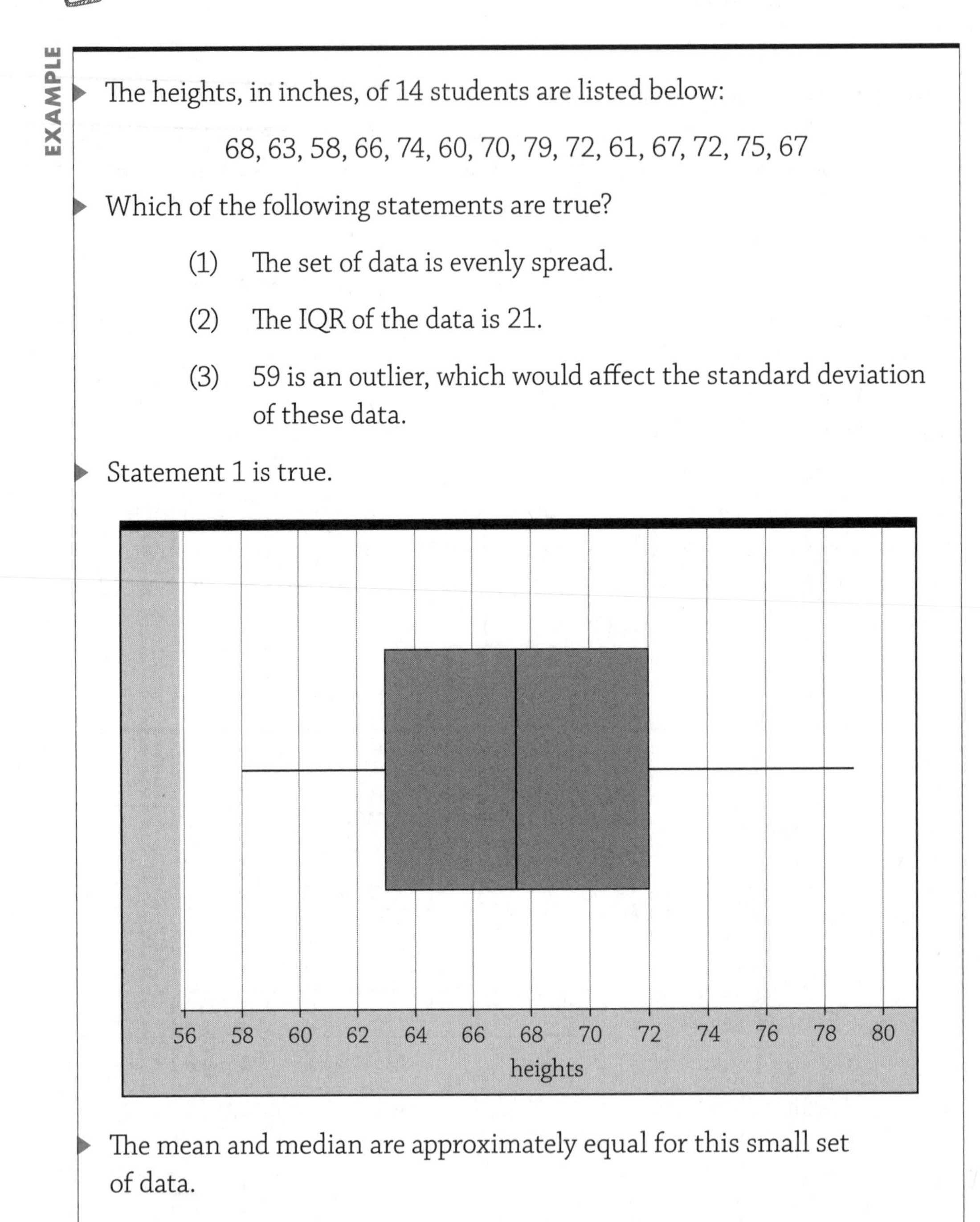

The mean and median are approximately equal for this small set of data.

Statement 2 is false. The range of the data is 21. The IQR is 9.

Statement 3 is false. This data set does not have any outliers.

Here is another example asking you to examine data. This time, you are given the results of the statistical computations.

EXAMPLE

▶ The students in Mrs. Francks's 5th and 9th period Algebra classes took the same test. The results of the scores are shown in the following table:

	$\bar{x}$	σ	n	min	Q_1	median	Q_3	max
5th Period	78.1	9.7	21	57	68	75	87	96
9th Period	78.3	8.4	22	59	67	76	89	94

▶ Based on these data, which class has the larger spread of test scores? Explain how you arrived at your conclusion.

▶ The 5th period class has the larger spread in test result as indicated by the larger standard deviation of the two groups.

In the next example, you are asked to draw a conclusion based on a graphical representation of the data.

EXAMPLE

▶ Laura was conducting a survey of sports participation between the local lacrosse and soccer programs. The ages for the participants are displayed in the accompanying histograms.

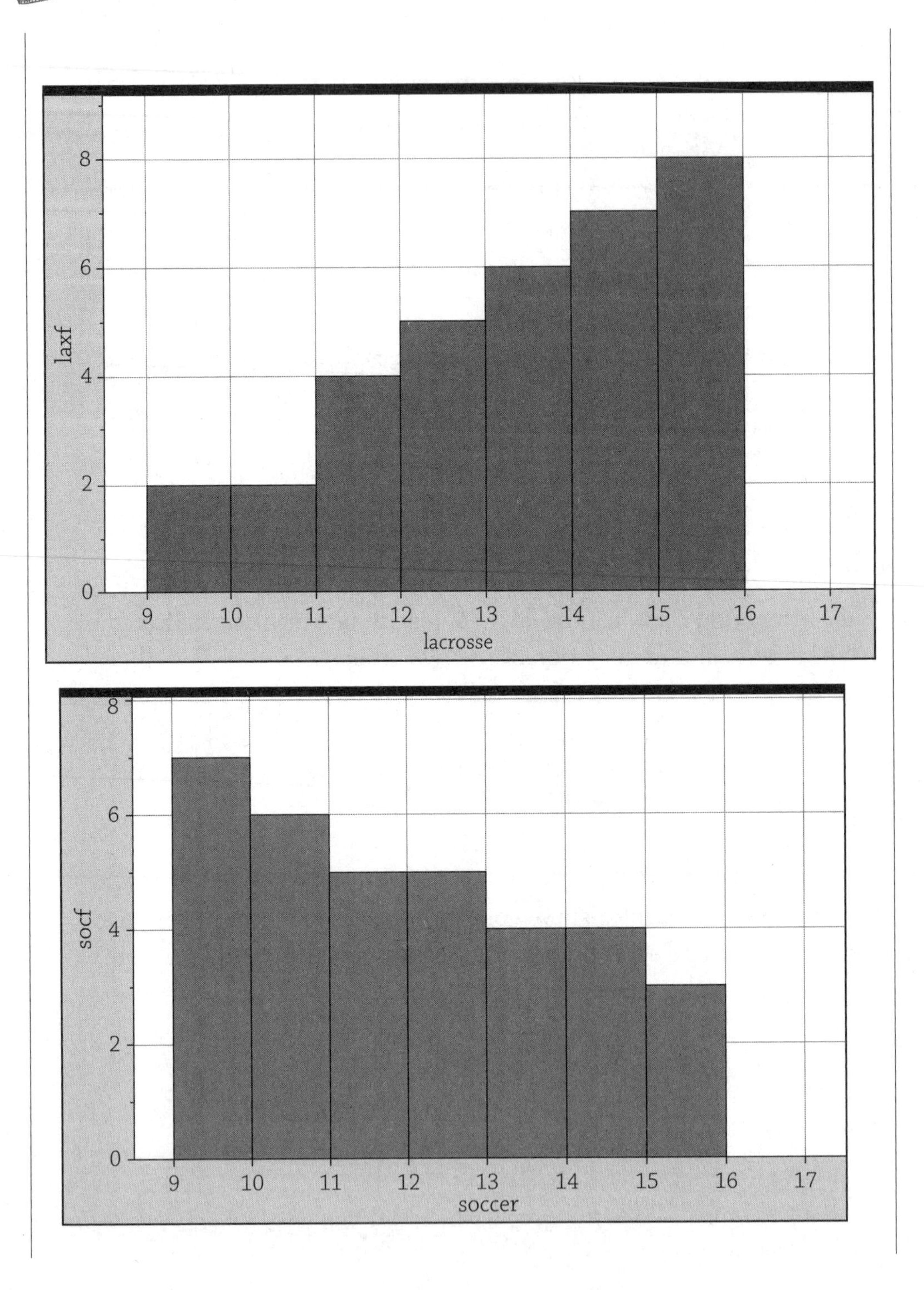
laxf
0
2
4
6
8
9
10
11
12
13
14
15
16
17
lacrosse
socf
0
2
4
6
8
9
10
11
12
13
14
15
16
17
soccer

- What conclusion can Laura draw about the median age of those who play each sport?
- There are 34 players in each sport. The median age will the average of the 17th and 18th player in each sport. The median age for lacrosse players is 13, and for soccer, the median age is 11.

Scatter Plots and Lines of Fit

Gathering data for the purpose of determining what has happened is worthwhile, but this is just a small part of what the study of statistics entails. A basic tenet in the study of mathematics is the search for patterns. Communities that gather census information hope to use this information to predict revenue income through taxes and fees collected but also to determine their expenses as they maintain infrastructure and provide other important services to their communities. The baseball manager who decides he needs a pinch hitter will choose from the players available the one with the best batting average against the pitcher who is on the mound for the opposing team.

We will begin by looking at those variables that are related by a linear relationship. For example, the graph below, called a **scatter plot**, shows the number of calories in a sandwich from Windy's on the horizontal axis and the corresponding number of calories from fat on the vertical axis.

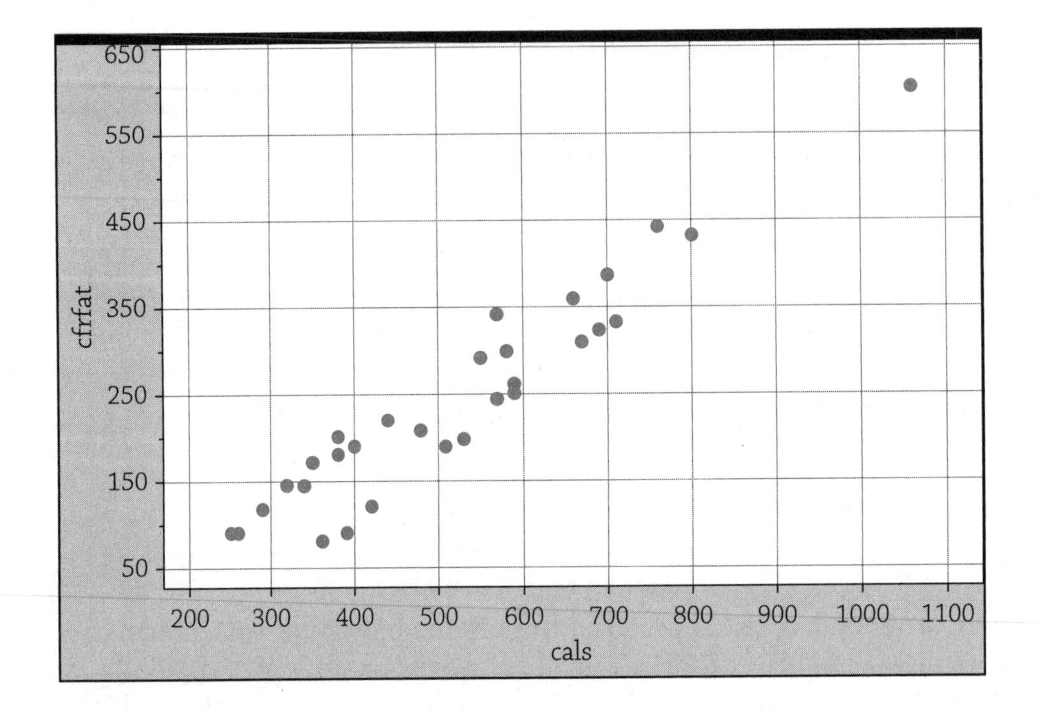

Unlike the vast majority of work you've done in math class, these points do not lie exactly on a line but resemble a linear pattern. (Think of using a very wide marker to draw the line through these points.) Our goal is to determine the "best" line to draw so that the differences between the actual data points plotted and the corresponding points on our line are as small as possible.

One such line might be

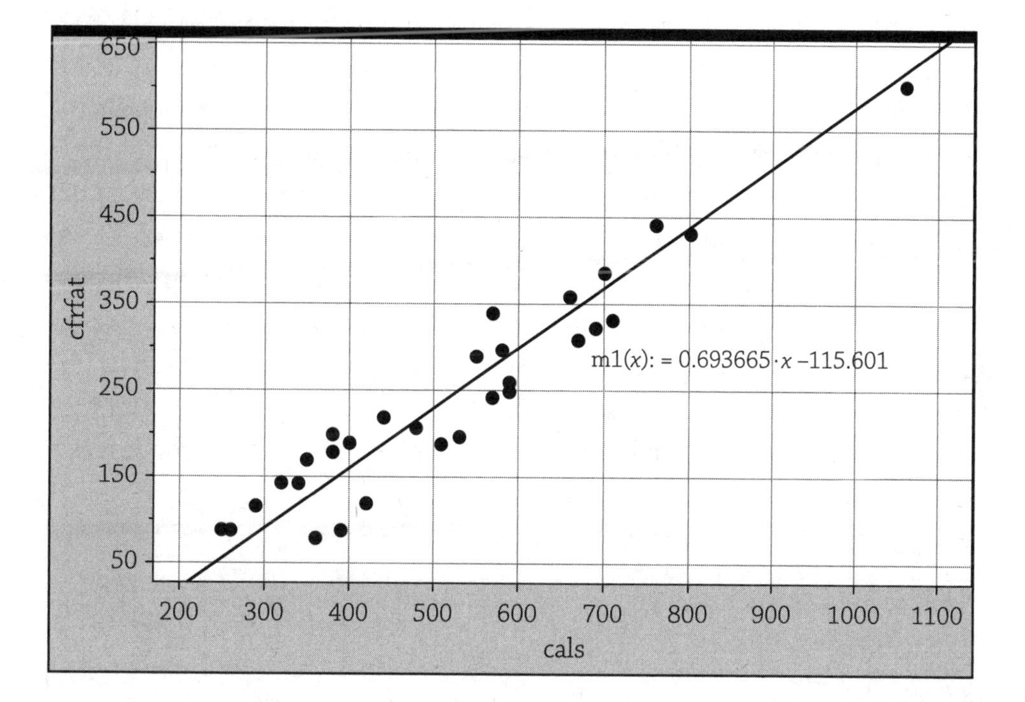

There are an infinite number of lines that we can try. The question is, "What is the equation of the best line to fit the data?" Well, you know that if we are asking the question, there is a good chance we also have an answer. Go to the Statistics menu on your calculator. Select Stat Calculations and then Linear Regression (mx + b). Enter the name of the list that is serving as the input values, then the list for the output values. You can change the location for where the regression equation will be stored.

RegEqn	m*x + b
m	0.625339
b	−80.5929
r^2	0.919138
r	0.958717
Resid	{−26.891...

There are five pieces of information in the result. First, we have the slope and vertical intercept for the line, Calories from Fat $= 0.625(\text{Calories}) - 80.593$. The slope tells us that for every extra calorie in the sandwich, there will be an additional 0.625 calories due to fat. The y-intercept in this case tells us that when the sandwich has 0 calories (is that possible?), there will be -80.593 calories due to fat. Well, that's not anything that is realistic. The reality is that a Windy's sandwich cannot have 0 calories and the negative number works when the input value is realistic. Let's look at that a different way. The domain for this equation is not the set of all real numbers. The domain comes from the caloric content of a Windy's sandwich, which we can estimate from the graph to be somewhere between 200 and 1100 calories.

Here is what the scatter plot and the line of best fit look like together:

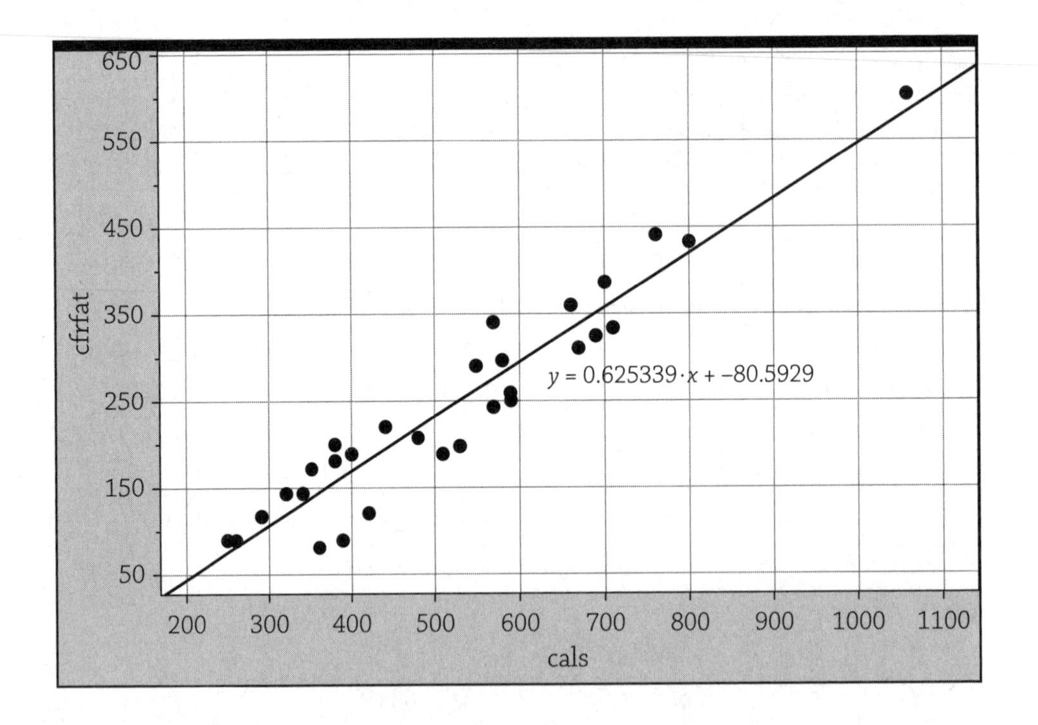

The value labeled r is the Pearson Correlation coefficient and is very often just referred to as the r-value. The r-value can take on values in the range from -1 to 1. A value of 1indicates a perfect relations for a line that has

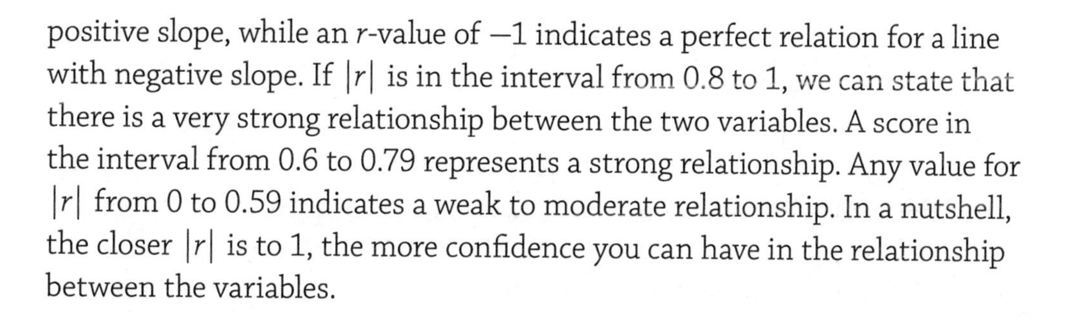

positive slope, while an r-value of -1 indicates a perfect relation for a line with negative slope. If $|r|$ is in the interval from 0.8 to 1, we can state that there is a very strong relationship between the two variables. A score in the interval from 0.6 to 0.79 represents a strong relationship. Any value for $|r|$ from 0 to 0.59 indicates a weak to moderate relationship. In a nutshell, the closer $|r|$ is to 1, the more confidence you can have in the relationship between the variables.

IRL

Regression equations are used to predict other values from the quantities being observed. BUT the only way they can legitimately be used is to use a value from the input variable that is within the domain of the input data to compute an estimated value for the output. In this example, we can use a hypothetical, or real, Windy's sandwich with a calorie content in the interval 200 to 1,100 to estimate the number of calories from fat for that sandwich. We cannot use a calorie content of more than 1,100 or less than 200, nor can we give a value for the number of calories from fat and solve the equation to determine the number of calories in the sandwich.

EXAMPLE

Estimate the number of calories from fat in a Windy's sandwich if the sandwich contains 600 calories.

Answer: Calories from fat $= 0.625339(600) - 80.5929 = 294.6$. (Full disclosure: I used the regression equation stored in the calculator. That is, for the TI-83/84, go to Y=, choose Y1, and finish the line with Y1(600) to get the correct result.)

FYI - The term tagged as r^2 is called the coefficient of determination. It represents the percent of the response variable that can be directly attributed to the input value. In this case, $r^2 = 0.919$, so we can conclude that 91.9% of the calories from fat come directly from the number of calories in the sandwich. (Things like the bun, condiments, etc. can account for the remaining calories.) You are NOT responsible to know this information unless your teacher says differently.

The last item from the calculation for the line of best fit is Resid, which is short for residual. One of the data points is (690, 324). (I have the data and can see it. You need to accept that I wrote the ordered pair correctly.) The predicted number of calories from fat is $f(690) = 350.1$. The residual for this value of 690 is $324 - 350.1 = -26.1$. The calculator automatically stores the residuals in a list called Resid each time it performs a regression. The graph of the residuals uses the same list for the inputs that was used in the regression and the resids as the output list.

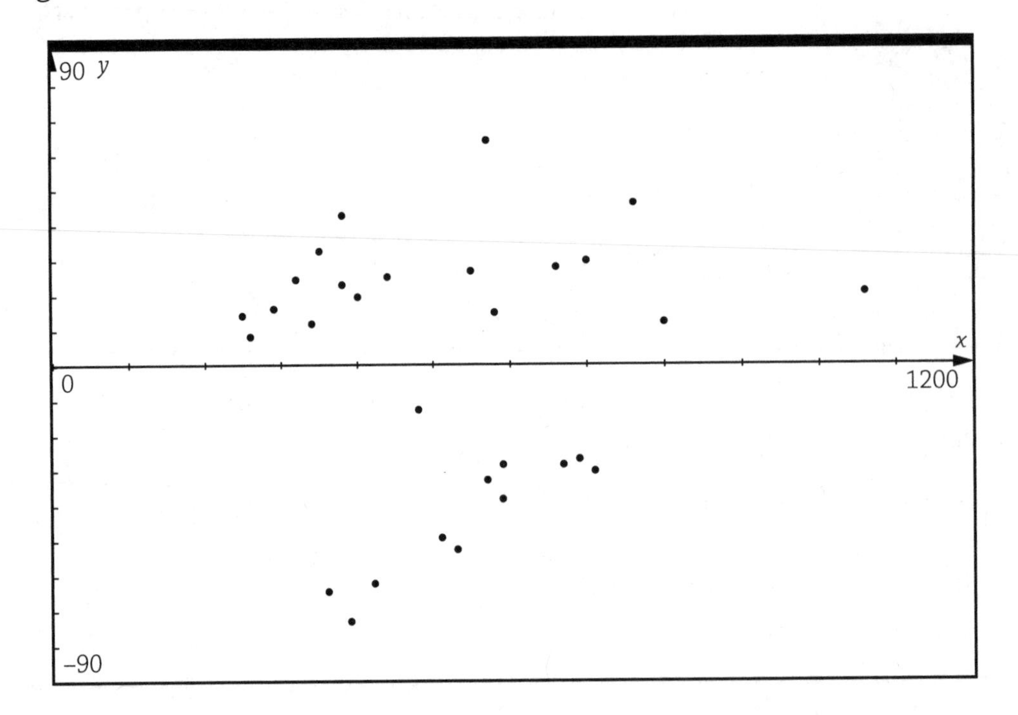

An indication that the equation determined from the regression calculation is the best it can be is if the graph of the residuals is completely random. Between examination of the r-value and the graph of the residuals, you can determine if the regression equation is a good fit.

Why do we go through this? Besides linear regression, there is also quadratic regression, power regression, and exponential regression to be considered.

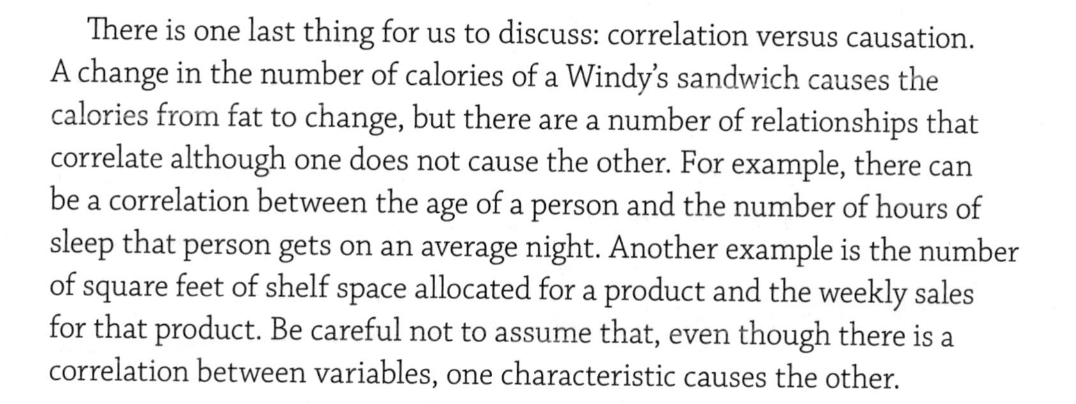

There is one last thing for us to discuss: correlation versus causation. A change in the number of calories of a Windy's sandwich causes the calories from fat to change, but there are a number of relationships that correlate although one does not cause the other. For example, there can be a correlation between the age of a person and the number of hours of sleep that person gets on an average night. Another example is the number of square feet of shelf space allocated for a product and the weekly sales for that product. Be careful not to assume that, even though there is a correlation between variables, one characteristic causes the other.

EXERCISES

EXERCISE 13-1

Use the data set 12, 12, 14, 15, 16, 18, 21, 25, 29, 31 to answer questions 1 to 6. (Assume data is a sample.)

1. What is the mode of the data set?
2. What is the median of the data set?
3. What is the mean of the data set?
4. What is the range of the data set?
5. What is the IQR of the data set?
6. What is the standard deviation for the data set?

The data presented in the table displays the number of hours per week students spent on HW. Use this data to answer questions 7 to 12. (Assume data is a population.)

Hours	Frequency
10	20
12	23
15	40
17	37
19	30
20	20
22	17
25	6

7. What is the mode of the data set?
8. What is the median of the data set?

9. What is the mean of the data set?

10. What is the range of the data set?

11. What is the IQR of the data set?

12. What is the standard deviation for the data set?

EXERCISE 13-2

1. Determine the value of the IQR for the data represented by the graph.

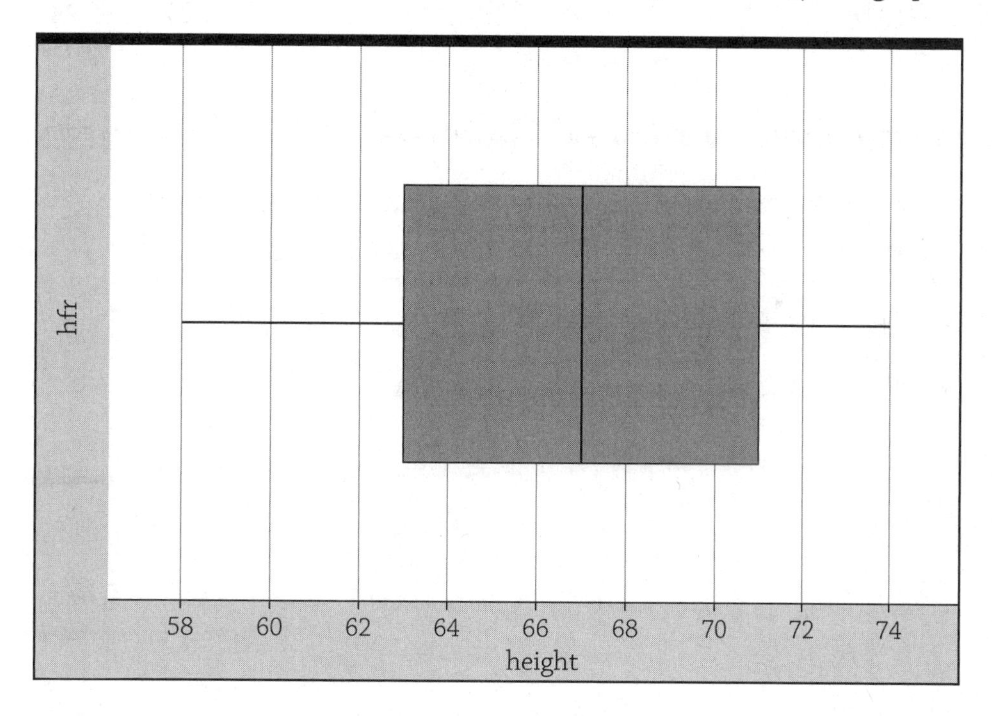

2. A student reads in a book that the equation representing the average number of hours per sleep changes with age and the equation given is avgsleep $= -0.1$*age $+ 8.3$. Interpret the meaning of the slope of the line.

3. The correlation coefficient given for the sleep study discussed in the preceding exercise is -0.87. Interpret the meaning of the correlation coefficient.

Alex works part-time at the local coffee shop. He kept a record of the number of cups of hot chocolate he sold to students and the high temperature for each day in order to complete a project for his statistics class. The data is displayed.

	Monday	Tuesday	Wednesday	Thursday	Friday	Saturday
High Temperature	51	56	48	45	38	52
Cups of Hot Chocolate Sold	23	19	27	28	32	24

Use this data to answer questions 4 to 6. Round answers to the nearest hundredth.

4. Determine the equation of the line of best fit that relates high temperature to the number of cups of hot chocolate sold.

5. Interpret the meaning of the y-intercept to this equation.

6. Predict the number of cups of hot chocolate that will be sold if the high temperature for the day is 40°.

Answer Key

1 Putting Numbers Together and Taking Them Apart

EXERCISE 1-1

1. $$5x^2+6y-5z = 5(-4)^2+6(7)-5(12) = 5(16)+42-60$$
$$= 80+42-60 = 62$$

2. $$(4x+3y)(4z+9x) = (4(-4)+3(7))(4(12)+9(-4))$$
$$= (-16+21)(48-36) = (5)(12)=60$$

3. $$\frac{-3x^2+7y+2z}{x(y-z)} = \frac{-3(-4)^2+7(7)+2(12)}{-4(7-12)} = \frac{-3(16)+49+24}{-4(-5)}$$
$$= \frac{-48+49+24}{20} = \frac{25}{20}=\frac{5}{4}$$

4. $6xyz = 6(-4)(7)(12)=-2016$

5. $$3y(2x-3z)^2 = 3(7)(2(-4)-3(12))^2 = 21(-8-36)^2$$
$$= 21(-44)^2=21(1936)=40{,}656$$

EXERCISE 1-2

1. $18p-12q = 18\left(\frac{2}{3}\right)-12\left(\frac{5}{6}\right)=12-10=2$

2. $72p+64r = 72\left(\frac{2}{3}\right)+64\left(\frac{-3}{4}\right)=48-48=0$

3. $pqr = \left(\frac{\cancel{2}}{\cancel{3}}\right)\left(\frac{5}{6}\right)\left(\frac{\cancel{-3}\ -1}{\cancel{4}\ 2}\right) = \frac{-5}{12}$

4. $12(p+r)-72q^2 = 12\left(\frac{2}{3}+\frac{-3}{4}\right)-72\left(\frac{5}{6}\right)^2 = 12\left(\frac{2}{3}\right)+12\left(\frac{-3}{4}\right)-72\left(\frac{25}{36}\right)$

$= 8-9-50=-51$

5. $\frac{5p+3r}{4p-q} = \frac{5\left(\frac{2}{3}\right)+3\left(\frac{-3}{4}\right)}{4\left(\frac{2}{3}\right)-\left(\frac{5}{6}\right)} = \frac{\frac{10}{3}-\frac{9}{4}}{\frac{8}{3}-\frac{5}{6}} = \frac{\left(\frac{10}{3}-\frac{9}{4}\right)}{\left(\frac{8}{3}-\frac{5}{6}\right)}\frac{12}{12} = \frac{12\left(\frac{10}{3}\right)-12\left(\frac{9}{4}\right)}{12\left(\frac{8}{3}\right)-12\left(\frac{5}{6}\right)}$

$= \frac{40-27}{32-10} = \frac{13}{22}$

EXERCISE 1-3

1. $3y+4z-5w+9w-5y+2z$

$= (3-5)y+(4+2)z+(9-5)w = -2y+6z+4w$

2. $12ab-9bc+8ab-6bc = (12+8)ab+(-9-6)bc = 20ab-15bc$

3. $9p^2-8p+4+10p^2-11p-5$

$= (9+10)p^2+(-8-11)p+(4-5) = 19p^2-19p-1$

4. $(9p^2-8p+4)-(10p^2-11p-5) = 9p^2-8p+4-10p^2+11p+5$

$= (9-10)p^2+(-8+11)p+(4+5) = -1p^2+3p+9 = -p^2+3p+9$

5. $(3x^3+5x^2-4x+6)+(4x^3-7x^2-8x+9)$

$= (3+4)x^3+(5-7)x^2+(-4-8)x+(6+9) = 7x^3-2x^2-12x+15$

6. $(3x^3+5x^2-4x+6)-(4x^3-7x^2-8x+9)$

$= 3x^3+5x^2-4x+6-4x^3+7x^2+8x-9$

$= (3-4)x^3+(5+7)x^2+(-4+8)x+(6-9) = -x^3+12x^2+4x-3$

7. $4(3x^3+5x^2-4x+6)+5(4x^3-7x^2-8x+9)$
   $=12x^3+20x^2-16x+24+20x^3-35x^2-40x+45$
   $=(12+20)x^3+(20-35)x^2+(-16-40)x+(24+45)$
   $=32x^3-15x^2-56x+69$

8. $5(3x^3+5x^2-4x+6)-2(4x^3-7x^2-8x+9)$
   $=15x^3+25x^2-20x+30-8x^3+14x^2+16x-18$
   $=(15-8)x^3+(25+14)x^2+(-20+16)x+(30-18)$
   $=7x^3+39x^2-4x+12$

9. $12ab^2c-9abc^2+8ab^2c-6a^2bc$
   $=(12+8)ab^2c-9abc^2-6a^2bc=20ab^2c-9abc^2-6a^2bc$

10. $3xyz+4xyz-5wxy+9wxy-5wxyz+2wxyz$
    $=(3+4)xyz+(-5+9)wxy+(-5+2)wxyz=7xyz+4wxy-3wxyz$

EXERCISE 1-4

1. $4x-3=21 \Rightarrow 4x=24 \Rightarrow x=6$
2. $4x+3=21 \Rightarrow 4x=18 \Rightarrow x=4.5$
3. $82-7y=54 \Rightarrow -7y=-28 \Rightarrow y=4$
4. $0.4z+2.9=8.6 \Rightarrow 0.4z=5.7 \Rightarrow z=14.25$
5. $\frac{2}{3}w+19=27 \Rightarrow \frac{2}{3}w=8 \Rightarrow w=12$

Relations and Functions

EXERCISE 2-1

1. $G=\{A(0,0), B(5,1), C(4,-3), D(2,-2), E(-1,-2), F(-4,0), G(-3,4), H(0,3)\}$

EXERCISE 2-2

1. {0, 5, 4, 2, –1, –4, –3}
2. {0, 1, –3, –2, 3, 4}

EXERCISE 2-3

1. P = {Salt Lake, Germany), (Turin, Germany), (Vancouver, United States), (Sochi, Russia), (PyeongChang, Norway)}
2. {Salt Lake, Turin, Vancouver, Sochi, PyeongChang}
3. {Germany, United States, Russia, Norway}
4. There was a unique winner in the medal count at each of the 21st-century Olympic Games.
5. Q is not a function because there is the likelihood that a student in class can be reached on a cell phone as well as a land line.

EXERCISE 2-4

1. $f(3)=(3)^2+9=9+9=18$
2. $f(-3)=(-3)^2+9=9+9=18$
3. $g(8)=\frac{8+4}{8-2}=\frac{12}{6}=2$
4. $g(-4)=\frac{-4+4}{-4-2}=\frac{0}{-6}=0$
5. $h(2) = 11 - 7(2) = -3$
6. $h(-3)=11-7(-3)=11+21=32$
7. $f(2)+g(3)=((2)^2+9)+\left(\frac{3+4}{3-2}\right)=(4+9)+\left(\frac{7}{1}\right)=20$
8. $\frac{f(4)\times g(5)}{h(2)}=\frac{((4)^2+9)\left(\frac{5+4}{5-2}\right)}{11-7(2)}=\frac{(16+9)\left(\frac{9}{3}\right)}{11-14}=\frac{(25)(3)}{-3}=-25$

9. Alex is not correct because each input value has just one output value. (It is very likely that Alex was confused because the output value was the same for both these expressions.)

Linear Equations

EXERCISE 3-1

1. $13x - 19 = 9x + 37 \Rightarrow 4x = 56 \Rightarrow x = 14$
2. $8w + 27 = 5w - 15 \Rightarrow 3w = -42 \Rightarrow w = -14$
3. $10t + 42 = 86 - 12t \Rightarrow 22t = 44 \Rightarrow t = 2$
4. $35 - 7p = 5p + 11 \Rightarrow 24 = 12p \Rightarrow p = 2$
5. $92w + 149 = 47w - 26 \Rightarrow 45w = -175 \Rightarrow w = \dfrac{-35}{9}$
6. $14 + 3(5x - 10) = 4(x - 9) + 3x \Rightarrow 14 + 15x - 30 = 4x - 36 + 3x$
   $\Rightarrow 15x - 16 = 7x - 36 \Rightarrow 8x = -20$
   $\Rightarrow x = -2.5$
7. $74 - 2(17 - 7x) = 11x + 82 \Rightarrow 74 - 34 + 14x = 11x + 82$
   $\Rightarrow 40 + 14x = 11x + 82 \Rightarrow 3x = 42$
   $\Rightarrow x = 14$
8. $45v - 7(6v + 5) = 13v - 2(4v + 9) \Rightarrow 45v - 42v - 35 = 13v - 8v - 18$
   $\Rightarrow 3v - 35 = 5c - 18 \Rightarrow -17 = 2v$
   $\Rightarrow v = -8.5$
9. $120p - 1340 = 480 - 3(95 - 20p) \Rightarrow 120p - 1340 = 480 - 285 + 60p$
   $\Rightarrow 120p - 1340 = 195 + 60p \Rightarrow 60p = 1535 \Rightarrow p = \dfrac{307}{12}$
10. $\dfrac{2}{3}x + \dfrac{5}{8} = \dfrac{5}{12}x - \dfrac{3}{4} \Rightarrow 24\left(\dfrac{2}{3}x + \dfrac{5}{8}\right) = 24\left(\dfrac{5}{12}x - \dfrac{3}{4}\right)$
    $\Rightarrow 24\left(\dfrac{2}{3}x\right) + 24\left(\dfrac{5}{8}\right) = 24\left(\dfrac{5}{12}x\right) - 24\left(\dfrac{3}{4}\right)$
    $\Rightarrow 16x + 15 = 10x - 18 \Rightarrow 6x = -33 \Rightarrow x = \dfrac{-11}{2}$

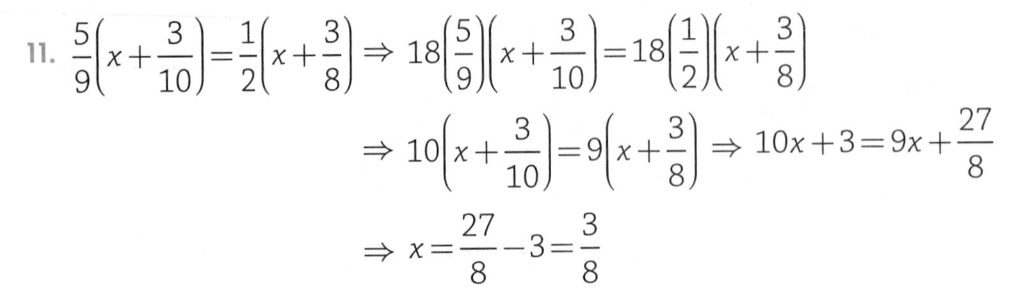

11. $\frac{5}{9}\left(x+\frac{3}{10}\right)=\frac{1}{2}\left(x+\frac{3}{8}\right) \Rightarrow 18\left(\frac{5}{9}\right)\left(x+\frac{3}{10}\right)=18\left(\frac{1}{2}\right)\left(x+\frac{3}{8}\right)$

$$\Rightarrow 10\left(x+\frac{3}{10}\right)=9\left(x+\frac{3}{8}\right) \Rightarrow 10x+3=9x+\frac{27}{8}$$

$$\Rightarrow x=\frac{27}{8}-3=\frac{3}{8}$$

12. $\frac{5}{3}(2x-5)+4=\frac{7}{6}(3x+7)-5 \Rightarrow 6\left(\frac{5}{3}(2x-5)+4\right)=6\left(\frac{7}{6}(3x+7)-5\right)$

$$\Rightarrow 10(2x-5)+24=7(3x+7)-30$$

$$\Rightarrow 20x-50+24=21x+49-30$$

$$\Rightarrow 20x-26=21x+19 \Rightarrow x=-45$$

13. $0.05(7x-19)=0.3x+1.5 \Rightarrow 100(0.05)(7x-19)=100(0.3x+1.5)$

$$\Rightarrow 5(7x-19)=30x+150 \Rightarrow 35x-95=30x+150$$

$$\Rightarrow 5x=245 \Rightarrow x=49$$

14. $\frac{12}{35}=\frac{x}{105} \Rightarrow 35x=1260 \Rightarrow x=36$

15. $\frac{2x+5}{10}=\frac{3x-4}{8} \Rightarrow 8(2x+5)=10(3x-4) \Rightarrow 16x+40=30x-40$

$$\Rightarrow 80=14x \Rightarrow x=\frac{40}{7}$$

16. $\frac{20}{x-4}=\frac{25}{x+2} \Rightarrow 25(x-4)=20(x+2) \Rightarrow 25x-100=20x+40$

$$\Rightarrow 5x=140 \Rightarrow x=28$$

EXERCISE 3-2

1. $5x+4y=30 \Rightarrow 4y=30-5x \Rightarrow y=\frac{30-5x}{4}$

2. $A=\frac{h}{2}(a+b) \Rightarrow 2A=h(a+b) \Rightarrow h=\frac{2A}{a+b}$

3. $ax+by=cx+dy \Rightarrow ax-cx=dy-by \Rightarrow (a-c)x=dy-by$

$$\Rightarrow x=\frac{dy-by}{a-c}=\frac{(d-b)y}{a-c}$$

EXERCISE 3-3

1. Sale price $= 0.85 \times 0.75 \times \$240 = \$153$
2. Final Price $= \$153 \times 1.07 = \$163.71$
3. $0.85 \times 1.07 \times \text{Retail} = 582.08 \Rightarrow \text{Retail} = \dfrac{582.08}{0.85 \times 1.07} = \$640$

EXERCISE 3-4

1.

Score	Frequency	Total
98	5	490
94	7	658
91	10	910
85	16	1360
78	18	1404
74	7	518
68	3	204

$$\text{Average} = \frac{\text{Sum of total}}{\text{Sum of frequency}} = \frac{5544}{66} = 84$$

2.

Score	Frequency	Total
95	4	380
89	6	534
87	x	$87x$
83	10	830
75	5	375

Solve the equation: $\dfrac{380+534+87x+830+375}{4+6+x+10+5} = 85.25$

$$\Rightarrow \frac{2119+87x}{x+25} = 85.25 \Rightarrow 2119+87x = 85.25(x+25)$$

$$\Rightarrow 2119+87x = 85.25x+2131.5 \Rightarrow 1.75x = 12.25 \Rightarrow x = 7$$

EXERCISE 3-5

	Math	English	Science	Social Studies	Total
Boys	35	30	40	30	135
Girls	40	55	30	30	155
Total	75	85	70	60	290

1. $\frac{75}{290} = 25.9\%$
2. $\frac{40}{155} = 25.8\%$
3. $\frac{40}{70} = 57.1\%$
4. $\frac{60}{145} = 41.4\%$
5. $\frac{60}{135} = 44.4\%$

Equations of Linear Functions

EXERCISE 4-1

1. $m = \frac{0-12}{3-0} = \frac{-12}{3} = -4$
2. $m = \frac{-1-(-5)}{4-(-2)} = \frac{4}{6} = \frac{2}{3}$
3. $m = \frac{12-(-2)}{9-9} = \frac{14}{0}$; undefined
4. $m = \frac{12-12}{13-(-5)} = \frac{0}{18} = 0$

5. $m=\frac{-6-5}{3-(-8)}=\frac{-11}{11}=-1$

6. $m=\frac{4-(-1)}{5-(-1)}=\frac{5}{6}$

7. $m=\frac{3-(-3)}{7-(-7)}=\frac{6}{14}=\frac{3}{7}$

8. Using model $y = mx + b$, the slope is -3.

9. Using the model $y-y_1=m(x-x_1)$, the slope is $\frac{-2}{3}$.

10. Using the model $Ax + By = C$, the slope is $\frac{-1}{2}$.

11. Parallel lines have equal slopes so the $m=\frac{12-4}{7-(-3)}=\frac{8}{10}=\frac{4}{5}$.

12. The slopes of perpendicular lines are negative reciprocals. The slope of the line through $(-3, 4)$ and $(7, 12)$ is $\frac{4}{5}$, so the slope of the line perpendicular to it is $\frac{-5}{4}$.

13. The slope of the line $5x - 7y = 35$ is $\frac{5}{7}$, so the slope of the line perpendicular to it is $\frac{-7}{5}$.

EXERCISE 4-2

1. Using the data (12, \$140) and (15, \$173), the slope of the line is

$$\frac{\$173-\$140}{15-12}=\frac{\$33}{3}=\$11/\text{hour}.$$

2. Using the data (80, \$87) and (110, \$106.50), the slope of the line is

$$\frac{\$106.50-\$87}{110-80}=\frac{\$19.50}{30}=\$0.65/\text{mile}.$$

EXERCISE 4-3

1. Because the slope is \$11/hour, the equation for her pay, p, must be of the form $p = 11h + b$ where h represents the number of hours she works. Using one of the ordered pairs from the problem to determine the value of b, we get $140 = 11(12) + b$, which gives $b = 8$. The equation of the line is $p = 11h + 8$.
2. Molly is paid \$8 if she works 0 hours. This is the extra amount she is paid to cover transportation costs.

EXERCISE 4-4

1. The slope is \$0.65/mile, so the amount, a, Alice is reimbursed for driving d miles is given by the equation $a = 0.65d + b$. Using the ordered pair (80, 87), we determine that $87 = 0.65(80) + b$ and $b = 35$. Therefore, $a = 0.65d + 35$.
2. $a = 0.65(205) + 35 = 168.25$

EXERCISE 4-5

1. slope $= 2 \Rightarrow y = 2x + b$. Use one of the ordered pairs to show that $b = -1$. The equation is $y = 2x - 1$.
2. $y - 3 = 2(x - 2)$ or $y - 9 = 2(x - 5)$
3. Slope is $\frac{3}{7}$ so the equation is $y-2=\frac{3}{7}(x+4)$ or $y-5=\frac{3}{7}(x-3)$.
4. Use the point-slope response to create the equation in standard form:

	$y-2=\frac{3}{7}(x+4)$	$y-5=\frac{3}{7}(x-3)$
Multiply by 7:	$7(y-2)=3(x+4)$	$7(y-5)=3(x-3)$
Distribute:	$7y-14=3x+12$	$7y-35=3x-9$

Bring variables to the left side and constants to the right side of the equation:

$-3x + 7y = 26$ $-3x + 7y = 26$

5. Slope $= \frac{-1}{5}$; point-slope: $y-10=\frac{-1}{5}(x-8)$ or $y-12=\frac{-1}{5}(x+2)$;

   slope-intercept: $y=\frac{-1}{5}x+\frac{58}{5}$; Standard: $x + 5y = 58$

6. Slope: $\frac{15}{4}$; point-slope: $y+4=\frac{15}{4}(x+7)$ or $y-11=\frac{15}{4}(x+3)$;

   slope-intercept: $y=\frac{15}{4}x+\frac{89}{4}$; standard form: $15x - 4y = -89$

7. Slope $= \frac{-5}{2}$; $-2=\frac{-5}{2}(4)+b \Rightarrow b=8$; Equation: $y=\frac{-5}{2}x+8$

8. Slope $= \frac{7}{3}$; $-3=\frac{7}{3}(5)+b \Rightarrow b=\frac{-44}{3}$; Equation: $y=\frac{7}{3}x-\frac{44}{3}$

9. $5x + 7y = C \Rightarrow 5(-2) + 7(-6) = C \Rightarrow C = -52$;
   Equation: $5x + 7y = -52$.

10. $10x - 11y = C \Rightarrow 10(4) - 11(-7) = C \Rightarrow C = 117$;
    Equation $10x - 11y = 117$

11. Point is (2, 0); $7x + 8y = C \Rightarrow 7(2) + 8(0) = C \Rightarrow C = 14$;
    Equation $7x + 8y = 14$

12. Slope $= \frac{-5}{3}$; point (0, 4); Equation: $y=\frac{-5}{3}x+4$.

13. Slope $= \frac{7}{5}$; Equation: point-slope: $y-5=\frac{7}{5}(x-3)$;

    slope-intercept: $y=\frac{7}{5}x+\frac{4}{5}$

14. $3x - 8y = C \Rightarrow 3(-10) - 8(6) = C \Rightarrow C = -78$; Equation: $3x - 8y = -78$

15. $3x + 8y = C \Rightarrow 3(-10) + 8(6) = C \Rightarrow C = 18$; Equation: $3x + 8y = 18$

EXERCISE 4-6

1. Solve for y: $x = 5y + 2 \Rightarrow x - 2 = 5y \Rightarrow y = \frac{x-2}{5} = \frac{1}{5}x - \frac{2}{5}$
2. Solve for y: $x = \frac{5}{11}y - 10 \Rightarrow x + 10 = \frac{5}{11}y \Rightarrow y = \frac{11}{5}(x+10) = \frac{11}{5}x + 22$
3. The graph of the function $y = f(x)$ is shown below. Sketch the graph of the inverse function on the same set of axes. Identify the coordinates of the three points on the line: (–5, 3), (–1, 1) , and (3, –1).

   Interchange the x- and y-coordinates: (3, –5), (1, –1), and (–1, 3). Plot these points and draw the line containing them.

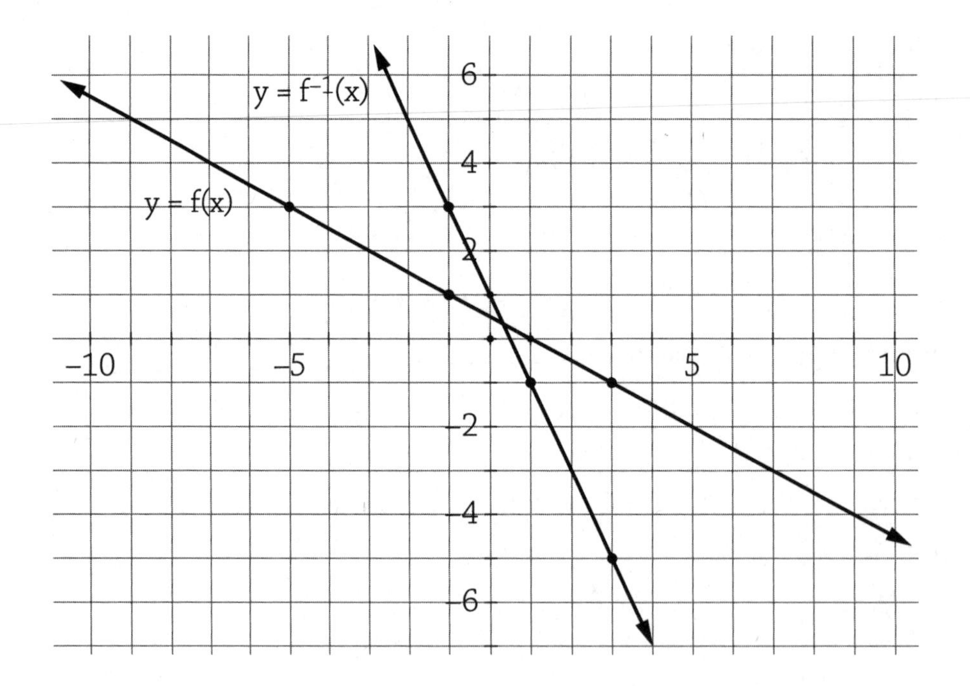

Applications of Linear Functions

EXERCISE 5-1

1. n: number of nickels

   $n + 15$: number of dimes

   $3n + 5$: number of quarters

   $n + n + 15 + 3n + 5 = 165 \Rightarrow 5n + 20 = 165 \Rightarrow 5n = 145 \Rightarrow n = 29$

   Cameron has 29 nickels, 44 dimes, and 92 quarters in his piggy bank.

2. Cameron has $.05(29) + .10(44) + .25(92) = \$28.85$ in his piggy bank.

EXERCISE 5-2

1. s: the number of singles

   $2s + 20$: the number of \$5 bills

   $2s + 20 + 50 = 2s + 70$: the number of \$10 bills

   $s + 2s + 20 + 2s + 70 = 240 \Rightarrow 5s + 90 = 240 \Rightarrow 5s = 150 \Rightarrow s = 30$

   Charlie had 30 singles, 80 five-dollar bills, and 130 ten-dollar bills in his drawer.

2. Charlie had $\$1(30) + \$5(80) + \$10(130) + \$12.35 = \$1742.35$ in his drawer at closing. (His manager might want to do a better job of putting some of that money is a safe during business hours.)

EXERCISE 5-3

1. w: width of the rectangle

   $2w - 1$: length of the rectangle

   $2w + 2(2w - 1) = 52 \Rightarrow 2w + 4w - 2 = 52$

   $\Rightarrow 6w = 54 \Rightarrow w = 9$

   The dimensions of the rectangle are 9 feet $\Rightarrow$ 17 feet.

$2w - 1$

w

2. L: length of the rectangle

$\frac{1}{2}L-2$: width of the rectangle

$2\left(\frac{1}{2}L-2\right)+2L=734 \Rightarrow L-4+2L=734$

$\Rightarrow 3L=738 \Rightarrow L=246$

The dimensions of the rectangle are 121 cm × 246 cm.

3. L: the length of a leg of the triangle

$2L-3$: the length of the base of the triangle

$L+L+2L-3=53 \Rightarrow 4L=56 \Rightarrow L=14$

The dimensions of the triangle are 14 inches × 14 inches × 25 inches.

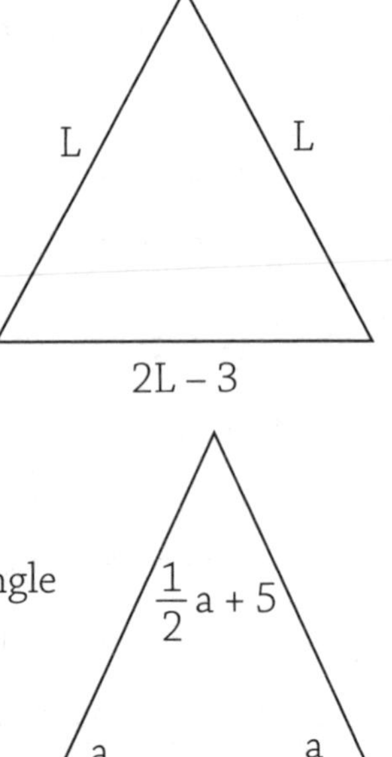

4. a: the measure of a base angle of the triangle

$\frac{1}{2}a+5$: the measure of the vertex angle of the triangle

$a+a+\frac{1}{2}a+5=180 \Rightarrow 2.5a=175 \Rightarrow a=70$

The measures of the angles of the triangle are 70°, 70°, and 40°.

5. a: the measure of the smallest angle of the quadrilateral

$3a$: the measure of the largest angle of the quadrilateral

$3a-10$: the measure of the fourth angle of the quadrilateral

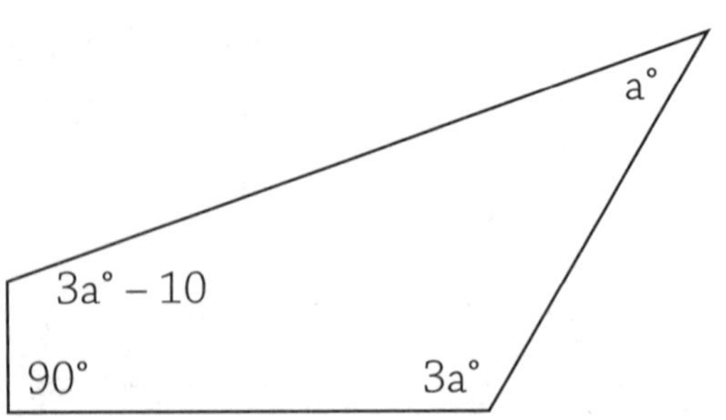

$90+a+3a+3a-10=360$

$\Rightarrow 7a+80=360 \Rightarrow 7a=280 \Rightarrow a=40$

The measures of the four angles are 90, 40, 120, and 110.

6. t: time each vehicle has traveled

   $45t + 55t = 300 \Rightarrow 100t = 300 \Rightarrow t = 3$

   The two vehicles are 300 miles apart at 4 p.m.

7. t: time Joanne travels when she passes John

   $25\left(t + \frac{1}{2}\right) = 40t \Rightarrow 25t + 12.5 = 40t \Rightarrow 12.5 = 15t$

   $\Rightarrow t = \frac{5}{6}$ hr $= 50$ minutes

   Joanne passes John at 10:20 a.m.

8. t: time it takes for the trucks to reach each other

   $65t + 60t = 1000 \Rightarrow 125t = 1000 \Rightarrow t = 8$

   The two trucks will pass each other at 4 p.m.

9. $\frac{7}{20} = \frac{d}{35} \Rightarrow 20d = 245 \Rightarrow d = 12.25$ inches

10. $\frac{1.4}{280} = \frac{v}{285} \Rightarrow 280v = 399 \Rightarrow v = 1.425$ cc

11. $d = kt^2 \Rightarrow 19.6 = k(2)^2 \Rightarrow k = 4.9 \Rightarrow d = 4.9(3)^2 \Rightarrow 44.1$ m

EXERCISE 5-4

1. Common difference $= 23 \Rightarrow 81 = 23(1) + b \Rightarrow b = 58$

   $\Rightarrow a = 23n + 58$

2. Common difference $= -6 \Rightarrow 99 = -6(1) + b \Rightarrow b = 105$

   $\Rightarrow a = -6n + 105$

3. Common difference $= 27 \Rightarrow 123 = 27(1) + b \Rightarrow b = 96$

   $\Rightarrow a = 27n + 96$

4. Common difference $= -13 \Rightarrow 42 = -13(1) + b \Rightarrow b = 55$

   $\Rightarrow a = -13n + 55$

EXERCISE 5-5

1. $a_{30} = 23(30) + 58 = 748$
2. $a_{50} = -6(50) + 105 = -195$
3. $a_{90} = 27(90) + 96 = 2526$
4. $a_{15} = -13(15) + 55 = -140$
5. Use the points (18, 44) and (33, 164) to determine the common difference is $\frac{164-44}{33-18} = \frac{120}{15} = 8$. $a_{18} = 44 \Rightarrow 44 = 8(18) + b$

   $\Rightarrow 44 = 144 + b$

   $\Rightarrow b = -100 \Rightarrow a_n = 8n - 100 \Rightarrow a_{200} = 8(200) - 100 = 1500$
6. Use the points (7, 74) and (12, 44) to determine the common difference $\frac{44-74}{12-7} = \frac{-30}{5} = -6$. $a_7 = 74 \Rightarrow 74 = -6(7) + b \Rightarrow b = 116$

   $\Rightarrow a_n = -6n + 116 \Rightarrow a_{25} = -6(25) + 116 = -34$
7. $a_1 = 8 - 100 = -92$ and $a_{250} = 8(250) - 100 = 1900$

   $\Rightarrow S_{250} = \frac{250}{2}(-92+1900) = 226{,}000$
8. $a_1 = -6 + 116 = 110$ and $a_{120} = -6(120) + 116 = -604$

   $\Rightarrow S_{120} = \frac{120}{2}(110-604) = -29{,}640$
9. There are 35 seats in the first row and 97 seats in the 32nd row (see the section on arithmetic sequences where this number was determined), so the number of seats in the center section is $S_{32} = \frac{32}{2}(35+97) = 2112$.

EXERCISE 5-6

1. Geometric: Those points 4 units from 7 on the number line are 3 and 11.

   Algebraic: $x - 7 = -4$ or $x - 7 = 4 \Rightarrow$ Add 7 to get $x = 3, 11$

2. Geometric: Those points 2 units from -3 on the number line are -5 and -1.

   Algebraic: $x + 3 = -2$ or $x + 3 = 2 \Rightarrow$ Subtract 3 to get $x = -5, -1$

3. Factor: $|2|\ |x - 3| = 10 \Rightarrow |x - 3| = 5$

   Geometric: Those points 5 units from 3 on the number line are -2 and 8.

   Algebraic: $2x - 6 = -10$ or $2x - 6 = 10 \Rightarrow$ Add 6: $2x = -4$ or $2x = 16 \Rightarrow$ Divide: $x = -2, 8$

4. Factor: $|3|\ |x + 3| = 15 \Rightarrow |x + 3| = 5$

   Geometric: Those points 5 units from -3 on the number line are -8 and 2.

   Algebraic: $3x + 9 = -15$ or $3x + 9 = 15 \Rightarrow$ Subtract 9: $3x = -24$ or $3x = 6 \Rightarrow$ Divide: $x = -8, 2$

5. Factor: $|2|\ |x - 3.5| = 11 \Rightarrow |x - 3.5| = 5.5$

   Geometric: Those points 5.5 units from 3.5 on the number line are -2 and 9.

   Algebraic: $2x - 7 = -11$ or $2x - 7 = 11 \Rightarrow$ Add 7: $2x = -4$ or $2x = 18 \Rightarrow$ Divide: $x = -2, 9$

6. Factor: $|4|\ |x + 1.25| = 13 \Rightarrow |x + 1.25| = 3.25$

   Geometric: Those points 3.25 units from -1.25 on the number line are -4.5 and 2.

   Algebraic: $4x + 5 = -13$ or $4x + 5 = 13 \Rightarrow$ Subtract 5: $4x = -18$ or $4x = 8 \Rightarrow$ Divide: $x = -4.5, 2$

7. Factor: $|-3|\ |x - 3| = 12 \Rightarrow |x - 3| = 4$

   Geometric: Those points 4 units from 3 on the number line are -1 and 7.

   Algebraic: $9 - 3x = -12$ or $9 - 3x = 12 \Rightarrow$ Subtract 9: $-3x = -21$ or $-3x = 3 \Rightarrow x = 7, -1$.

8. Factor: $|-5|\ |x - 1.6| = 10 \Rightarrow |x - 1.6| = 2$

   Geometric: Those points 2 units from 1.6 on the number line are -0.4 and 3.6.

Algebraic: $8 - 5x = -10$ or $8 - 5x = 10 \Rightarrow$ Subtract 8: $-5x = -18$ or $-5x = 2 \Rightarrow$ Divide: $x = 3.6, -0.4$.

9. Subtract 10: $|3x + 2| = 5 \Rightarrow$ Factor: $|3|\left|x + \frac{2}{3}\right| = 5 \Rightarrow \left|x + \frac{2}{3}\right| = \frac{5}{3}$

   Geometric: Those points $\frac{5}{3}$ units from $\frac{-2}{3}$ on the number line are $\frac{-7}{3}$ and 1.

   Algebraic: $3x + 2 = -5$ or $3x + 2 = 5 \Rightarrow$ Subtract 2: $3x = -7$ or $3x = 3 \Rightarrow x = \frac{-7}{3}, 1$.

Linear Inequalities

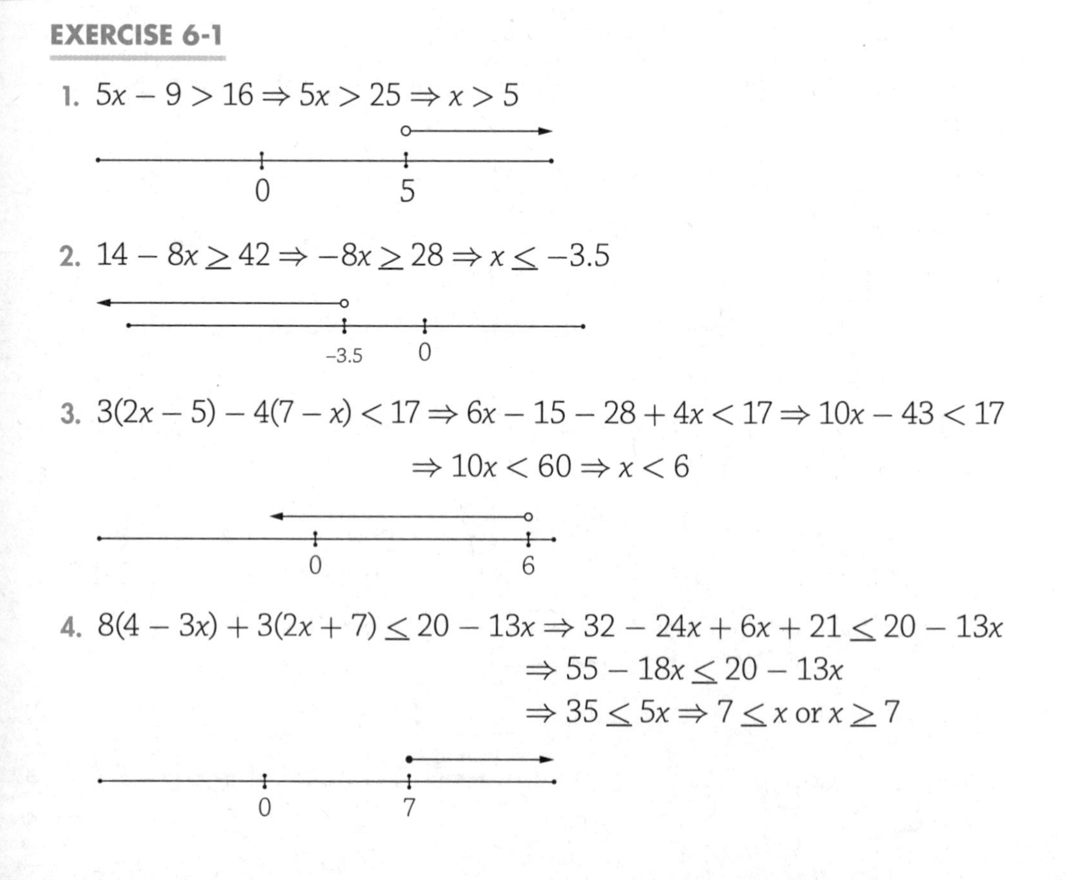

EXERCISE 6-1

1. $5x - 9 > 16 \Rightarrow 5x > 25 \Rightarrow x > 5$

2. $14 - 8x \geq 42 \Rightarrow -8x \geq 28 \Rightarrow x \leq -3.5$

3. $3(2x - 5) - 4(7 - x) < 17 \Rightarrow 6x - 15 - 28 + 4x < 17 \Rightarrow 10x - 43 < 17$
   $\Rightarrow 10x < 60 \Rightarrow x < 6$

4. $8(4 - 3x) + 3(2x + 7) \leq 20 - 13x \Rightarrow 32 - 24x + 6x + 21 \leq 20 - 13x$
   $\Rightarrow 55 - 18x \leq 20 - 13x$
   $\Rightarrow 35 \leq 5x \Rightarrow 7 \leq x$ or $x \geq 7$

5. $11 < 2x - 7 \leq 29 \Rightarrow 18 < 2x \leq 36 \Rightarrow 9 < x \leq 18$

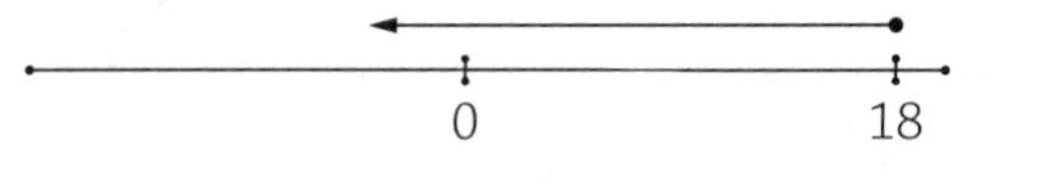

6. $32 < 4 - 7x < 53 \Rightarrow 28 < -7x < 49 \Rightarrow -4 > x > -7$ or $-7 < x < -4$

7. $7x + 19 \leq 33$ or $12x - 35 > 49 \Rightarrow 7x \leq 14$ or $12x > 84 \Rightarrow x \leq 2$ or $x > 7$

8. $17 - 3x < 26$ or $14 + 8x > 5x - 7 \Rightarrow -3x < 9$ or $3x > -21 \Rightarrow x > -3$ or $x > -7 \Rightarrow x > -7$

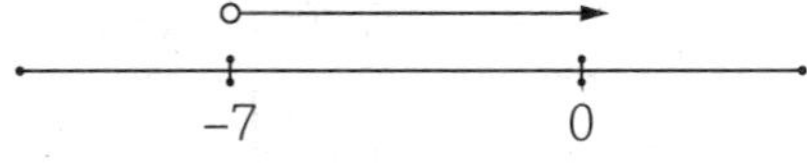

9. $3\frac{2}{3} \leq \frac{5}{9}x - 4\frac{1}{2} < 4\frac{5}{12} \Rightarrow$ Write as improper fractions: $\frac{11}{3} \leq \frac{5}{9}x - \frac{9}{2} < \frac{53}{12}$

$\Rightarrow$ Multiply by the least common denominator (72):

$$72\left(\frac{11}{3}\right) \leq 72\left(\frac{5}{9}x - \frac{9}{2}\right) < 72\left(\frac{53}{12}\right)$$

$\Rightarrow 24(11) \leq 8(5x) - 36(9) < 6(53)$

$\Rightarrow$ Multiply: $264 \leq 40x - 324 < 318$

$\Rightarrow$ Add 324: $588 \leq 40x < 642$

$\Rightarrow$ Divide by 40: $14.7 \leq x < 16.05$

10. $810{,}000 > 20{,}000x + 197{,}000$ or $15{,}000x - 210{,}000 > 60{,}000$

$\Rightarrow$ Isolate terms in x: $613{,}000 > 20{,}000x$ or $15{,}000x > 270{,}000$

$\Rightarrow$ Divide: $30.65 > x$ or $x > 18 \Rightarrow x < 30.65$ or $x > 18$

$\Rightarrow$ Set of real numbers

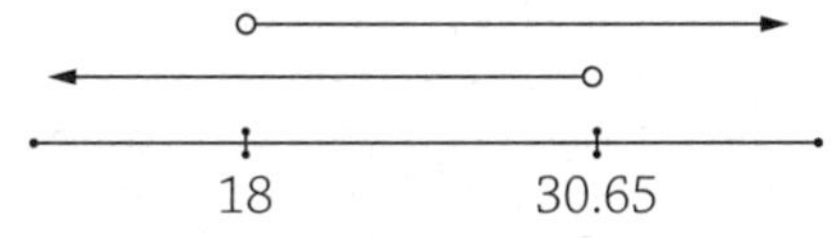

EXERCISE 6-2

1. Geometrically: The set of points at most 5 units from 9 is $4 \leq x \leq 14$.

   Algebraically: $|x-9| \leq 5 \Rightarrow -5 \leq x-9 \leq 5 \Rightarrow$ add 9: $4 \leq x \leq 14$

2. Geometrically: The set of points greater than 7 units from -3 is $x < -10$ or $x > 4$

   Algebraically: $|x+3| > 7 \Rightarrow x+3 < -7$ or $x+3 > 7 \Rightarrow x < -10$ or $x > 4$

3. Geometrically: $|2x-5| \geq 13 \Rightarrow |2|\ |x-2.5| \geq 13 \Rightarrow |x-2.5| \geq 6.5 \Rightarrow$

   The set of points at least 6.5 units from 2.5 are $x \leq -4$ or $x \geq 9$.

   Algebraically: $|2x-5| \geq 13 \Rightarrow 2x-5 \leq -13$ or $2x-5 \geq 13 \Rightarrow 2x \leq -8$ or $2x \geq 18 \Rightarrow x \leq -4$ or $x \geq 9$

4. Geometrically: $|3x+10| < 14 \Rightarrow |3|\left|x+\frac{10}{3}\right| < 14 \Rightarrow \left|x+\frac{10}{3}\right| < \frac{14}{3} \Rightarrow$

   The set of points less than $\frac{14}{3}$ units from $\frac{-10}{3}$ is $\frac{-24}{3} < x < \frac{4}{3}$

   $\Rightarrow -8 < x < \frac{4}{3}$.

   Algebraically: $|3x+10| < 14 \Rightarrow -14 < 3x+10 < 14 \Rightarrow -24 < 3x < 4$

   $\Rightarrow -8 < x < \frac{4}{3}$

EXERCISE 6-3

1. $|z-12| \leq 0.25$

2. Midway between -8 and 0 is -4. The points graphed are no more than 4 units from -4, so $|x+4| < 4$.

EXERCISE 6-4

1. $y > -4x - 2$

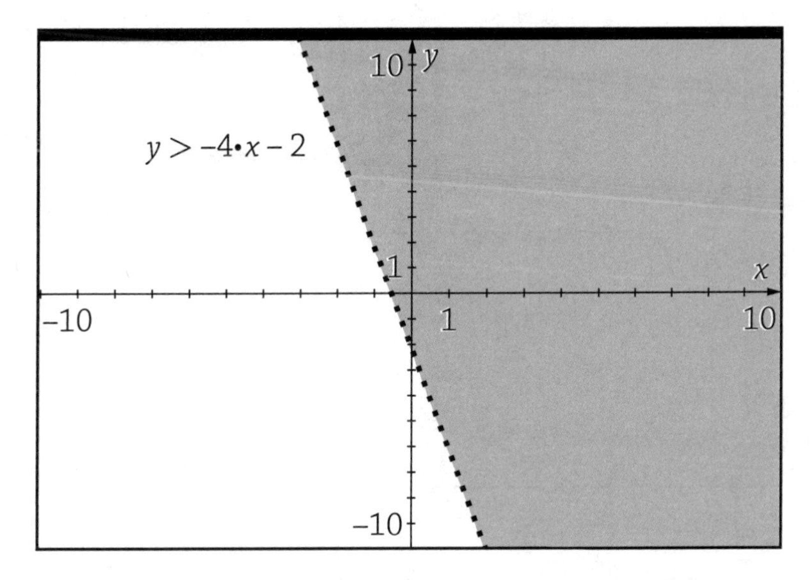

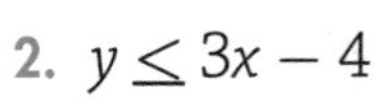

2. $y \leq 3x - 4$

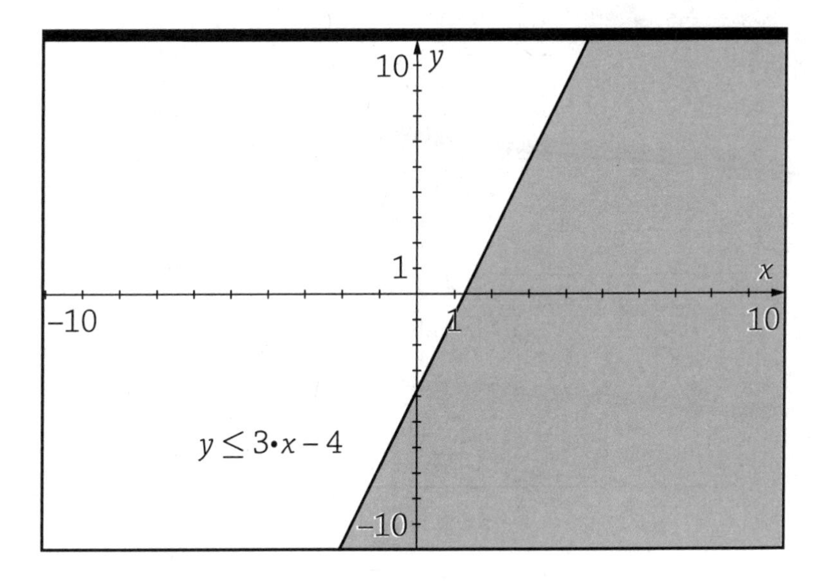

3. $2x + 5y \geq 20$

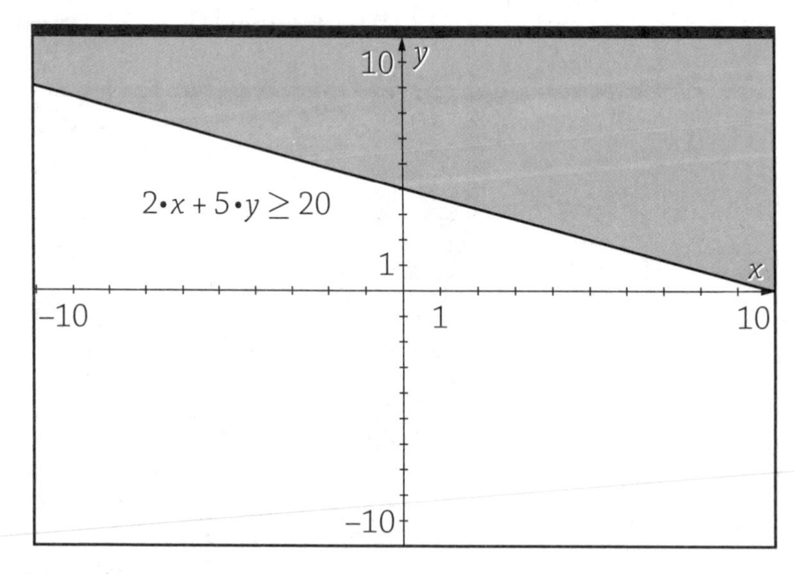

4. $3x - 2y < 12$

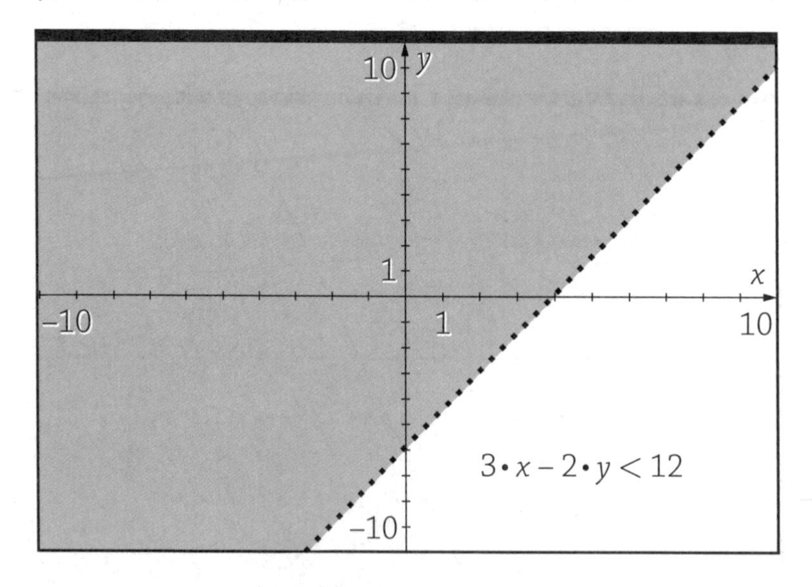

Systems of Linear Equations and Inequalities

EXERCISE 7-1

1.

2.

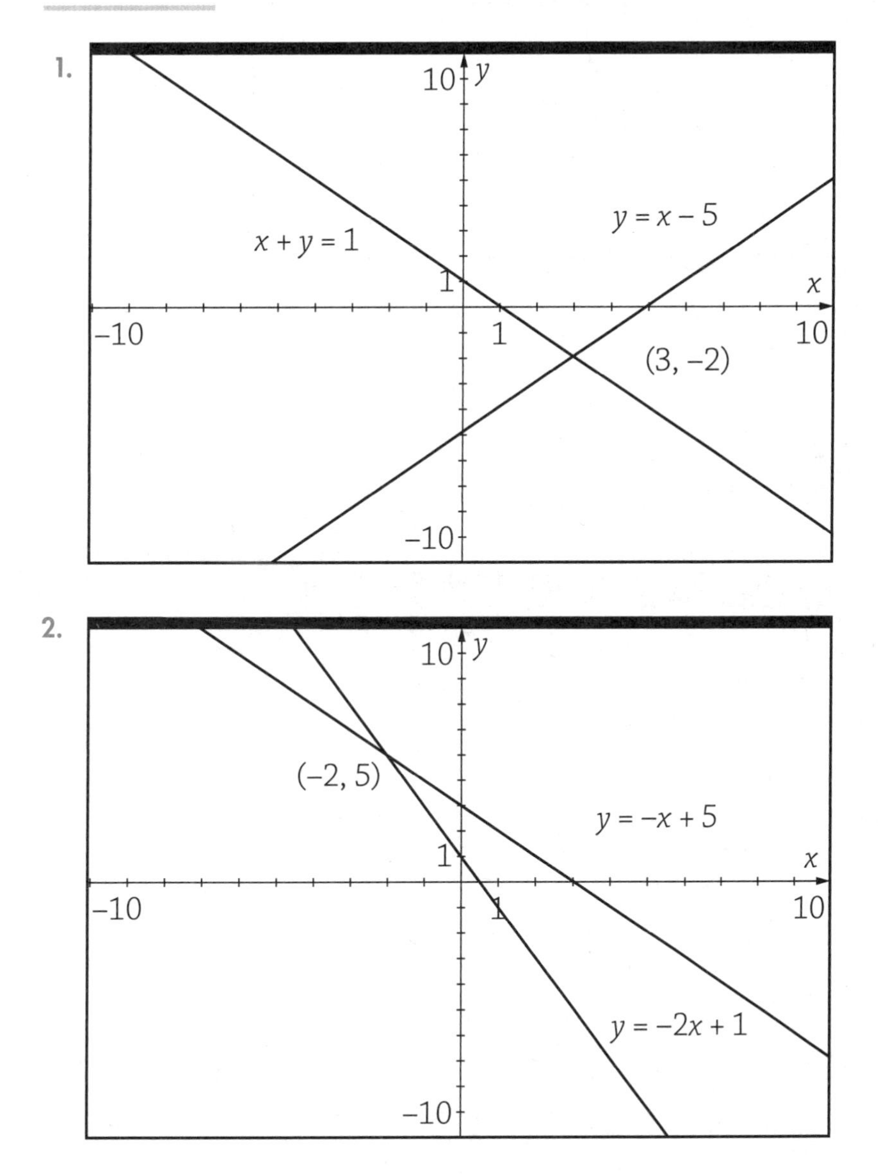

3.

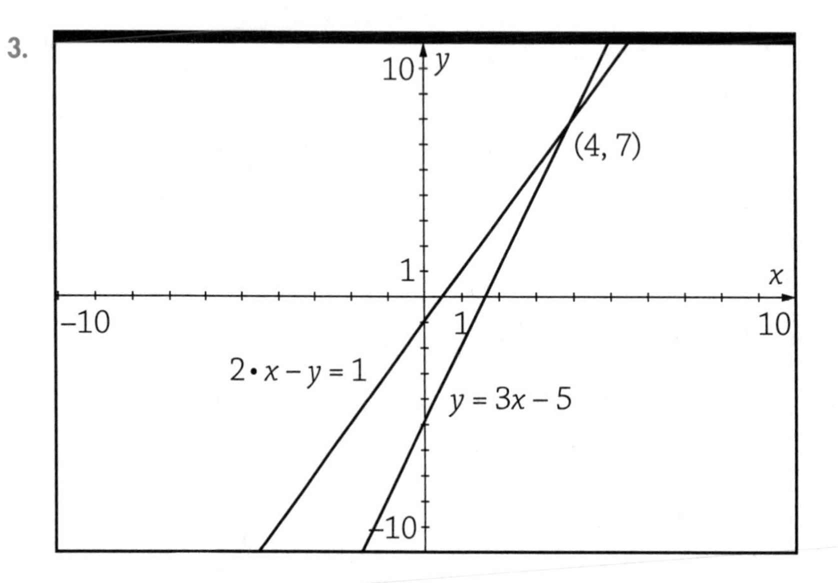

EXERCISE 7-2

1. $(-4, 3)$

$y = 3x + 15 \Rightarrow -x - 1 = 3x + 15 \Rightarrow -16 = 4x \Rightarrow x = -4$
$\Rightarrow y = 4 - 1 = 3$

2. $(-2, 9)$

$3x + 2y = 12 \Rightarrow 3x + 2(-2x + 5) = 12 \Rightarrow 3x - 4x + 10 = 12$
$\Rightarrow -x = 2 \Rightarrow x = -2 \Rightarrow y = -2(-2) + 5 = 9$

3. $(4, 9)$

$7x - 3y = 1 \Rightarrow 7x - 3(2x + 1) = 1 \Rightarrow 7x - 6x - 3 = 1 \Rightarrow x = 4$
$\Rightarrow y = 2(4) + 1 = 9$

EXERCISE 7-3

1. $(2, 3)$

$4x + 3y = 17 \Rightarrow 4x + 3y = 17$

$3(5x - y = 7) \Rightarrow 15x - 3y = 21$

Add: $19x = 38 \Rightarrow x = 2 \Rightarrow 4(2) + 3y = 17 \Rightarrow 3y = 9 \Rightarrow y = 3$

2. (7, 12)

$$\begin{aligned} 5(5x+3y=71) \\ -3(2x+5y=74) \end{aligned} \Rightarrow \begin{aligned} 25x+15y=355 \\ -6x-15y=-222 \end{aligned}$$

Add: $19x = 133 \Rightarrow x = 7 \Rightarrow 5(7) + 3y = 71 \Rightarrow 35 + 3y = 71 \Rightarrow 3y = 36$
$\Rightarrow y = 12$

3. (11, 7)

$$\begin{aligned} -3(5x-8y=-1) \\ 5(3x+5y=68) \end{aligned} \Rightarrow \begin{aligned} -15x+24y=3 \\ 15x+25y=340 \end{aligned}$$

Add: $49y = 343 \Rightarrow y = 7 \Rightarrow 3x + 5(7) = 68 \Rightarrow 3x + 35 = 68 \Rightarrow 3x = 33$
$\Rightarrow x = 11$

4. No solution

$$\begin{aligned} 6x+9y=27 \\ -3(2x+3y=10) \end{aligned} \Rightarrow \begin{aligned} 6x+9y=27 \\ -6x-9y=-30 \end{aligned}$$

Add: $0 = -3 \Rightarrow$ No solution

EXERCISE 7-4

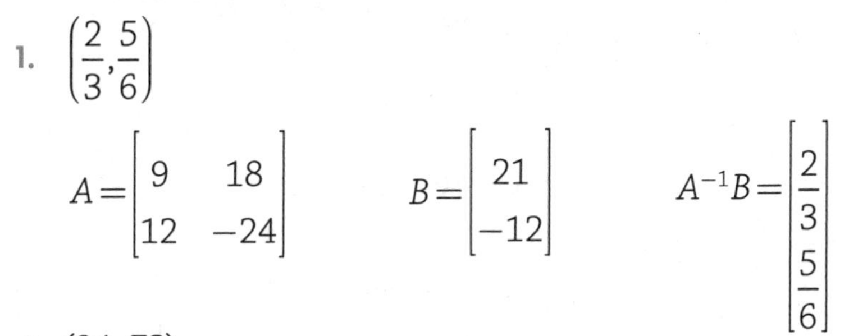

1. $\left(\frac{2}{3}, \frac{5}{6}\right)$

$$A=\begin{bmatrix} 9 & 18 \\ 12 & -24 \end{bmatrix} \qquad B=\begin{bmatrix} 21 \\ -12 \end{bmatrix} \qquad A^{-1}B=\begin{bmatrix} \frac{2}{3} \\ \frac{5}{6} \end{bmatrix}$$

2. (84, 73)

$$A=\begin{bmatrix} 3 & -2 \\ 5 & 7 \end{bmatrix} \qquad B=\begin{bmatrix} 106 \\ 931 \end{bmatrix} \qquad A^{-1}B=\begin{bmatrix} 84 \\ 73 \end{bmatrix}$$

3. (–47, 57)

$$A=\begin{bmatrix}4 & 7\\5 & -2\end{bmatrix} \qquad B=\begin{bmatrix}-211\\-349\end{bmatrix} \qquad A^{-1}B=\begin{bmatrix}-47\\57\end{bmatrix}$$

EXERCISE 7-5

1. (8, 7)
2. (7, 2)
3. (12, –11)
4. (4.5, 8)
5. (–1.03, 2.97)
6. Infinite solutions—same line
7. (16, –23)

EXERCISE 7-6

1. d: number of dimes; q: number of quarters

   $$\begin{matrix} d+q=200 \\ 10d+25q=3740 \end{matrix} \Rightarrow \begin{matrix} -10(d+q=200) \\ 10d+25q=3740 \end{matrix} \Rightarrow \begin{matrix} -10d-10q=-2000 \\ 10d+25q=3740 \end{matrix}$$

   Add: $15q = 1740 \Rightarrow q = 116$

   Cameron and Carson have 84 dimes and 116 quarters.

2. x: number of \$1 coins; y: number of \$2 coins

   $$\begin{matrix} x+y=59 \\ x+2y=96 \end{matrix} \Rightarrow \begin{matrix} -(x+y=59) \\ x+2y=96 \end{matrix} \Rightarrow \begin{matrix} -x-y=-59 \\ x+2y=96 \end{matrix}$$

   Add: $y = 37$

   Diane and Chris have thirty-seven \$2 coins and twenty-two \$1 coins.

3. r: rate in still water; c: rate of the current

   $$\begin{matrix} 2(r+c)=18 \\ 6(r-c)=18 \end{matrix} \Rightarrow \begin{matrix} r+c=9 \\ r-c=3 \end{matrix}$$

Add: $2r = 12 \Rightarrow r = 6$

The boat's speed in still water is 6 mph, and the speed of the current is 3 mph.

4. a: plane's speed on eastbound flight; w: speed of the jet stream

$$\begin{array}{ll} \text{Eastbound}: 15(a+w)=7200 & a+w=480 \\ & \Rightarrow \\ \text{Westbound}: 15(a+240-w)=7200 & a-w=240 \end{array}$$

Add: $2a = 720 \Rightarrow a = 360$

The jet stream has a speed of 120 mph. The pilot flies east at 360 mph and west at 600 mph to maintain the schedule.

5. w: number of ml of water added; a: number of ml of 70% acid solution

$$\begin{array}{l} a+w=105 \\ .70a=.40(a+w) \end{array} \Rightarrow \text{Substitute } .70a=.40(105) \Rightarrow .70a=42 \Rightarrow a=60$$

6. Add 45 ml water to 60 ml of a 70% acid solution to make 105 ml of a 40% acid solution.

EXERCISE 7-7

1.

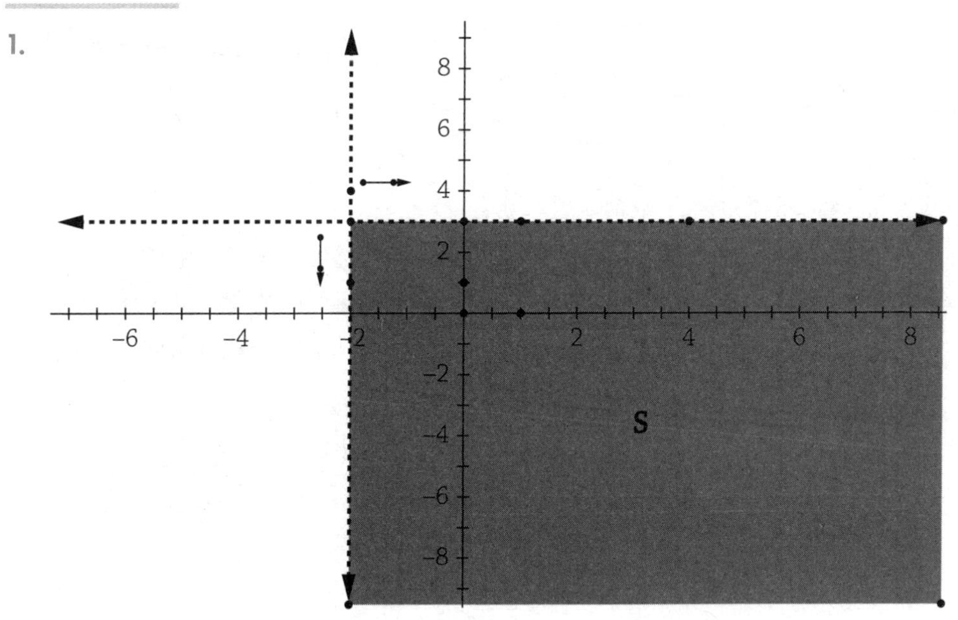

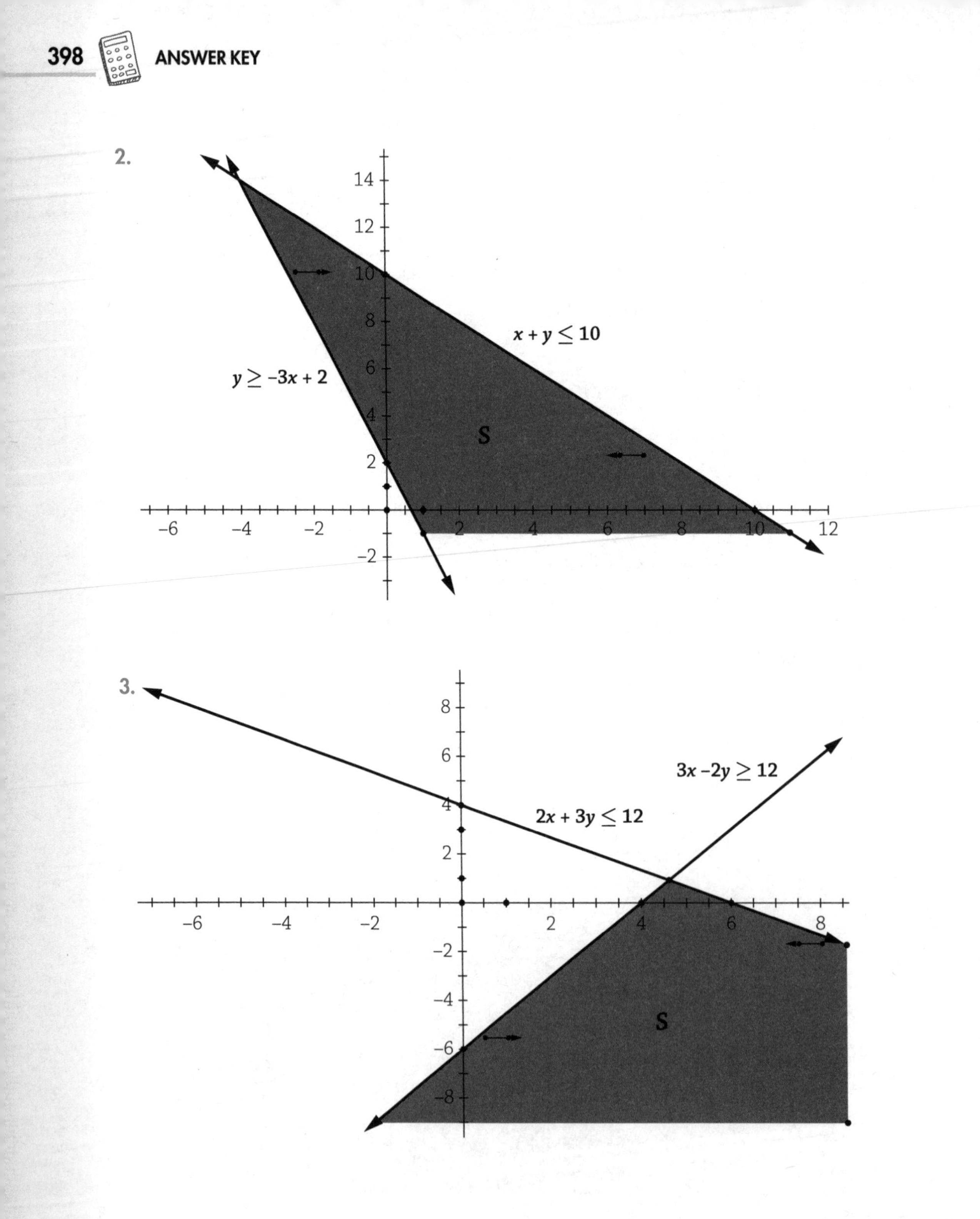

2.
$x + y \leq 10$
$y \geq -3x + 2$
S
3.
$3x - 2y \geq 12$
$2x + 3y \leq 12$
S

4.

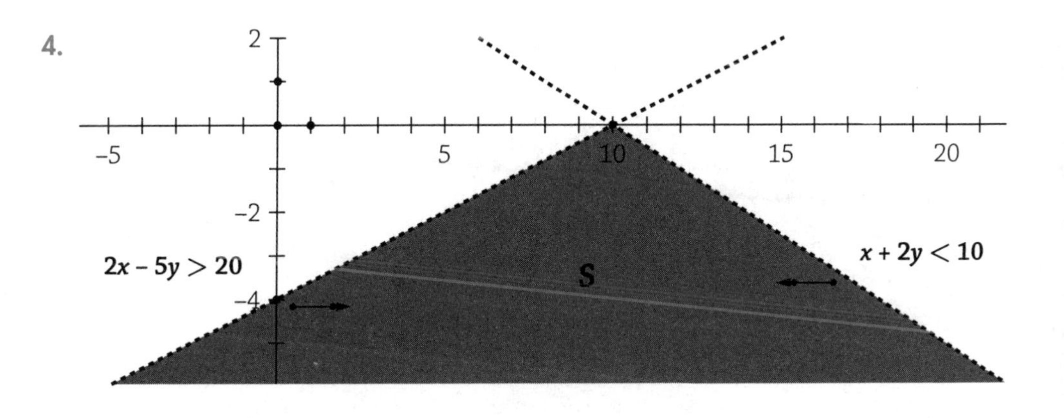

Exponents and Exponential Functions

EXERCISE 8-1

1. $(4x^2)(19x^7) \Rightarrow (4\times 19)(x^2)(x^7) \Rightarrow 76x^9$

2. $500x^7y^7z^9$

3. $(6m^4)^3 \Rightarrow 6^3(m^4)^3 \Rightarrow 216m^{12}$

4. $(-4a^3b^2c^4)^3 \Rightarrow (-4)^3(a^3)^3(b^2)^3(c^4)^3 \Rightarrow -64a^9b^6c^{12}$

5. $(7w^3)^2(4w^5)^3 \Rightarrow (7^2)(4^3)(w^3)^2(w^5)^3 \Rightarrow (49)(64)(w^6)(w^{15}) \Rightarrow 3136w^{21}$

6. $\dfrac{(6a^3)^2}{9a^5} \Rightarrow \dfrac{6^2(a^3)^2}{9a^5} \Rightarrow \dfrac{36a^6}{9a^5} \Rightarrow 4a$

7. $\dfrac{(15r^6p^4)^2}{(5r^4p^2)^3} \Rightarrow \dfrac{15^2(r^6)^2(p^4)^2}{5^3(r^4)^3(p^2)^3} \Rightarrow \dfrac{225r^{12}p^8}{125r^{12}p^6} \Rightarrow \dfrac{9p^2}{5}$

8. $\left(\dfrac{96k^{-5}g^4}{120k^7g^{-2}}\right)^3 \Rightarrow \left(\dfrac{4g^6}{5k^{12}}\right)^3 \Rightarrow \dfrac{4^3(g^6)^3}{5^3(k^{12})^3} \Rightarrow \dfrac{64g^{18}}{125k^{36}}$

9. $\left(\frac{45c^{-3}d^{5}}{30c^{-5}d^{-3}}\right)^{3} \Rightarrow \left(\frac{3c^{2}d^{8}}{2}\right)^{3} \Rightarrow \frac{27c^{6}d^{24}}{8}$

10. $\left(\frac{12w^{3}v^{-4}}{15w^{-2}v^{3}}\right)^{2}\left(\frac{5w^{-4}v^{4}}{6w^{5}v^{-2}}\right)^{3} \Rightarrow \left(\frac{4w^{5}}{5v^{7}}\right)^{2}\left(\frac{5v^{6}}{6w^{9}}\right)^{3} \Rightarrow \left(\frac{\cancel{16}^{2}w^{10}}{\cancel{25}v^{14}}\right)\left(\frac{\cancel{125}^{5}v^{18}}{\cancel{216}^{27}w^{27}}\right) \Rightarrow \frac{10v^{4}}{27w^{17}}$

11. $27^{\frac{-2}{3}} \Rightarrow \left(27^{\frac{1}{3}}\right)^{-2} \Rightarrow 3^{-2} \Rightarrow \frac{1}{9}$

12. $\left(\frac{25b^{-4}}{c^{10}}\right)^{\frac{-3}{2}} \Rightarrow \left(\frac{25}{b^{4}c^{10}}\right)^{\frac{-3}{2}} \Rightarrow \left(\frac{b^{4}c^{10}}{25}\right)^{\frac{3}{2}} \Rightarrow \frac{\left(b^{4}\right)^{\frac{3}{2}}\left(c^{10}\right)^{\frac{3}{2}}}{\left(25^{\frac{1}{2}}\right)^{3}} \Rightarrow \frac{b^{6}c^{15}}{5^{3}} \Rightarrow \frac{b^{6}c^{15}}{125}$

EXERCISE 8-2

1. 2.73×10^{19}
2. 1.35×10^{-13}
3. 8.73×10^{14}
4. 2.1×10^{27}
5. 9.1×10^{-17}
6. 4.6×10^{3}
7. 7.9×10^{7}
8. 7.51×10^{13}

EXERCISE 8-3

1. $2^{x+4}=256 \Rightarrow 2^{x+4}=2^{8} \Rightarrow x+4=8 \Rightarrow x=4$
2. $3^{x+2}=81 \Rightarrow 3^{x+2}=3^{4} \Rightarrow x+2=4 \Rightarrow x=2$
3. $2^{5x-10}=8^{x} \Rightarrow 2^{5x-10}=2^{3x} \Rightarrow 5x-10=3x \Rightarrow -10=-2x \Rightarrow x=5$
4. $4^{x-3}=16^{x+1} \Rightarrow 4^{x-3}=\left(4^{2}\right)^{x-3} \Rightarrow 4^{x-3}=4^{2x-6} \Rightarrow x-3=2x-6 \Rightarrow 3=x$
5. $25^{2x+1}=125^{4-x} \Rightarrow \left(5^{2}\right)^{2x+1}=\left(5^{3}\right)^{4-x} \Rightarrow 5^{4x+2}=5^{12-3x}$

   $\Rightarrow 4x+2=12-3x \Rightarrow 7x=10 \Rightarrow x=\frac{10}{7}$

6. $5^x = 18$. There is no common base for 5 and 18, so solve by graphing.

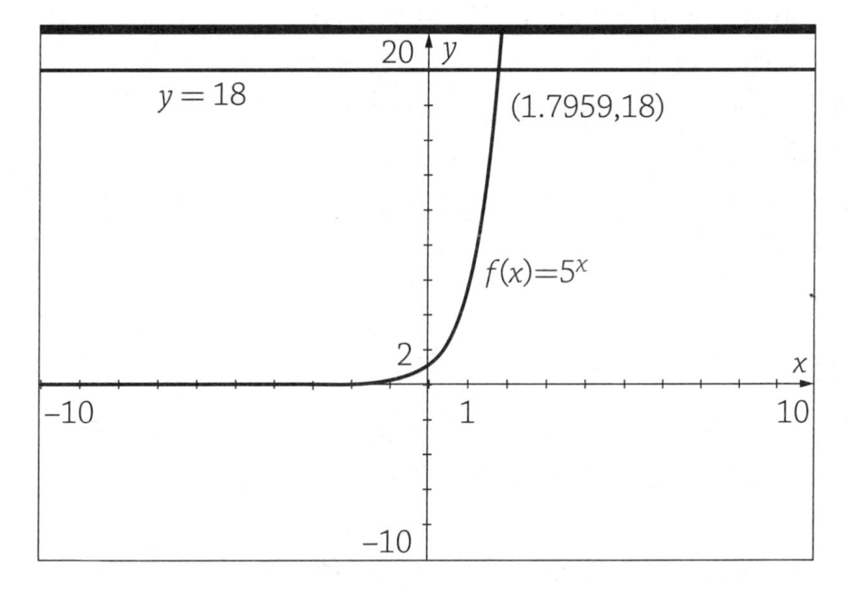

$x = 1.7959$ (to the nearest ten-thousandth)

7. $3^{x+2} - 7 = 20 \Rightarrow 3^{x+2} = 27 \Rightarrow 3^{x+2} = 3^3 \Rightarrow x + 2 = 3 \Rightarrow x = 1$

EXERCISE 8-4

1. The common ratio is $\frac{3}{2}$, so the defining function is $f(n) = 8\left(\frac{3}{2}\right)^{n-1}$.

   The 15th term is $f(15) = 8\left(\frac{3}{2}\right)^{14} = \frac{4,782,969}{2,048} = 2,335.43$

2. The common ratio is $\frac{-1}{2}$ so the defining function

   is $f(n) = 204,800\left(\frac{-1}{2}\right)^{n-1}$. The 20th term is $f(20) = 204,800\left(\frac{-1}{2}\right)^{19}$

   $= \frac{-25}{64} = -0.390625.$

3. $f(4) = 96 = ab^3$ and $f(9) = 3072 = ab^8$. Rewrite ab^8 as ab^3b^5 and substitute 96 for ab^3 to get $3072 = 96b^5$. Divide by 96 to get $32 = b^5$. Solve the equation to find $b = 2$ and $a = 12$. The 17th term is $12(2)^{16} = 786{,}432$.

4. $f(2) = ab = 5400$ and $f(5) = ab^4 = 1600$. This becomes $abb^3 = 1600$. Substitute 5400 for ab to get the equation $5400b^3 = 1600$. Solve: $b^3 = \frac{1600}{5400} = \frac{8}{27}$ so that $b = \frac{2}{3}$ and $a = 8100$. The 10th term of the sequence is $8100\left(\frac{2}{3}\right)^9 = \frac{51{,}200}{3}$.

5. Multiply the previous term in the sequence by 3. The terms of the sequence are: 8, 24, 72, 216, 648, 1944, 5832, 17,496, 52,488, 157,464. The tenth term is 157,464.

6. Multiply the previous term in the sequence by $\frac{3}{2}$. The terms of the sequence are: 24, 36, 54, 81, $\frac{243}{2}$, $\frac{729}{4}$, $\frac{2187}{8}$, $\frac{6561}{16}$. The eighth term is $\frac{6561}{16}$.

7. Multiply the previous term by 2 and then subtract 20. There terms of the sequence are 40, 60, 100, 180, 340, 660, 1300, and 2580. The seventh term in the sequence is 1300.

8. The first two terms are 2 and 3. The next terms are found by multiplying the previous term by 5 and subtracting twice the term prior to that. The terms in the sequence are 2, 3, $5(3) - 2(2) = 11$, $5(11) - 2(3) = 49$, $5(49) - 2(11) = 223$. The fifth term is 223.

EXERCISE 8-5

1. The amount of money that will be in the account is

$$3000\left(1 + \frac{0.016}{4}\right)^{4\times10} = 3000(1.004)^{40} = \$3519.41.$$

2. The 2020 population for Canada is predicted to be $34.1(1.012)^{10} = 38.42$ million people.
3. The percent of the original dosage of I^{123} left in the bloodstream after 24 hours is $100(2)^{\frac{-24}{13.22}} = 28.41\%$.
4. $T(5) = 31.95$ degrees.

Quadratic Expressions and Equations

EXERCISE 9-1

1. $12x^2+6x-9+21x^2-17x+91=33x^2-11x+82$
2. $-21x^2+17x+39+11x^2-25x-13=-10x^2-8x+26$
3. $(42x^2+62x+52)+(33x^2-37x+41)=75x^2+25x+93$
4. $(42x^2+62x+52)-(33x^2-37x+41)=42x^2+62x+52-33x^2+37x-41$
   $=9x^2+29x+11$
5. $12x^2+6x-9-(21x^2-17x+91)=12x^2+6x-9-21x^2+17x-91$
   $=-9x^2+23x-100$
6. $11x^2-25x-13-(-21x^2+17x+39)=11x^2-25x-13+21x^2-17x-39$
   $=32x^2-42x-52$
7. $-8(3x^2-7x-2)=(-8)(3x^2)+(-8)(-7x)+(-8)(-2)$
   $=-24x^2+56x+16$
8. $3x(2x^3+5x-7)=3x(2x^3)+3x(5x)+3x(-7)=6x^4+15x^2-21x$
9. $9x^2y(7x^2+5xy-8y^2)=9x^2y(7x^2)+9x^2y(5xy)+9x^2y(-8y^2)$
   $=63x^4y+45x^3y-72x^2y^3$

10. $(x+6)(x+4)=x^2+4x+6x+24=x^2+10x+24$

11. $(x+3)(x-4)=x^2-4x+3x-12=x^2-x-12$

12. $(x-8)(x-5)=x^2-5x-8x+40=x^2-13x+40$

13. $(x+8)(x-7)=x^2-7x+8x-56=x^2+x-56$

14. $(4x+3)(x-7)=4x^2-28x+3x-21=4x^2-25x-21$

15. $(x+3)(5x-7)=5x^2-7x+15x-21=5x^2+8x-21$

16. $(x-3)(2x-1)=2x^2-x-6x+3=2x^2-7x+3$

17. $(3x+4)(x-2)=3x^2-6x+4x-8=3x^2-2x-8$

18. $(4x+3)(3x-2)=12x^2-8x+9x-6=12x^2+x-6$

19. $(4x-3)(5x+4)=20x^2+16x-15x-12=20x^2+x-12$

20. $(4x-3)(4x+3)=(4x)^2-(3)^2=16x^2-9$

21. $(3x+4y)(3x-4y)=(3x)^2-(4y)^2=9x^2-16y^2$

22. $(3x+4)^2=(3x)^2+2(3x)(4)+(4)^2=9x^2+24x+16$

23. $(4x-3)^2=(4x)^2-2(4x)(3)+(3)^2=16x^2-24x+9$

24. $(2x+5y)^2=(2x)^2+2(2x)(5y)+(5y)^2=4x^2+20xy+25y^2$

25. $(3w-7p)^2=(3w)^2-2(3w)(7p)+(7p)^2=9w^2-42wp+49p^2$

26. $(2x+3)(x^2-3x-18)=(2x+3)(x^2)+(2x+3)(-3x)+(2x+3)(-18)$

$=2x(x^2)+3(x^2)+2x(-3x)+3(-3x)+2x(-18)+3(-18)$

$=2x^3+3x^2-6x^2-9x-36x-54=2x^3-3x^2-45x-54$

27. $(2x-5)(x^2-4x-12)=(2x-5)(x^2)+(2x-5)(-4x)+(2x-5)(-12)$

$= 2x(x^2)-5(x^2)+2x(-4x)-5(-4x)+2x(-12)-2(-12)$

$= 2x^3-5x^2-8x^2+20x-24x+24=2x^3-13x^2-4x+24$

28. $(x+3)(3x^2-4x-2)=(x+3)(3x^2)+(x+3)(-4x)+(x+3)(-2)$

$= x(3x^2)+3(3x^2)+x(-4x)+3(-4x)+x(-2)+3(-2)$

$= 3x^3+9x^2-4x^2-12x-2x-6=3x^3+5x^2-14x-6$

29. $(x+y)(3x^2-4xy-5y^2)=(x+y)(3x^2)+(x+y)(-4xy)+(x+y)(-5y^2)$

$= (x+y)(3x^2)+(x+y)(-4xy)+(x+y)(-5y^2)$

$= x(3x^2)+y(3x^2)+x(-4xy)+y(-4xy)+x(-5y^2)+y(-5y^2)$

$= 3x^3+3x^2y-4x^2y-4xy^2-5xy^2-5y^3=3x^3-x^2y-9xy^2-5y^3$

30. $(x^2-3)(3x^2+7x+5)=(x^2-3)(3x^2)+(x^2-3)(7x)+(x^2-3)(5)$

$= x^2(3x^2)-3(3x^2)+x^2(7x)-3(7x)+x^2(5)-3(5)$

$= 3x^4-9x^2+7x^3-21x+5x^2-15=3x^4+7x^3-4x^2-21x-15$

EXERCISE 9-2

1. $5x^3-10x^2=5x^2(x-2)$
2. $12y^3-36y^4=12y^3(1-3y)$
3. $z^2+9z+14=(z+2)(z+7)$
4. $p^2+29p+100=(p+4)(p+25)$
5. $r^2-20r+75=(r-5)(r-15)$
6. $m^2-4m-32=(m-8)(m+4)$

7. $g^2 - 4g - 45 = (g - 9)(g + 5)$

8. $k^2 + 8k - 48 = (k + 12)(k - 4)$

9. $4t^2 - 25 = (2t - 5)(2t + 5)$

10. $64v^2 - 169f^2 = (8v - 13f)(8v + 13f)$

11. $36d^2 + 108de + 81e^2 = 9(4d^2 + 12de + 9e^2) = 9(2d + 3e)^2$

12. $36q^2 - 180q + 225 = 9(4q^2 - 20q + 25) = 9(2q - 5)^2$

13. $2x^2 + 11x + 15 = (2x + 5)(x + 3)$

14. $7x^2 + 13x - 24 = (7x - 8)(x + 3)$

15. $11x^2 - 20x - 39 = (11x + 13)(x - 3)$

16. $6x^2 - 7x - 20 = (3x + 4)(2x - 5)$

EXERCISE 9-3

1. $x^2 - 5x - 24 = 0 \Rightarrow (x - 8)(x + 3) = 0 \Rightarrow x = 8, -3$
2. $y^2 - 12y + 20 = 0 \Rightarrow (y - 2)(y - 10) = 0 \Rightarrow y = 2, 10$
3. $4d^2 - 49 = 0 \Rightarrow (2d - 7)(2d + 7) = 0 \Rightarrow d = \pm\frac{7}{2}$
4. $v^2 - 16v + 64 = 0 \Rightarrow (v - 8)^2 = 0 \Rightarrow v = 8$

Quadratic Functions and Equations

EXERCISE 10-1

1. $x^2 - 4x + 2 \Rightarrow (x^2 - 4x + (2)^2) - 4 + 2 \Rightarrow (x - 2)^2 - 2$

2. $p^2 + 10p + 7 \Rightarrow (p^2 + 10p + (5)^2) - 25 + 7 \Rightarrow (p + 5)^2 - 18$

3. $w^2+7w-9 \Rightarrow \left(w^2+7w+\left(\frac{7}{2}\right)^2\right)-\left(\frac{7}{2}\right)^2-9 \Rightarrow \left(w+\frac{7}{2}\right)^2-\frac{85}{4}$

4. $2m^2-12m+5 \Rightarrow 2(m^2-6m+(3)^2)-2(3)^2+5 \Rightarrow 2(m-3)^2-13$

5. $3r^2+8r+4 \Rightarrow 3\left(r^2+\frac{8}{3}r+\left(\frac{4}{3}\right)^2\right)-3\left(\frac{4}{3}\right)^2+4 \Rightarrow 3\left(r+\frac{4}{3}\right)^2-\frac{4}{3}$

EXERCISE 10-2

1. $y=x^2-10x+3$; Axis of symmetry: $x=\frac{-(-10)}{2(1)}=5$; Vertex: (5, –22)

2. $y=-x^2+8x+3$; Axis of symmetry: $x=\frac{-8}{2(-1)}=4$; Vertex: (4, 19)

3. $d(x)=4x^2-24x-5$; Axis of symmetry: $x=\frac{-(-24)}{2(4)}=3$; Vertex: (3, –41)

4. $v(x)=-3x^2-12x+2$; Axis of symmetry: $x=\frac{-(-12)}{2(-3)}=-2$;

   Vertex: (–2, 14)

5. $y=(x-3)^2+1$; Axis of symmetry: $x=\frac{-(-24)}{2(4)}=3$; Vertex: (3, –41)

6. $y=(x+5)^2-3$; Axis of symmetry: $x=-5$; Vertex: (–5, –3)

7. $k(x)=-5(x-2)^2+9$; Axis of symmetry: $x=2$; Vertex: (2, 9)

8. $n(x)=\frac{2}{3}(x+1)^2+2$; Axis of symmetry: $x=-1$; Vertex: (–1, 2)

9. $y \geq -41$

10. $y \leq 9$

EXERCISE 10-3

1. $x^2 - 7x - 12 = 0 \Rightarrow x = \dfrac{7 \pm \sqrt{(-7)^2 - 4(1)(-12)}}{2(1)} \Rightarrow x = \dfrac{7 \pm \sqrt{97}}{2}$

2. $3f^2 - 7f - 4 = 0 \Rightarrow f = \dfrac{7 \pm \sqrt{(-7)^2 - 4(3)(-4)}}{2(3)} \Rightarrow f = \dfrac{7 \pm \sqrt{97}}{6}$

3. $8m^2 + 7m - 2 = 0 \Rightarrow m = \dfrac{-7 \pm \sqrt{7^2 - 4(8)(-2)}}{2(8)} \Rightarrow m = \dfrac{-7 \pm \sqrt{113}}{16}$

4. $-4x^2 + 5x + 3 = 0 \Rightarrow x = \dfrac{-5 \pm \sqrt{5^2 - 4(-4)(3)}}{2(-4)} \Rightarrow x = \dfrac{-5 \pm \sqrt{73}}{-8}$

5. $\dfrac{1}{2}x^2 + 10x + 9 = 0 \Rightarrow x = \dfrac{-10 \pm \sqrt{10^2 - 4\left(\dfrac{1}{2}\right)(9)}}{2\left(\dfrac{1}{2}\right)} \Rightarrow x = \dfrac{-10 \pm \sqrt{82}}{1} = -10 \pm \sqrt{82}$

6. $\dfrac{-3}{4}y^2 + 8x + 2 = 0 \Rightarrow y = \dfrac{-8 \pm \sqrt{(8)^2 - 4\left(\dfrac{-3}{4}\right)(2)}}{2\left(\dfrac{-3}{4}\right)} \Rightarrow y = \dfrac{-8 \pm \sqrt{70}}{\dfrac{-3}{2}} = \dfrac{-16 \pm 2\sqrt{70}}{-3}$

EXERCISE 10-4

1. $12x^2 - 17x - 5 = 0 \Rightarrow x = \dfrac{17 \pm \sqrt{17^2 - 4(12)(-5)}}{2(12)} = \dfrac{17 \pm \sqrt{529}}{24}$

$$= \frac{17 \pm 23}{24} = \frac{17 + 23}{24}, \frac{17 - 23}{24} = \frac{5}{3}, \frac{-1}{4}$$

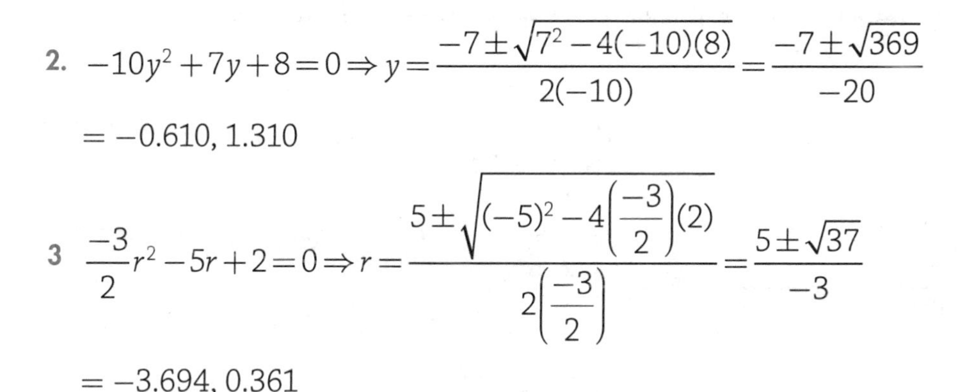

2. $-10y^2+7y+8=0 \Rightarrow y=\dfrac{-7\pm\sqrt{7^2-4(-10)(8)}}{2(-10)}=\dfrac{-7\pm\sqrt{369}}{-20}$

$=-0.610, 1.310$

3 $\dfrac{-3}{2}r^2-5r+2=0 \Rightarrow r=\dfrac{5\pm\sqrt{(-5)^2-4\left(\dfrac{-3}{2}\right)(2)}}{2\left(\dfrac{-3}{2}\right)}=\dfrac{5\pm\sqrt{37}}{-3}$

$=-3.694, 0.361$

EXERCISE 10-5

1. $p(x)=\sqrt{x+2}+1$: Move the graph of $y=\sqrt{x}$ left 2 units and up 1 unit.
2. $g(x)=2|x-3|-1$: Move the graph of $y=|x|$ 3 units to the right, stretch it form the x-axis by a factor of 2, and down 1 unit.
3. $d(x)=\dfrac{1}{x-1}+2$: Move the graph of $y=\dfrac{1}{x}$ right 1 unit and up 2 units.
4.

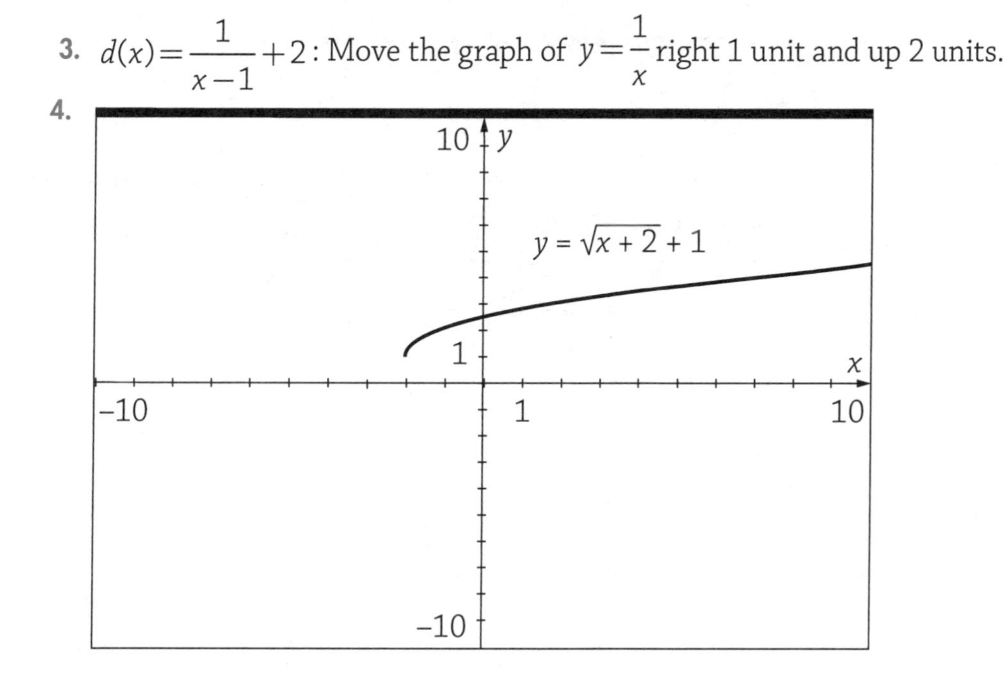

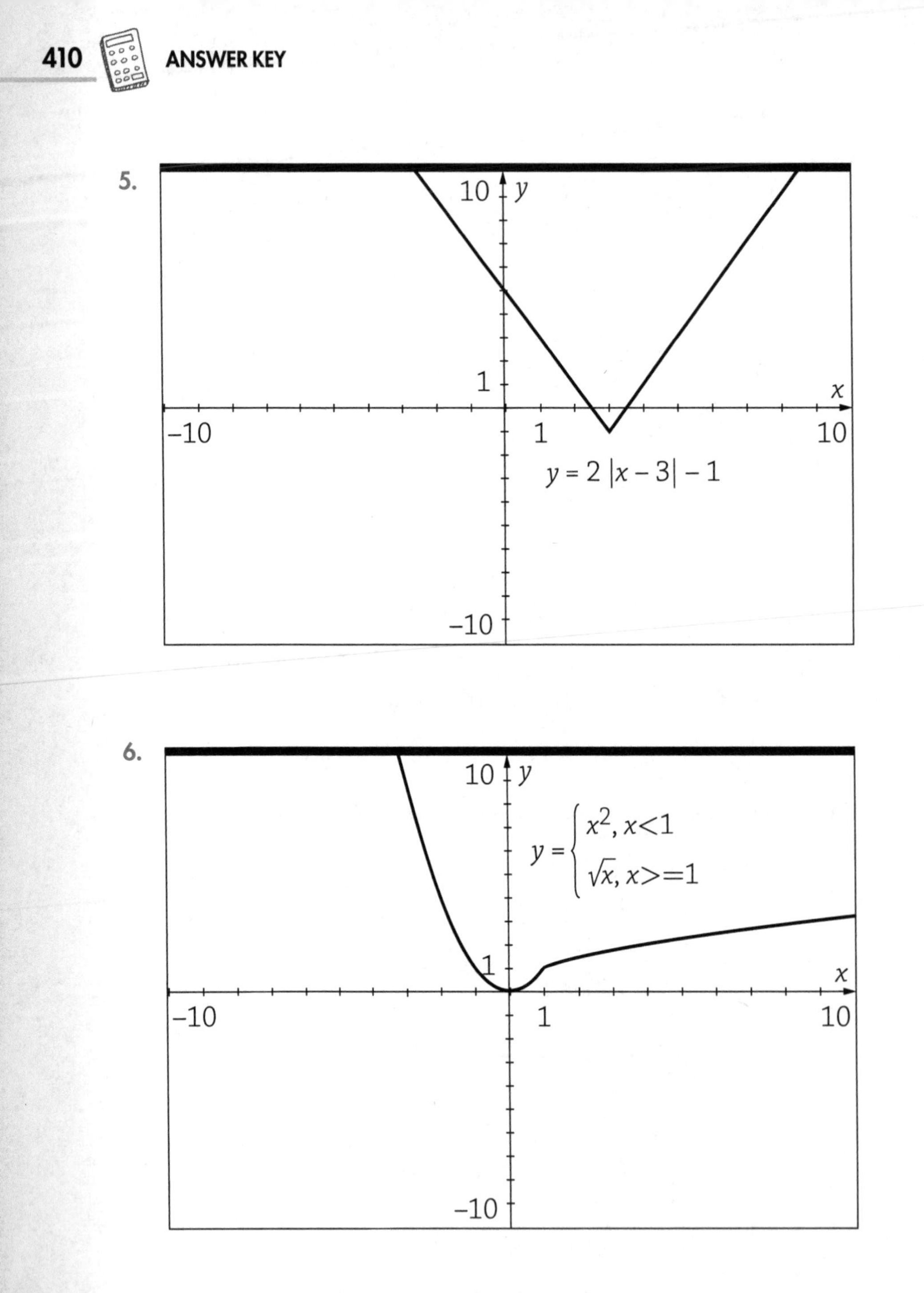
5.
10
y
1
x
–10
1
10
y = 2 |x – 3| – 1
–10
6.
10
y
y =
x², x<1
√x, x>=1
1
x
–10
1
10
–10

11

Rational Expressions and Equations

EXERCISE 11-1

1. $x(x-4)\neq 0 \Rightarrow x\neq 0,4$
2. $(x-6)(x+3)\neq 0 \Rightarrow x\neq -3,6$
3. All real numbers (x^2+4 never equals 0)

EXERCISE 11-2

1. $(0.3)(5.2)=3.9l \Rightarrow l=0.4$ m
2. $(100)(5)=120p \Rightarrow P=4.1667$ Pascals
3. $(1200)(4)^2=I(10)^2 \Rightarrow I=192$ foot-candles

EXERCISE 11-3

1. $x=\dfrac{-3}{2}$ (Set denominator equal to zero.)
2. $x=1.2$ (Set numerator equal to zero.)

EXERCISE 11-4

1. $x=\dfrac{-5}{2},1$
2. $x=-1, 6$

EXERCISE 11-5

1. $\dfrac{4x-x^2}{x^2-16}=\dfrac{x\cancel{(4-x)}-1}{\cancel{(x-4)}(x+4)}=\dfrac{-x}{x+4}$

2. $\dfrac{2x^2-x-1}{x^2-6x+5}=\dfrac{(2x+1)\cancel{(x-1)}}{(x-5)\cancel{(x-1)}}=\dfrac{2x+1}{x-5}$

3. $$\frac{4x-12}{6x+18}\times\frac{9x^2-81}{x^2-4x+3}\Rightarrow\frac{\cancel{4}^{2}(x-3)}{\cancel{6}^{3}\cancel{(x+3)}}\times\frac{\cancel{9}^{3}\cancel{(x+3)}\cancel{(x-3)}}{\cancel{(x-3)}(x-1)}\Rightarrow\frac{6(x-3)}{x-1}$$

4. $$\frac{x^2+10x+25}{x^2+7x+10}\times\frac{x^2-4x-12}{2x^2+11x+5}\Rightarrow\frac{\cancel{(x+5)}\cancel{x}}{\cancel{(x+2)}\cancel{(x+5)}}\times\frac{(x-6)\cancel{(x+2)}}{(2x+1)\cancel{(x+5)}}\Rightarrow\frac{x-6}{2x+1}$$

5. $$\frac{4x^2-1}{2x^2+x-1}\div\frac{2x^2+3x+1}{x^2+5x+4}\Rightarrow\frac{4x^2-1}{2x^2+x-1}\times\frac{x^2+5x+4}{2x^2+3x+1}$$
$$\Rightarrow\frac{\cancel{(2x-1)}\cancel{(2x+1)}}{\cancel{(2x-1)}\cancel{(x+1)}}\times\frac{\cancel{(x+1)}(x+4)}{\cancel{(2x+1)}(x+1)}\Rightarrow\frac{x+4}{x+1}$$

6. $$\frac{x^2-3x-28}{2x^2-2x-40}\div\frac{6x^2-66x+168}{18x-72}\Rightarrow\frac{x^2-3x-28}{2x^2-2x-40}\times\frac{18x-72}{6x^2-66x+1}$$
$$\Rightarrow\frac{(x-7)(x+4)}{2(x^2-x-20)}\times\frac{\cancel{18}^{3}(x-4)}{\cancel{6}(x^2-11x+28)}$$
$$\Rightarrow\frac{\cancel{(x-7)}\cancel{(x+4)}}{2(x-5)\cancel{(x+4)}}\times\frac{3\cancel{(x-4)}}{\cancel{(x-7)}\cancel{(x-4)}}\Rightarrow\frac{3}{2(x-5)}$$

7. $$\frac{\left(\frac{2}{3}+\frac{5}{6}\right)}{\left(\frac{11}{12}-\frac{3}{4}\right)}\frac{12}{12}=\frac{8+10}{11-9}=\frac{18}{2}=9$$

8. $$\frac{\left(\frac{x}{y}-\frac{y}{x}\right)}{\left(\frac{1}{y}-\frac{1}{x}\right)}\frac{xy}{xy}=\frac{x^2-y^2}{x-y}=\frac{(x+y)\cancel{(x-y)}}{\cancel{x-y}}=x+y$$

9. $$\frac{\left(\frac{2}{x+1}-\frac{3}{x-1}\right)}{\left(1-\frac{24}{x^2-1}\right)}\frac{(x+1)(x-1)}{(x+1)(x-1)}=\frac{2(x-1)-3(x+1)}{x^2-1-24}=\frac{2x-2-3x-3}{x^2-25}$$

$$=\frac{-x-5}{(x-5)(x+5)}=\frac{-\cancel{(x+5)}}{(x-5)\cancel{(x+5)}}=\frac{-1}{x-5}$$

10. $$\frac{\left(\frac{x+1}{x-2}-\frac{x-1}{x+2}\right)}{\left(1+\frac{4}{x^2-4}\right)}\frac{(x-2)(x+2)}{(x-2)(x+2)}=\frac{(x+1)(x+2)-(x-1)(x-2)}{x^2-4+4}$$

$$=\frac{(x^2+3x+2)-(x^2-3x+2)}{x^2}=\frac{6x}{x^2}=\frac{6}{x}$$

EXERCISE 11-6

1. $$\frac{5}{3x-15}+\frac{7}{2x-10}\Rightarrow\frac{5}{3(x-5)}+\frac{7}{2(x-5)}\Rightarrow\frac{5}{3(x-5)}\left(\frac{2}{2}\right)+\frac{7}{2(x-5)}\left(\frac{3}{3}\right)$$

$$\Rightarrow\frac{10+21}{6(x-5)}=\frac{31}{6(x-5)}$$

2. $$\frac{8}{x-4}-\frac{4}{x+3}\Rightarrow\left(\frac{8}{x-4}\right)\left(\frac{x+3}{x+3}\right)-\left(\frac{4}{x+3}\right)\left(\frac{x-4}{x-4}\right)\Rightarrow\frac{8(x+3)-4(x-4)}{(x+3)(x-4)}$$

$$\Rightarrow\frac{8x+24-4x+16}{(x+3)(x-4)}=\frac{4x+40}{(x+3)(x-4)}$$

3. $$\frac{x+1}{x-1}-\frac{x-4}{x+4}\Rightarrow\left(\frac{x+1}{x-1}\right)\left(\frac{x+4}{x+4}\right)-\left(\frac{x-4}{x+4}\right)\left(\frac{x-1}{x-1}\right)\Rightarrow\frac{(x^2+5x+4)-(x^2-5x+4)}{(x+4)(x-1)}$$

$$\Rightarrow\frac{x^2+5x+4-x^2+5x-4}{(x+4)(x-1)}=\frac{10x}{(x+4)(x-1)}$$

4. $$\frac{x+5}{2x-2}-\frac{x-3}{2x+8}\Rightarrow\left(\frac{x+5}{2(x-1)}\right)\left(\frac{x+4}{x+4}\right)-\left(\frac{x-3}{2(x+4)}\right)\left(\frac{x-1}{x-1}\right)$$

$$\Rightarrow\frac{(x^2+9x+20)-(x^2-4x+3)}{2(x-1)(x+4)}$$

$$\Rightarrow\frac{x^2+9x+20-x^2+4x-3}{2(x-1)(x+4)}=\frac{13x+17}{2(x-1)(x+4)}$$

5. $$\frac{2x+5}{x-3}-\frac{3x-4}{x-4}\Rightarrow\left(\frac{2x+5}{x-3}\right)\left(\frac{x-4}{x-4}\right)-\left(\frac{3x-4}{x-4}\right)\left(\frac{x-3}{x-3}\right)$$

$$\Rightarrow\frac{(2x^2-3x-20)-(3x^2-13x+12)}{(x-3)(x-4)}$$

$$\Rightarrow\frac{2x^2-3x-20-3x^2+13x-12}{(x-3)(x-4)}=\frac{-x^2+10x-32}{(x-3)(x-4)}$$

6. $$\frac{x+1}{x^2-2x}+\frac{x-4}{x^2+5x}\Rightarrow\left(\frac{x+1}{x(x-2)}\right)\left(\frac{x+5}{x+5}\right)+\left(\frac{x-4}{x(x+5)}\right)\left(\frac{x-2}{x-2}\right)$$

$$\Rightarrow\frac{(x^2+6x+5)+(x^2-6x+8)}{x(x-2)(x+5)}$$

$$\Rightarrow\frac{2x^2+13}{x(x-2)(x+5)}$$

7. $$\frac{7}{x^2-x-2}+\frac{5}{x^2-5x+6}\Rightarrow\left(\frac{7}{(x-2)(x+1)}\right)\left(\frac{x-3}{x-3}\right)+\left(\frac{5}{(x-2)(x-3)}\right)\left(\frac{x+1}{x+1}\right)$$

$$\Rightarrow\frac{7(x-3)+5(x+1)}{(x+1)(x-2)(x-3)}$$

$$\Rightarrow\frac{7x-21+5x+5}{(x+1)(x-2)(x-3)}=\frac{12x-16}{(x+1)(x-2)(x-3)}$$

8. $$\frac{x+7}{x^2-x-2}-\frac{x-5}{x^2-5x+6}\Rightarrow\left(\frac{x+7}{(x-2)(x+1)}\right)\left(\frac{x-3}{x-3}\right)-\left(\frac{x-5}{(x-2)(x-3)}\right)\left(\frac{x+1}{x+1}\right)$$

$$\Rightarrow\frac{(x+7)(x-3)-(x-5)(x+1)}{(x+1)(x-2)(x-3)}\Rightarrow\frac{(x^2+4x-21)-(x^2-4x-5)}{(x+1)(x-2)(x-3)}$$

$$\Rightarrow\frac{x^2+4x-21-x^2+4x+5}{(x+1)(x-2)(x-3)}$$

$$\Rightarrow\frac{8x-16}{(x+1)(x-2)(x-3)}=\frac{8(x-2)}{(x+1)(x-2)(x-3)}=\frac{8}{(x+1)(x-3)}$$

9. $$\frac{2x+1}{x^2-10x+25}-\frac{x+4}{x^2-25}\Rightarrow\left(\frac{2x+1}{(x-5)^2}\right)\left(\frac{x+5}{x+5}\right)-\left(\frac{x+4}{(x-5)(x+5)}\right)\left(\frac{x-5}{x-5}\right)$$

$$\Rightarrow\frac{(2x+1)(x+5)-(x+4)(x-5)}{(x-5)^2(x+5)}$$

$$\Rightarrow\frac{(2x^2+11x+5)-(x^2-x-20)}{(x-5)^2(x+5)}\Rightarrow\frac{2x^2+11x+5-x^2+x+20}{(x-5)^2(x+5)}$$

$$=\frac{x^2+12x+25}{(x-5)^2(x+5)}$$

10. $$\frac{2x+5}{x^2-10x}-\frac{x+1}{x^2-20x+100}\Rightarrow\left(\frac{2x+5}{x(x-10)}\right)\left(\frac{x-10}{x-10}\right)-\left(\frac{x+1}{(x-10)^2}\right)\frac{x}{x}$$

$$\Rightarrow\frac{(2x+5)(x-10)-x(x+1)}{x(x-10)^2}$$

$$\Rightarrow\frac{(2x^2-15x-50)-(x^2+x)}{x(x-10)^2}\Rightarrow\frac{2x^2-15x-50-x^2-x}{x(x-10)^2}=\frac{x^2-16x-50}{x(x-10)^2}$$

EXERCISE 11-7

1. $x = -9, 14$

$$\frac{4}{x+10} + \frac{10}{x+6} = \frac{2}{3} \Rightarrow \left(\frac{4}{x+10}\right)3(x+10)(x+6) + \left(\frac{10}{x+6}\right)3(x+10)(x+6)$$
$$= \left(\frac{2}{3}\right)3(x+10)(x+6)$$

$$\Rightarrow 12(x+6) + 30(x+10) = 2(x^2 + 16x + 60) \Rightarrow 12x + 72 + 30x + 300$$
$$= 2x^2 + 32x + 120$$

$$\Rightarrow 42x + 372 = 2x^2 + 32x + 120 \Rightarrow 2x^2 - 10x - 252 = 0 \Rightarrow 2(x^2 - 5x - 126)$$
$$= 0 \Rightarrow 2(x-14)(x+9) = 0$$

2. $p = -2, 9$

$$\frac{3}{p-4} - \frac{1}{2} = \frac{1}{n+1} \Rightarrow \left(\frac{3}{p-4}\right)2(p-4)(p+1) - \left(\frac{1}{2}\right)2(p-4)(p+1)$$
$$= \left(\frac{1}{p+1}\right)2(p-4)(p+1)$$

$$\Rightarrow 6(p+1) - (p^2 - 3p - 4)(p+1) = 2(p-4) \Rightarrow 6p + 6 - p^2 + 3p + 4$$
$$= 2p - 8 \Rightarrow -p^2 + 9p + 10 = 2p - 8$$

$$\Rightarrow -p^2 + 7p + 18 = 0 \Rightarrow -(p^2 - 7p - 18) = 0 \Rightarrow -(p-9)(p+2) = 0$$

3. $g = 11.$

$$\frac{2}{g-6} - \frac{5}{g+4} = \frac{1}{g+4} \Rightarrow \left(\frac{2}{g-6}\right)(g+4)(g-6) - \left(\frac{5}{g+4}\right)(g+4)(g-6)$$
$$= \left(\frac{1}{g+4}\right)(g+4)(g-6)$$

$\Rightarrow 2(g+4)-5(g-6)=1(g-6) \Rightarrow 2g+8-5g+30=g-6$

$\Rightarrow -3g+38=g-6 \Rightarrow 44=4g$

4. $m=-7, \frac{-75}{7}$

$\frac{m+8}{m+10}+\frac{m+6}{m+12}=\frac{2}{15} \Rightarrow \left(\frac{m+8}{m+10}\right)15(m+10)(m+12)$

$+\left(\frac{m+6}{m+12}\right)15(m+10)(m+12)=\left(\frac{2}{15}\right)15(m+10)(m+12)$

$\Rightarrow 15(m+8)(m+12)+15(m+10)(m+6)=2(m+10)(m+12)$

$\Rightarrow 15((m^2+20m+96)+(m^2+16m+60))=2(m^2+22m+120)$

$\Rightarrow 15(2m^2+36m+156)=2(m^2+22m+120)$

$\Rightarrow 30m^2+540m+2340=2m^2+44m+240$

$\Rightarrow 28m^2+496m+2100=0$

$\Rightarrow 7m^2+124m+525=0$

$\Rightarrow (7m+75)(m+7)=0$

5. $n=\pm 15$

$\frac{n}{n+5}-\frac{n-10}{n-5}=\frac{1}{4} \Rightarrow \left(\frac{n}{n+5}\right)4(n+5)(n-5)-\left(\frac{n-10}{n-5}\right)4(n+5)(n-5)$

$=\left(\frac{1}{4}\right)4(n+5)(n-5)$

$\Rightarrow 4n(n-5)-4(n+5)(n-10)=(n+5)(n-5)$

$$\Rightarrow 4n^2 - 20n - 4(n^2 - 5n - 50) = n^2 - 25$$

$$\Rightarrow 4n^2 - 20n - 4n^2 + 20n + 200 = n^2 - 25$$

$$\Rightarrow 200 = n^2 - 25 \Rightarrow 225 = n^2$$

6. $r = \frac{-3}{2}, 10$

$$\frac{r-2}{r} + \frac{r-4}{r+2} = \frac{13}{r} \Rightarrow \left(\frac{r-2}{r}\right) r(r+2) + \left(\frac{r-4}{r+2}\right) r(r+2) = \left(\frac{13}{r}\right) r(r+2)$$

$$\Rightarrow (r-2)(r+2) + (r-4)r = 13(r+2)$$

$$\Rightarrow r^2 - 4 + r^2 - 4r = 13r + 26$$

$$\Rightarrow 2r^2 - 17r - 30 = 0 \Rightarrow (2r+3)(r-10) = 0$$

EXERCISE 11-8

1. $3x - 5$

$$\begin{array}{r} 3x-5 \\ x+6 \overline{) 3x^2 + 13x - 30} \\ \underline{3x^2 + 18x} \quad\quad \\ -5x - 30 \\ \underline{-5x - 30} \end{array}$$

2. $2x^2 + 7x + 6$

$$\begin{array}{r} 2x^2 + 7x + 6 \\ 2x-3 \overline{) 4x^3 + 8x^2 - 9x - 18} \\ \underline{4x^3 - 6x^2} \quad\quad\quad\quad \\ 14x^2 - 9x - 18 \\ \underline{14x^2 - 21x} \quad\quad \\ 12x - 18 \\ \underline{12x - 18} \end{array}$$

3. $x^3+6x^2+12x+16$

$$\begin{array}{r l}
 & x^3+6x^2+12x+16 \\
x-2\overline{)} & x^4+4x^3+0x^2-8x-32 \\
 & \underline{x^4-2x^3} \\
 & \quad 6x^3+0x^2-8x-32 \\
 & \quad \underline{6x^3-12x^2} \\
 & \qquad 12x^2-8x-32 \\
 & \qquad \underline{12x^2-24x} \\
 & \qquad\quad 16x-32 \\
 & \qquad\quad \underline{16x-32}
\end{array}$$

4. $x^2+4+\dfrac{1}{x-5}$

$$\begin{array}{r l}
 & x^2 \qquad +4 \\
x-5\overline{)} & x^3-5x^2+4x-19 \\
 & \underline{x^3-5x^2} \\
 & \qquad 4x-19 \\
 & \qquad \underline{4x-20} \\
 & \qquad\quad 1
\end{array}$$

Radical Functions and Geometry

EXERCISE 12-1

1. $\sqrt{24}=\sqrt{4}\sqrt{6}=2\sqrt{6}$

2. $\sqrt{54}=\sqrt{9}\sqrt{6}=3\sqrt{6}$

3. $\sqrt{162}=\sqrt{81}\sqrt{2}=9\sqrt{2}$

4. $\sqrt{450}=\sqrt{225}\sqrt{2}=15\sqrt{2}$

5. $7\sqrt{24}+4\sqrt{54}=7\sqrt{4}\sqrt{6}+4\sqrt{9}\sqrt{6}=7(2\sqrt{6})+4(3\sqrt{6})$
$=14\sqrt{6}+12\sqrt{6}=26\sqrt{6}$

6. $4\sqrt{98}-2\sqrt{162}=4\sqrt{49}\sqrt{2}-2\sqrt{81}\sqrt{2}=4(7\sqrt{2})-2(9\sqrt{2})$
$=28\sqrt{2}-18\sqrt{2}=10\sqrt{2}$

7. $4\sqrt{600a}-7\sqrt{150a}=4(\sqrt{100}\sqrt{6a})-7(\sqrt{25}\sqrt{6a})$
$=4(10\sqrt{6a})-7(5\sqrt{6a})=40\sqrt{6a}-35\sqrt{6a}=5\sqrt{6a}$

8. $(4\sqrt{6})(5\sqrt{3})=20\sqrt{18}=20\sqrt{9}\sqrt{2}=20(3\sqrt{2})=60\sqrt{2}$

9. $(4\sqrt{15})(5\sqrt{30})=(4\sqrt{15})(5\sqrt{15}\sqrt{2})=20(15\sqrt{2})=300\sqrt{2}$

10. $(8\sqrt{2ab})(3\sqrt{32ac})=24\sqrt{64a^2bc}=24\sqrt{64a^2}\sqrt{bc}=24(8a\sqrt{bc})=192a\sqrt{bc}$

11. $(5\sqrt{7})^2=25(7)=175$

12. $(-6\sqrt{5})^2=36(5)=180$

13. $(4\sqrt{5r})^2=16(5r)=80r$

14. $\dfrac{5\sqrt{128}}{4\sqrt{50}}=\dfrac{5\sqrt{64}\sqrt{2}}{4\sqrt{25}\sqrt{2}}=\dfrac{40\sqrt{2}}{20\sqrt{2}}=2$

15. $(18+3\sqrt{27})+(-7+2\sqrt{108})=(18+3(\sqrt{9}\sqrt{3}))+(-7+2(\sqrt{36}\sqrt{3}))$
$=(18+9\sqrt{3})+(-7+12\sqrt{3})=11+21\sqrt{3}$

16. $(18+3\sqrt{27})-(-7+2\sqrt{108})=(18+9\sqrt{3})-(-7+12\sqrt{3})=18+9\sqrt{3}+7-12\sqrt{3}$
$=25-3\sqrt{3}$

17. $5(8+5\sqrt{216})+7(4-8\sqrt{54})=5(8+5(\sqrt{36}\sqrt{6}))+7(4-8(\sqrt{9}\sqrt{6}))$

$= 5(8+30\sqrt{6})+7(4-24\sqrt{6})=40+150\sqrt{6}+28-168\sqrt{6}=68-18\sqrt{6}$

18. $5(8+5\sqrt{216})-7(4-8\sqrt{54})=40+150\sqrt{6}-28+168\sqrt{6}=12+318\sqrt{6}$

19. $(8+2\sqrt{3})(7-4\sqrt{3})=56-32\sqrt{3}+14\sqrt{3}-8(3)=56-18\sqrt{3}-24=32-18\sqrt{3}$

20. $(9+3\sqrt{12})(2-4\sqrt{48})=(9+3(\sqrt{4}\sqrt{3}))(2-4(\sqrt{16}\sqrt{3}))$
$=(9+6\sqrt{3})(2-16\sqrt{3})=18-144\sqrt{3}+12\sqrt{3}-96(3)$

$=18-132\sqrt{3}-288=-270-132\sqrt{3}$

21. $(8+2\sqrt{3})(8-2\sqrt{3})=8^2-(2\sqrt{3})^2=64-12=52$

22. $(7-4\sqrt{3})^2=7^2-2(7)(4\sqrt{3})+(4\sqrt{3})^2=49-56\sqrt{3}+48=97-56\sqrt{3}$

23. $\left(\frac{2+\sqrt{5}}{3-\sqrt{5}}\right)\left(\frac{3+\sqrt{5}}{3+\sqrt{5}}\right)=\frac{6+2\sqrt{5}+3\sqrt{5}+5}{9-5}=\frac{11+5\sqrt{5}}{4}$

24. $\left(\frac{4+\sqrt{5}}{4-\sqrt{5}}\right)\left(\frac{4+\sqrt{5}}{4+\sqrt{5}}\right)=\frac{16+2(4)(\sqrt{5})+(\sqrt{5})^2}{16-5}=\frac{16+8\sqrt{5}+5}{11}=\frac{21+8\sqrt{5}}{11}$

25. $\frac{7-2\sqrt{8}}{3-\sqrt{18}}=\left(\frac{7-4\sqrt{2}}{3-3\sqrt{2}}\right)\left(\frac{3+3\sqrt{2}}{3+3\sqrt{2}}\right)=\frac{21+21\sqrt{2}-12\sqrt{2}-24}{9-18}$
$=\frac{-3+9\sqrt{2}}{-9}=\frac{-3(1-3\sqrt{2})}{-9}=\frac{1-3\sqrt{2}}{3}$

EXERCISE 12-2

1. $x^2-6x+7=0 \Rightarrow x=\dfrac{6\pm\sqrt{(-6)^2-4(1)(7)}}{2(1)} \Rightarrow x=\dfrac{6\pm\sqrt{36-28}}{2}$

$\Rightarrow x=\dfrac{6\pm\sqrt{8}}{2} \Rightarrow x=\dfrac{6\pm2\sqrt{2}}{2}=3\pm\sqrt{2}$

2. $x^2-4x-41=0 \Rightarrow x=\dfrac{4\pm\sqrt{(-4)^2-4(1)(-41)}}{2(1)} \Rightarrow x=\dfrac{4\pm\sqrt{16+164}}{2}=\dfrac{4\pm\sqrt{180}}{2}$

$=\dfrac{4\pm\sqrt{(9)(4)(5)}}{2}=\dfrac{2\pm6\sqrt{5}}{2}=2\pm3\sqrt{5}$

3. $x^2+4x-3=0 \Rightarrow x=\dfrac{-4\pm\sqrt{4^2-4(1)(-3)}}{2(1)}=\dfrac{-4\pm\sqrt{16+12}}{2}$

$=\dfrac{-4\pm\sqrt{28}}{2}=\dfrac{-4\pm\sqrt{(4)(7)}}{2}=\dfrac{-4\pm2\sqrt{7}}{2}=-2\pm\sqrt{7}$

$=\dfrac{-4\pm2\sqrt{7}}{2}=-2\pm\sqrt{7}$

4. $16x^2+40x+19=0 \Rightarrow x=\dfrac{-40\pm\sqrt{40^2-4(16)(19)}}{2(16)}=\dfrac{-40\pm\sqrt{1600-1216}}{32}$

$=\dfrac{-40\pm\sqrt{384}}{32}=\dfrac{-40\pm\sqrt{(4)(96)}}{32}$

$=\dfrac{-40\pm2\sqrt{(16)(6)}}{32}=\dfrac{-40\pm8\sqrt{6}}{32}=\dfrac{8\left(-5\pm\sqrt{6}\right)}{32}=\dfrac{-5\pm\sqrt{6}}{4}$

5. $49x^2-14x-1=0 \Rightarrow x=\dfrac{14\pm\sqrt{(-14)^2-4(49)(-1)}}{2(49)}=\dfrac{14\pm\sqrt{196+196}}{98}$

$=\dfrac{14\pm\sqrt{(196)(2)}}{98}=\dfrac{14\pm14\sqrt{2}}{98}=\dfrac{14\left(1\pm\sqrt{2}\right)}{98}=\dfrac{1\pm\sqrt{2}}{7}$

EXERCISE 12-3

1. $\sqrt{2x+11}=5 \Rightarrow \left(\sqrt{2x+11}\right)^2=5^2 \Rightarrow 2x+11=25 \Rightarrow 2x=14 \Rightarrow x=7$

2. $2\sqrt{x+25}-3=9 \Rightarrow 2\sqrt{x+25}=12 \Rightarrow \sqrt{x+25}=6 \Rightarrow \left(\sqrt{x+25}\right)^2=$
   $\Rightarrow x+25=36 \Rightarrow x=11$

3. $\sqrt{x^2+9}+x=1 \Rightarrow \sqrt{x^2+9}=1-x \Rightarrow \left(\sqrt{x^2+9}\right)^2=(1-x)^2$
   $\Rightarrow x^2+9=x^2-2x+1 \Rightarrow -2x=8 \Rightarrow x=-4$

4. $\sqrt{3x+1}=x-3 \Rightarrow \left(\sqrt{3x+1}\right)^2=(x-3)^2 \Rightarrow 3x+1=x^2-6x+9$
   $\Rightarrow x^2-9x+8=0 \Rightarrow (x-1)(x-8)=0$
   $\Rightarrow x=1,8$
   Reject $x=1$ because $\sqrt{4} \neq -2$. Therefore, $x=8$.

5. $\sqrt{12x+13}-x=4 \Rightarrow \sqrt{12x+13}=x+4 \Rightarrow \left(\sqrt{12x+13}\right)^2=(x+4)^2$
   $\Rightarrow 12x+13=x^2+8x+16 \Rightarrow x^2-4x+3=0$
   $\Rightarrow (x-1)(x-3)=0 \Rightarrow x=1,3$
   Check to show that both solutions work.

EXERCISE 12-4

1. The legs are 20(3) and 20(4), so the hypotenuse must be $20(5)=100$.
2. The hypotenuse is 3(25) and one leg is 3(5), so the other leg must be $3(24)=72$.
3. This is not one of the Pythagorean triples that we discussed so use the Pythagorean theorem.
   $x^2+60^2=61^2 \Rightarrow x^2=61^2-60^2=121$ so $x=11$.
4. $x^2+10^2=24^2 \Rightarrow x^2=24^2-10^2=476 \Rightarrow x=\sqrt{476}=\sqrt{4}\sqrt{124}$
   $=2\sqrt{4}\sqrt{31}=4\sqrt{31}$

5. $20^2+25^2=x^2 \Rightarrow x^2=400+625=1025 \Rightarrow x=\sqrt{1025}=\sqrt{25}\sqrt{41}=5\sqrt{41}$

6. $x=7$

$(x-2)^2+(x+5)^2=(x+6)^2 \Rightarrow x=-1,7$. Reject $x=-1$.

EXERCISE 12-5

1. $\sin(32)=\dfrac{x}{200} \Rightarrow x=200\sin(32)=105.98$

2. $\tan(62)=\dfrac{80}{x} \Rightarrow x\tan(62)=80 \Rightarrow x=\dfrac{80}{\tan(62)}=42.54$

3. $\cos(21)=\dfrac{74}{x} \Rightarrow x\cos(21)=74 \Rightarrow x=\dfrac{74}{\cos(21)}=79.26$

4. $\sin(70)=\dfrac{x}{83.2} \Rightarrow x=83.2\sin(70)=78.18$

5. $\tan(50)=\dfrac{x}{215} \Rightarrow x=215\tan(50)=256.23$

EXERCISE 12-6

1. $\cos(\theta)=\dfrac{25}{28} \Rightarrow \theta=\cos^{-1}\left(\dfrac{25}{28}\right)=26.8°$

2. $\tan(\theta)=\dfrac{30}{17} \Rightarrow \theta=\tan^{-1}\left(\dfrac{30}{17}\right)=60.5°$

3. $\cos(\theta)=\dfrac{54}{72} \Rightarrow \theta=\cos^{-1}\left(\dfrac{54}{72}\right)=41.41°$

4. $\sin(\theta)=\dfrac{41}{52} \Rightarrow \theta=\sin^{-1}\left(\dfrac{41}{52}\right)=52.04°$

5. $\cos(\theta)=\dfrac{81}{112} \Rightarrow \theta=\cos^{-1}\left(\dfrac{81}{112}\right)=43.68°$

6. $\tan(\theta)=\dfrac{1}{12} \Rightarrow \theta=\tan^{-1}\left(\dfrac{1}{12}\right)=4.8°$

Probability and Statistics

EXERCISE 13-1

Results of One-Variable Stats

$\overline{X}$	19.3
ΣX	193.
ΣX^2	4157
$sx: = sn-...$	6.92901
$\sigma x: = \sigma n...$	6.57343
n	10.
Min$X$	12.
Q_1X	14.
Median X	17.
Q_3X	25.
Max$X$	31.

1. Mode = 12
2. Median = 17
3. Mean = 19.3
4. Range = 31 − 12 = 19
5. IQR = 25 − 14 = 11
6. Standard deviation $s = 6.93$

Results of One-Variable Stats

$\bar{X}$	16.5751
ΣX	3199.
ΣX^2	55813.
$sx := sn-...$	3.81141
$\sigma x := \sigma n...$	3.80153
n	193.
MinX	10.
Q_1X	15.
Median X	17.
Q_3X	19.
MaxX	25.

7. Mode = 15
8. Median = 17
9. Mean = 16.6
10. Range = 25 − 10 = 15
11. IQR = 19 − 15 = 4
12. Standard deviation $\sigma = 3.8$

EXERCISE 13-2

1. IQR = 71 − 63 = 8
2. The slope indicates that for each year a person, ages they will need 0.1 hours less sleep time each night.
3. The correlation coefficient −0.87 indicates that there is a very strong negative correlation between the variables.

4. Cups of hot chocolate $= -0.70$ (high temperature) $+ 59.30$
5. The y-intercept predicts that when the temperature is 0°, Alex will sell 59 cups of hot chocolate.
6. The prediction is that Alex will sell 31 cups of hot chocolate when the high temperature is 40°.

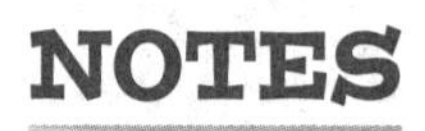

NOTES

W9-BNM-663

NUSD STATE PRESCHOOLS

unLovaBLe

Dan Yaccarino

Henry Holt and Company • New York

This book is for my buddy Alfred.

Henry Holt and Company, LLC, *Publishers since 1866*
115 West 18th Street, New York, New York 10011
www.henryholt.com

Henry Holt is a registered trademark of Henry Holt and Company, LLC
Copyright © 2001 by Dan Yaccarino. All rights reserved.
Distributed in Canada by H. B. Fenn and Company Ltd.

Library of Congress Cataloging-in-Publication Data
Yaccarino, Dan. Unlovable / Dan Yaccarino.
Summary: Alfred, a pug, is made to feel inferior by a cat, a parrot, and the other neighborhood dogs, until a new dog moves in next door and helps Alfred to realize he is just fine the way he is.
[1. Pug—Fiction. 2. Dogs—Fiction. 3. Self-perception—Fiction. 4. Friendship—Fiction.] I. Title.
PZ7.Y125Un 2001 [E]—dc21 00-47299

First published in hardcover in 2001 by Henry Holt and Company
First Owlet Edition—2004 / Designed by Dan Yaccarino and Donna Mark
The artist used gouache on watercolor paper to create the illustrations for this book.

ISBN 0-8050-6321-8 (hardcover)
3 5 7 9 10 8 6 4 2
ISBN 0-8050-7532-1 (paperback)
1 3 5 7 9 10 8 6 4 2

Printed in the United States of America on acid-free paper. ∞

Alfred was unlovable.

At least that's what the cat told him every chance he got.

"You've got the *ugliest* mug I've ever seen. *No one* could love you!"

Alfred tried his best to ignore the remarks, but it was difficult, especially since the cat had taught the parrot to say "Unlovable! *Squawk!* Unlovable!" whenever Alfred walked by.

The goldfish gurgled in agreement.

But what was it that made him unlovable?

His snoring?

The way that he ate?

His little curly tail?

None of the neighborhood dogs would have a thing to do with him.

"His mouth is too small to hold a ball," a big German shepherd sneered.

"His legs are way too short for running," snickered a greyhound.

A pampered poodle chuckled. "Did you see that face?"

“Beat it, shrimp,” growled a Doberman.

“You couldn’t even scare a mailman.”

Alfred didn't like staying in the house since the cat was always making fun of him, the parrot was always squawking "Unlovable!" and the goldfish was always gurgling in agreement. So Alfred spent most of his time alone in the backyard.

One day a new family moved in next door. Alfred tried to see if they had a dog who might be his friend, but he was too little to look over the fence.

As he was sniffing around, he heard something on the other side.

"Hello?" Alfred called.

"Hi," came the answer. "I'm Rex. I just moved in."

"My name is Alfred." And, without thinking, he blurted out, "I'm a golden retriever."

"Glad to meet you," Rex replied.

Alfred and Rex talked for hours. Alfred said he liked sleeping in the sun, dog food, and scratching. Rex did too. Rex said he hated baths and going to the vet. Alfred did too.

It began to get dark. Soon it was time for dinner, and they both went inside. That night Alfred thought about how much he liked Rex and how much they had in common. Then he thought about the fib he had told. Alfred was sure they'd be friends—as long as Rex never saw how unlovable he was.

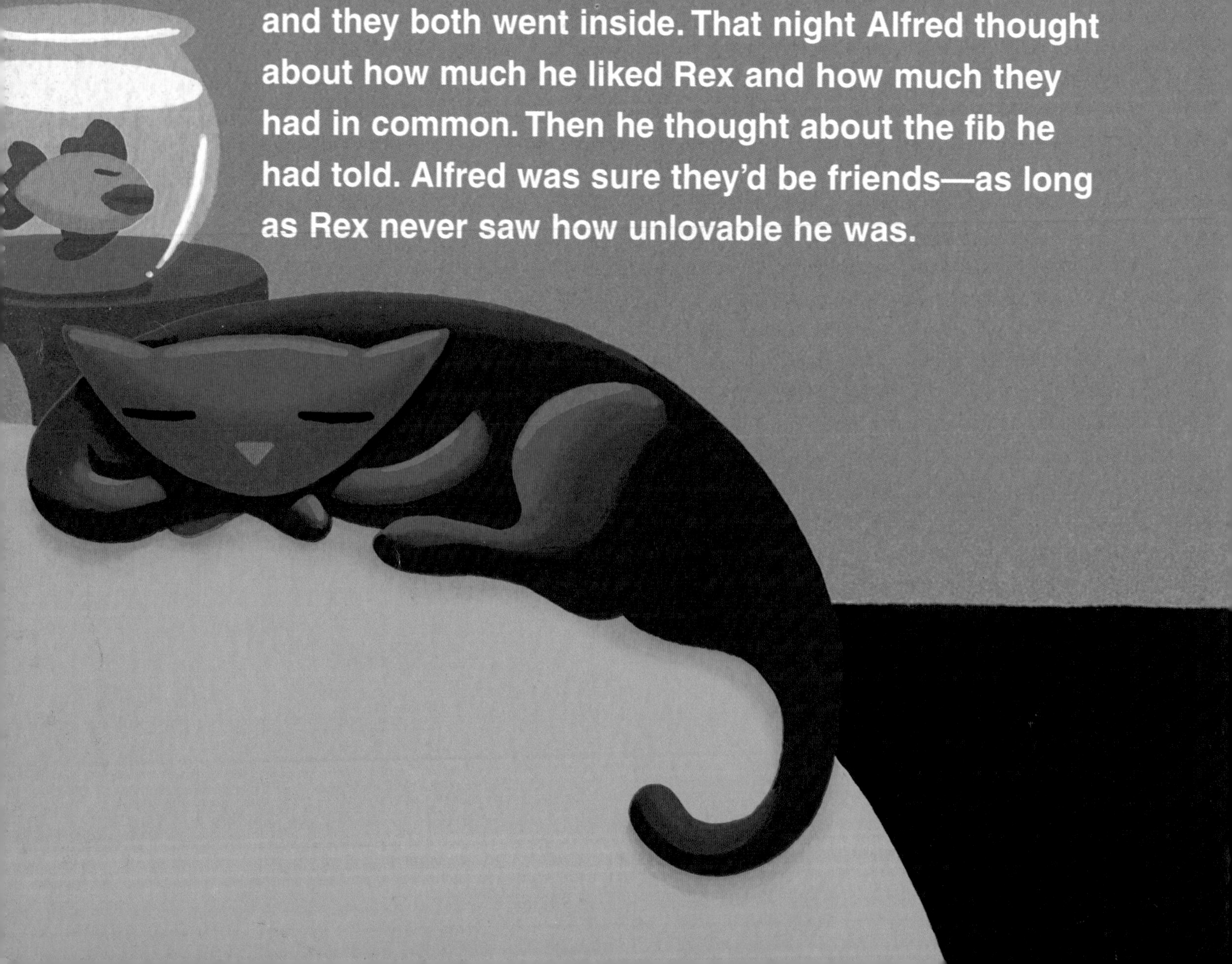

The next day when Alfred and Rex were chatting, a squirrel jumped onto the fence between them. They both barked at it. The squirrel took one look at Alfred and climbed up a tree.

Rex said, "You sure showed that squirrel who's boss, Alfred."

But Alfred was thinking, *If Rex ever sees me he'll run away too.*

"I'm going to dig a hole under the fence," Rex said one day. "Then I can squeeze through to your side, and we can meet."

Alfred heard Rex digging. When he heard Rex wiggle under the fence, he ran and hid behind a bush. Then he heard Rex call, "Alfred, where are you?"

Suddenly Rex poked his head into the bush where Alfred was hiding.

“Y-you look just like me!” gasped Alfred.

“Wow! This is great,” said Rex. “You’re not a golden retriever after all!”

Alfred had to laugh. Who cared what the others said? Rex was his friend, and Rex liked him just the way he was.

Together Alfred and Rex ran.

They jumped.

They played.

And Alfred never felt unlovable again.